AF848570

Earth and Environmental Sciences Library

Earth and Environmental Sciences Library (EESL) is a multidisciplinary book series focusing on innovative approaches and solid reviews to strengthen the role of the Earth and Environmental Sciences communities, while also providing sound guidance for stakeholders, decision-makers, policymakers, international organizations, and NGOs.

Topics of interest include oceanography, the marine environment, atmospheric sciences, hydrology and soil sciences, geophysics and geology, agriculture, environmental pollution, remote sensing, climate change, water resources, and natural resources management. In pursuit of these topics, the Earth Sciences and Environmental Sciences communities are invited to share their knowledge and expertise in the form of edited books, monographs, and conference proceedings.

More information about this series at https://link.springer.com/bookseries/16637

Shalini Yadav · Abdelazim M. Negm ·
Ram Narayan Yadava
Editors

Wastewater Assessment, Treatment, Reuse and Development in India

Springer

Editors
Shalini Yadav
Centre of Excellence in Advanced Water and Environmental Research
Rabindranath Tagore University
Bhopal, Madhya Pradesh, India

Ram Narayan Yadava
Research and International Affairs
Madhyanchal Professional University
Bhopal, Madhya Pradesh, India

Abdelazim M. Negm
Hydraulics and Water Resources in the Water and Water Structures Engineering Department
Faculty of Engineering
Zagazig University
Zagazig, Egypt

ISSN 2730-6674 ISSN 2730-6682 (electronic)
Earth and Environmental Sciences Library
ISBN 978-3-030-95785-8 ISBN 978-3-030-95786-5 (eBook)
https://doi.org/10.1007/978-3-030-95786-5

© Springer Nature Switzerland AG 2022
This work is subject to copyright. All rights are reserved by the Publisher, whether the whole or part of the material is concerned, specifically the rights of translation, reprinting, reuse of illustrations, recitation, broadcasting, reproduction on microfilms or in any other physical way, and transmission or information storage and retrieval, electronic adaptation, computer software, or by similar or dissimilar methodology now known or hereafter developed.
The use of general descriptive names, registered names, trademarks, service marks, etc. in this publication does not imply, even in the absence of a specific statement, that such names are exempt from the relevant protective laws and regulations and therefore free for general use.
The publisher, the authors and the editors are safe to assume that the advice and information in this book are believed to be true and accurate at the date of publication. Neither the publisher nor the authors or the editors give a warranty, expressed or implied, with respect to the material contained herein or for any errors or omissions that may have been made. The publisher remains neutral with regard to jurisdictional claims in published maps and institutional affiliations.

This Springer imprint is published by the registered company Springer Nature Switzerland AG
The registered company address is: Gewerbestrasse 11, 6330 Cham, Switzerland

Preface

Assessment, treatment, and reuse of polluted surface and groundwater and its management in India and developing countries are the backbone of sustainability for the functioning of healthy ecosystems, reliable agricultural productivity, maintenance of desirable environmental quality, and enjoyment of quality lifestyle. Scientists, Academicians, Researchers, Practicing Engineers, Consultants, Planners, Policy Makers, Economists and Social Scientists, Managers, and Leaders from around the world are sharing their knowledge, skills, experiences, and expertise through research findings and the case studies under different climatic environments.

The motivation of this volume titled "Wastewater Assessment, Treatment, Reuse and Development in India" is mainly the need to support India and the developing countries by providing necessary skilled information on the past, present, and future scenarios on the assessment, treatment, and reuse of polluted surface and groundwater and its management with the growing population, demands of water for domestic, agricultural, and industrial uses and decreasing of freshwater resources per capita in terms of quantity and its quality.

This book consists of 16 chapters in five parts dealing with various methodologies and technologies on assessment, treatment, and reuse of surface as well as groundwater. Part: I deals with the Monitoring and Assessment of Polluted Water consisting of three chapters dealing with the Coastal Reservoirs, Heavy Metal Accumulation, and Water Purification in Mitigating Global Water Demand. Whereas Part: II describes Wastewater Treatment and Reuse consisting of five chapters dealing with Potential for the use of Treated Wastewater in Industries, Understanding Sustainable Arsenic Mitigation Technology Application, Sequential Anaerobic-Aerobic Co-Treatment of Three Herbicides Mixture in Water, Wastewater Management and Treatment Technologies with Recycling and Reuse Issues Leading to Zero Liquid Discharge, and Wastewater and Its Impacts.

Part: III on Wastewater Remediation comprises five chapters and discusses the Advanced Bioremediation Techniques for Hydrocarbon Polluted Sites, Municipal Wastewater a Remedy for Water Stress Region, Biosorptive Removal of a Herbicide, Glyphosate Using Waste Activated Sludge of Tannery Industry, Wastewater Treatment with Special Reference to Activated Sludge Process, and Assessment of

Water Resources Development. Part: IV on Waste and Sustainability is consisting of two chapters which is focused on Sustainable Construction Practices for Residential Buildings to Reduce the Water Footprint; and Water Resources Availability and Its Teleconnection with Large Scale Climatic Oscillations Over a River Basin. The last Part: V is devoted to conclusions and recommendations of the book. The novelty of each chapter under all the above cited five parts is summarized in brief as follows:

Chapter entitled "Coastal Reservoirs: A Technology That Can Quench Indian Thirst" discusses the importance of coastal reservoirs that can adequately supply sufficient, high-quality, and affordable water without desalination with minimum environmental/social impacts. Further, the importance of coastal reservoirs for Indian capital coastal cities is discussed, and their layouts are presented.

Chapter entitled "Heavy Metal Accumulation in Dominant Edible Fish Species of River Kolab in Koraput District of Odisha, India" focused on the management policy of the effluents released from the industries to the river to trigger the alkaline condition of the water body so that the heavy metal species get compartmented in the sediment bed instead of entering in the body tissues of aquatic organisms. Further, it suggests that the dose of liming should be optimized depending on the pH of the wastewater generated from these industries.

Chapter entitled "Cake and Membrane Filtration in Mitigating Global Water Demand" presented the status of global water purification, preventing global pollution, challenges on infections, demand of increasing need of technology compared to green technology, and importance of JPCF and MBST Processes. It further discusses the importance of water clarification by using MBST, Cake Filtration, Algae Application in WWT (Waste Water Treatment), Zero Discharge Technology, and Cake Filtration-Based Technology in mitigating global water demand.

Chapter entitled "Potential for Use of Treated Waste Water for Industrial Reuse in India" focuses on the appropriate technology for the treatment of wastewater and safe disposal with minimum environmental and health impacts including important to understand the social and economic context of a society/community/city before a technology is implemented for the use of treated wastewater.

Chapter entitled "Understanding Sustainable Arsenic Mitigation Technology Application in the Indian Subcontinent" discusses the magnitude of arsenic contamination of groundwater in the Indian subcontinent and its associated public health crisis due to chronic consumption of arsenic-contaminated groundwater including the arsenic removal technologies that are currently available in the subcontinent, their mechanism of operation, and their efficacy of performance. Further, it pointed out that the technology application and its sustainability are greatly dependent on the chemistry of the groundwater and the operational life of the systems including monitoring and ability of the community to successfully manage the high-arsenic-containing waste generated from the processes.

Chapter entitled "Sequential Anaerobic-Aerobic Co-Treatment of Three Herbicides Mixture in Water: A Comprehensive Study on Biotransformation" dealt with the sequential anaerobic-aerobic treatment method to withstand the toxic condition of the herbicides. Further, it pointed out that the long-term treatment process carried out was able to remove the herbicides at high efficiency with a high biogas production rate and noted that the sludge stabilization ratio observed for the herbicide treatment reactor is within the required range.

Chapter entitled "Wastewater Management and Treatment Technologies with Recycling and Reuse Issues in India Leading to Zero Liquid Discharge (ZLD)" addressed the industrial and municipal wastewater treatment with its effective technological implementation through recycle and reuse with the focus on wastewater recycling opportunities to reduce freshwater consumption including wastewater management and treatment practices with respect to the current scenario and challenges/opportunities. Further, it suggested that Industries and municipalities must focus on optimum water resource utilization with reducing demand and minimizing water losses by implementing result-oriented cost-effective technology.

Chapter entitled "Wastewater and Its Impacts in India" focused on identifying strategies on how to bridge the glaring gap between the facilities created for treatment and the waste being generated, which is visible in the form of a huge quantity of wastewater being left untreated. Also, it informed that the current practice of treating the wastewater through a centralized wastewater system needs to be relooked to bring flexibility to the system and make it cost-effective by promoting the adoption of innovative sustainable treatment technologies; it advised that the treatment of wastewater is an essential process for the conservation of environment.

Chapter entitled "Nutrient Loading Impact on Remediation of Hydrocarbon Polluted Groundwater Using Constructed Wetland" dealt with the development of an effective and low-cost plant-based bioremediation process by optimizing the nutrient application in a hydrocarbon polluted domain. The findings of this study help in the management and remediation of hydrocarbon polluted site including to quantify the extent of shallow groundwater decontamination under real agricultural field-based nutrient loadings in the areas affected by the acute hydrocarbon pollution.

Chapter entitled "Municipal Wastewater—A Remedy for Water Stress in India" focused on the importance and details about the advanced secondary municipal wastewater treatment techniques and tertiary treatment methods to reuse municipal wastewater, and it suggested that a proper characterization of wastewater must be mandatory to select the suitable technology and design a proper STP including the selection of technologies, and design of STP must ensure the degradation of organics in wastewater to the maximum extent. It should have provision not to discharge any form of rejects from the wastewater to the environment.

Chapter entitled "Biosorptive Removal of a Herbicide, Glyphosate Using Waste Activated Sludge of Tannery Industry" has objectives to establish adsorption characteristics for glyphosate removal in the aqueous solution both in synthetic as well as in simulated municipal wastewater using generated metal-laden waste as biosorbent and to investigate different experimental approaches involving optimization of batch adsorption, and influence of different environmental parameters like biosorbent dosage, pH, initial glyphosate concentrations, and contact time. Further, it is suggested the simulated municipal wastewater of glyphosate provides a complete and sustainable solution for glyphosate removal.

Chapter entitled "Wastewater Treatment Processes with Special Reference to Activated Sludge Process in Indian Conditions for Water Use Sustainability" focused on the wastewater treatment which is one of the most important sanitation issues required for maintaining a healthy ecosystem, and various industrial, commercial, and domestic activities produce huge quantities of wastewater. Further, it pointed out that the activated sludge process is an important type of secondary treatment, which aims to clean the wastewater by separating it into effluent and sludge. The effluent can then be discharged into any water body while the sludge can be used as manure or incinerated as per the requirement.

Chapter entitled "Assessment of Water Resources in Development of Rajasthan" discusses to save the traditional and historic water resource bodies to ensure the availability of water resources with the increase in urbanization and its demand. Also, it is suggested for a proper legislation and regulation over the water resources exploitation, water conservation to the maximum extent, recycling and reuse of water, and recovery of cost from the beneficiaries must be imposed.

Chapter entitled "Sustainable Construction Practices for Residential Buildings to Reduce the Water Footprint" is dedicated to the environment-friendly construction practices by optimizing the use of water during several construction phases of a reinforced cement concrete framed multi-story residential building and discussed to assess and analyze the water footprint of a residential building during different construction stages and apply water conservation, recovery, and recycling methods.

Chapter entitled "Water Resources Availability and Its Teleconnection with Large Scale Climatic Oscillations Over Godavari River Basin" presents a brief discussion regarding the water availability over major river basins in India. In particular, the water availability and linkage between the monthly streamflow and climate oscillations like SST, IOD, and ISMI are analyzed. In addition, the water availability at different streamflow gauging points is assessed using the flow duration curve. This chapter also provides a comprehensive documentation regarding the surface and groundwater availability over different river basins in India which shall be helpful to the water manager for developing sustainable water resources planning and management in a specific River Basin.

Chapter entitled "Update, Conclusions and Recommendations for "Wastewater Assessment, Treatment, Reuse and Development in India"" closes the book by providing a summary and an update of the related studies to the book theme. More importantly, it presents the main conclusions and recommendations that we extracted from all the 16 chapters presented in the book in addition to the editors experiences.

A word of thank is a must for all contributors who made this book a reality including the authors, Springer editors, and the Springer team who worked in a very collaborative and supportive mode with us from the start of the book project until its publication and beyond. Special thanks are due to Andrey Kostianoy and Alexis Vizcaino.

The volume editor would be happy to receive any comments to further improve the future editions. Comments, feedbacks, and suggestions for further improvement or proposals for new chapters for the next editions are welcome and should be sent directly to the volume editors.

Bhopal, India Shalini Yadav
Zagazig, Egypt Abdelazim M. Negm
Bhopal, India Ram Narayan Yadava
April 2020

Contents

Contributors

Abhishek School of Environment and Society, Tokyo Institute of Technology, Yokohama, Japan

Kakoli Banerjee Department of Biodiversity & Conservation of Natural Resources, Central University of Odisha, Sunabeda, Koraput, India

Sarajit Basu Department of Chemical Engineering, IIT-Bombay, Bombay, India; Jadavpur University, Kolkata, India

Manisha Bhatkulkar Department of Zoology, Jawaharlal Nehru Arts, Commerce and Science College, Wadi, Nagpur, India

Ashim Kumar Bhattacharya Academy of Water Technology and Environ Management, Kolkata, India;
Environment Pollution Monitoring Cell, National Institute of Technological Teachers Training and Research, Salt Lake, Kolkata, India

N. K. Brahma Department of Chemical Engineering, IIT-Kharagpur, Kharagpur, West Bengal, India

Bharati Brahmacharimayum State Climate Change Cell Manipur, Directorate of Environment, Govt. of Manipur, Imphal East, India

Jew Das Department of Civil Engineering, National Institute of Technology Warangal, Warangal, India

Pranab Kumar Ghosh Department of Civil Engineering, Indian Institute of Technology Guwahati, Guwahati, India

Sourja Ghosh Water Technology Division, CSIR-Central Glass and Ceramic Research Institute, Kolkata, India

Pankaj Kumar Gupta Faculty of Environment, University of Waterloo, Waterloo, Canada

Saiful Islam Civil Engineering Department, College of Engineering, King Khalid University, Abha, Saudi Arabia

Nadim Reza Khandaker North South University, Dhaka, Bangladesh

Gopal Raj Khemundu Department of Biodiversity & Conservation of Natural Resources, Central University of Odisha, Sunabeda, Koraput, India

Sudesh Kumar Department of Chemistry, Banasthali Vidyapith, Rajasthan, India

Biswajita Mahapatra Department of Biodiversity & Conservation of Natural Resources, Central University of Odisha, Sunabeda, Koraput, India

A. Mahapatra Institute of Genetic Engineering, Badu, Kolkata, India

G. B. Mahesh Department of Civil Engineering, Madanapalle Institute of Technology and Science, Angallu, Madanapalle, India

Swachchha Majumdar Water Technology Division, CSIR-Central Glass and Ceramic Research Institute, Kolkata, India

Basavaraju Manu Department of Civil Engineering, Dakshina Kannada District, National Institute of Technology, Mangaluru, India

Abdelazim M. Negm Water and Water Structures Engineering Department, Faculty of Engineering, Zagazig University, Zagazig, Egypt

El-Sayed E. Omran Soil and Water Department, Faculty of Agriculture, Suez Canal University, Ismailia, Egypt;
Institute of African Research and Studies and Nile Basin Countries, Aswan University, Aswan, Egypt

Pratibha Department of Chemistry, Banasthali Vidyapith, Rajasthan, India

Mohammad Moshiur Rahman North South University, Dhaka, Bangladesh

Lata Ramrakhiani Water Technology Division, CSIR-Central Glass and Ceramic Research Institute, Kolkata, India

Rishi Rana Department of Civil Engineering, Jaypee University of Information Technology, Waknaghat, District Solan, India

S. N. Roy Ceramic Membrane Division, CSIR-CGCRI, Kolkata, India;
Adamas University, Kolkata, India

G. Sekaran Environmental Science and Engineering Division, SRMIST Ramapuram campus, Ramapuram, Chennai, India

Ram Karan Singh Department of Civil and Environmental Engineering, The NorthCap University, Gurgaon, India;
The ICFAI University Dehradun, Dehradun, India

Mansvee Singh Department of Applied Psychology, The University of Massachutes Amherst, Amherst, USA

Supriya Singh Department of Chemistry, Banasthali Vidyapith, Rajasthan, India

Vanshika Singh Department of Chemistry, Banasthali Vidyapith, Rajasthan, India

S. Swarnalatha Environmental Science and Engineering Division, CSIR – Central Leather Research Institute (CSIR – CLRI), Adyar, Chennai, India

Vineet Tirth Department of Mechanical Engineering, College of Engineering, King Khalid, University, Abha, Kingdom of Saudi Arabia;
Mechanical Engineering Department, College of Engineering, King Khalid University, Abha, Saudi Arabia

Amit Tirth Department of Public Health Dentistry, Kothiwal Dental College and Research Center, Moradabad, India

N. V. Umamahesh Department of Civil Engineering, National Institute of Technology Warangal, Warangal, India

N. Vinayagamoorthy Department of Environmental Health Engineering, Faculty of Public Health, Sri Ramachandra Institute of Higher Education & Research (Deemed to be University), Porur, Chennai, India

Ram Narayan Yadava Research and International Affairs, Madhyanchal Professional University, Ratibad, India

Brijesh Kumar Yadav Department of Hydrology, Indian Institute of Technology Roorkee, Uttarakhand, India

Shalini Yadav Centre of Excellence in Advanced Water and Environmental Research, Rabindranath Tagore University, Bhopal, India

Shu-Qing Yang Center for Coastal Reservoir Research, Eng, University of Wollongong, Wollongong, Australia

Monitoring and Assessment of Polluted Water

Coastal Reservoirs: A Technology That Can Quench Indian Thirst

Shu-Qing Yang

Abstract Indian is the second populous country in the world, and many places have suffered from water shortage problems due to its spatial non-uniformity of the rainfall patterns. Every year, this country receives about 4000 km^3/year rainfall, and its rivers drain 1900 km^3/year of runoff. Currently 450 km^3/year of surface water and 250 km^3/year of groundwater are used. The estimated ultimate water need is about 1180 km^3/year. In order to solve Indian water crisis, many solutions have been proposed like wastewater reuse, desalination and inter-basin water diversion etc. Different from these proposals, this chapter analyzes the opportunities and feasibilities of coastal reservoirs (CR) that are freshwater reservoirs in seawater to harvest the runoff lost to the sea. The preliminary results show that CR can effectively supply India sufficient, high-quality and affordable water with minimum environmental/social impacts. In this chapter, the coastal reservoirs for Indian capital coastal cities are discussed, and their layouts are presented.

Keywords Water resources · Coastal reservoirs · Inland dams · Water supply · Indian water availability

1 Introduction

With ever growing population, ongoing urbanization and lack in sustainable developments, the world is facing critical problems in terms of natural resources such as fresh water. With the increase in population and industries, water demand is being increased dramatically within the last few decades and expected to raise exponentially over the next years. Within a generation, demand in many countries is forecasted to exceed their supply capacity by an estimated 40% by 2050, and 4 billion people from 54 countries are expected to face water scarcity [1]. According to World Health Organization 2.1 billion people currently lack safe drinking water at home [2].

S.-Q. Yang (✉)
Center for Coastal Reservoir Research, Eng, University of Wollongong, Northfield Ave, Wollongong, NSW 2522, Australia
e-mail: shuqing@uow.edu.au

© Springer Nature Switzerland AG 2022

S. Yadav et al. (eds.), *Wastewater Assessment, Treatment, Reuse and Development in India*, Earth and Environmental Sciences Library,
https://doi.org/10.1007/978-3-030-95786-5_1

With India, being the second-most populous country in the world, its demand for water is expected to exceed all current supply capacity and the country is set to become water scarce country by 2025. Currently, around 70% of irrigation and 80% of domestic water comes from groundwater which is getting depleted rapidly [3]. To sustain against this scarcity India requires to step further on developing new strategies with alternative resources and water management systems. As India is the 9th country by total renewable water resources [4], developing such strategies at earliest stage can minimize water scarcity in future. With huge inland Himalayan and Perennial rivers flowing through almost all the states of India and being surrounded by the oceans from three sides, India is blessed with rich water resources. Its average rainfall in India is about 4,000 km^3/year, but most of India's rainfall comes over a 4-month period starting from June to September. India also experiences years of excess monsoons and floods, followed by below-average or late monsoons with droughts. Despite abundant rain in July–September, some regions face a shortage of freshwater, while some other parts of the country receive excessive rains resulting in floods. This geographical and time variance in availability of natural water versus the year-round demand for irrigation, drinking and industrial water creates a demand–supply gap that keeps worsening with rising population.

To overcome the water shortage problem caused by the uneven distribution over time and space, India has paid great attention to water resources development in inland regions, this includes the construction of 5000 middle to large dams and huge inter-basin water diversion projects like the inter-basin links to transfer water from Himalayan to Peninsular rivers. The over-developed water resources in inland regions may result in many socio-environmental issues. For example, one of the largest rivers in the world, the Ganga and Brahmaputra has become a seasonal river and the over-development also intensifies social conflicts between upper regions and lower regions, even worsens the inter-state relationships and also inter-country relationship (like India–Bangladesh, India–Pakistan, etc.). Some of the other measures adopted include use of groundwater, seawater desalination, wastewater reuse, rainwater harvesting, but these measures have not yielded good results as expected. To overcome water shortage in coastal areas such as Tamil Nadu, big industries have started to explore the possibilities of initializing desalination plants to fulfill the industrial demands. Currently there are over 100 small and large desalination plants running all over the country. However, these projects are of huge costs. Hence, other solutions in utilizing or conserving the abundant rainwater are needed.

Yang and his team members [1] realize that the world is not running out of water, but water is running out of our coastal rivers, our shortage is not water, but storage shortage. In the world, every year our runoff to the sea is about 40,000 km^3/year, far more than our predicted water usage in 2030, i.e., 4,200 km^3/year. The global water crisis is caused by the wrong assumption that people can only construct dams in mountainous areas, not in the sea. Coastal reservoirs are freshwater reservoirs whose dam is situated in a seawater environment. In 2017, the International Association for Coastal Reservoirs (IACRR) was established officially, which aims at the freshwater development from the sea without desalination (www.iacrr.org).

In the 1930s when USA engineers constructed their highest Hoover dam across Colorado River gorge, Dutch politicians and engineers jointly started to construct the 32 km long IJsselmeer dam in seawater to develop the freshwater from the Rhine River. At almost the same time, a group of dreamers in Australia made their attempts to enclose the outlets of Murray-Darling River, the largest rivers in Australia. They wanted to change the brackish Alexandrina Lake into a freshwater lake. Today, there are about 70 coastal reservoirs in the world like China, Korea, Japan, Netherlands, Russia, India, Singapore etc. These coastal reservoirs serve multiple purposes that change over time like flood disaster mitigation, seawater intrusion prevention, water supply and urban regeneration, due to the societal demands and priorities change. With the ever-growing pressure on coastal regions, and the use of coastal reservoirs for water resources development becomes intensified, and integrated and sustainable management of coastal reservoirs is a must. Among them, China's experience of coastal reservoirs is useful for India as both countries are similar in many aspects like population, climate and development (in Table 1). Hong Kong is the first city in the world using coastal reservoirs for its urban water supply, Shanghai is the largest city whose water supply fully depends on coastal reservoirs.

Different from the strategy of inland water resources development, this chapter discusses the strategy of coastal reservoirs which is a freshwater reservoir in seawater to harvest floodwater lost to the sea. According to this strategy, water from coastal reservoirs at river mouths can be used for many purposed, inter-basin water diversion could be achieved using pipeline/canal to link the coastal reservoirs. This chapter analyzes the feasibility of such coastal reservoirs and their impact on Indian water crisis.

2 Water Problems and Water Availability

India is located between 68° E to 97° E and 8° N to 37° N with a land area of 3.29 million km^2 [5] which is just about 1/3rd of the land area of China. India's geography is diverse. The topography of India consists of the lofty and high Himalayan mountains in the north which were historically formed by the collision between Indian and Eurasian plates and plain valleys of the Ganges, Indus, and Brahmaputra south of the Himalaya, and large peninsular hills and plateau, and some narrow coastal plains. With the diverse geology, the drainage system of India varies widely [6]. The major river basins in India are listed in Table 2.

The rainfall in this country mainly depends on the south-west monsoon. The average annual rainfall over the country is about 1250 mm, or it receives about 4125 km^3/year of rainwater, but the spatial variation is very high, from less than 100 mm in the western deserts, to 11,000 mm in parts of the north-east. The average annual runoff from Indian mainland is about 2436 km^3 [7]. Others estimate the total runoff is about 1,633–1,881 km^3, giving the runoff coefficient from 0.40–0.45, but the widely accepted runoff is 1869 km^3/year [8] as shown in Table 2. India has a long coastline of 7,600 km. India's utilizable surface water resources are estimated around

Table 1 Existing coastal reservoirs in China

Name	Catchment (km^2)	Dam length (m)	Capacity (million m^3)	Surface area (km^2)	Year completed	Province/River
Tuoshan Weir (b)	350	134			833	Zhejiang/Yin
Mulanpi (b)	219	110			1064	Fujian/Mulanxi
Sheyang tide gate (b)	4036	1900			1956	Jiansu/Sheyanghe
Xinyanggang tide gate (b)	2478	240			1957	Jiangsu/Mangsehe
Haihe tide gate (b)	32,700	2100			1967	Hebei/Ziyahe
Duliujian tide gate (b)	46,000	220			1967	Hebei/Hiaihe
Plover cove (c)	45.9	2000	230	12	1968	Hong Kong
Datang harbor (b, c)	Ningbo	70	46.7	4.79	1973	Zhejiang
Huchen harbor (b, c)	Ningbo	102	81.7	9.4	1973	Zhejiang/Zhongbu
High island (c)	200	750 + 431	281	6	1978	Hong Kong
Shanhusha (c)	55,600	2000	1.9	0.4	1979	Zhejiang/Qiangtang
Baogang/Shanghai (c)	1.8 milliom	3700	12	1.8	1985	China/Yangtze
Chen Hang/Shanghai (c)	1.8 milliom	4700	8.3	1.35	1992	China/Yangtze
Yuhuan (a, c)	166	1080	64.1	36.5	1998	China/Zhejiang
Reservatorio (c)		2000	4	0.33	2002	Macao
Bahe (c)	256	2622	10.5	11	2004	Shandong/Wanglian
Caoe (b, c, d)	5099	700	146	43.1	2007	Zhejiang/Caoejiang
Qingcaosha/Shanghai (c)	1.8 milliom	43,000	553	66	2011	China/Yangtze
Dongfengxisha (c)	1.8 milliom	12,000	9.8	3.74	2016	China/Yangtze
Yuguang island (c)	489	7000	6.7	13.6	Under study	Shangdong/Henghe/Baima/Jili
Dayawan (c)	62.7	1000	2	0.5	Planned	Shengzhen

a = land reclamation; b = seawater intrusion; c = water supply; d = flood defence.

Table 2 Surface water resources in major river basins of India

River	Catchment area	Average annual runoff	Storage completed	Exploitable surface water
	*1000 km^2	km^3/year	km^3	km^3/year
Indus	1165	73.3	14.5	46
Ganga	1080	525	37.4	250
Brahmaputra	651	585	1.09	24
Brahmani, Baitani and Subarnarekha	39 + 14 + 18.9	40.9		25
Mahanadi	141.6	66.9	8.9	50
East flowing rivers between Mahanadi and Pennar	89	22.5		13.1
Godavari	312.8	110.5	17.3	76.3
Krishna	258.9	78.1	32.2	58
Pennar	55.2	6.3	2.4	6.9
Cauvery	81.2	21.4	7.2	19
East flowing rivers between Pennar and Kanyakumari	102	16.5		16.7
Rajasthan desert		Negligible		
West flowing rivers of Kutch, Saurashtra	330	15.1		15.0
Sabarmati	21.7	3.8	1.3	1.9
Mahi	34.8	41	30.3	3.1
Narmada	97.4	45.6	11.7	34.5
Tapi	66	14.9		14.5
West flowing rivers from Tapi to Tadri	56	87.4		11.9
West flowing rivers from Tapi to Kanyakumari	56	113.5		24.3
Rivers drain to Bangladesh and Burma		31		
Total		1869.3		

690 km^3/year, and utilizable groundwater sources 432 km^3/year [7]. Currently the total water usage of India is about 710 km^3/year in 2010 [7], it is estimated by the National Commission (1999) that the annual withdrawal requirements, by 2025 and 2050 could be 1050 km^3 and 1300 km^3, respectively. India has a very severe water crisis and the water deficit could constrain its social-economic development. The worst drought in the 20th Century occurred in 1918 and 68% of India land was affected. The second worst drought covered 64% of its total area in 1987–1988. In 2000–2001, a drought affected 145 million people in 8 states.

The climate of India comprises arid desert in the west, glaciers in the north, and humid tropical southwest. The country has four seasons: winter (December, January and February), summer (March, April and May), a monsoon rainy season (June to September), and a post-monsoon period (October to November). Its west coasts like Kerala, Karnataka and Goa have the wet tropical weather, but its east coast has both wet and dry tropical climates. Humid sub-tropical weather dominates its northeast, but arid or semi-arid climate occurs in the northwest region like Gujarat, Punjab, Haryana and Maharashtra. In India, about 42% of rainfall is evaporated.

Population-wise India is the second most populous country in the world with a population of 1.34 billion in 2018 which is three times as the population of the United states of America and 16% of the world's population. The population is estimated to reach 1.6 billion and 2 billion in 2030 and 2050, respectively. However, India has only 4% of global available freshwater resources. Due to rapid growth of its population, the water availability per capita has been reduced significantly from 5177 m^3/person/year in 1951, to 1820 m^3/person/year in 2001 and estimated to be 1341 in 2025 and 1140 m^3/year/person in 2050 [25].

According to India's National Register of Large Dams (NRLD), India has currently 5254 large dams, about 50% of which were constructed during 1970–1989. Among these large dams, 2069 dams are in Maharashtra, 898 in Madhya Pradesh, 619 in Gujarat and 1168 dams are distributed in other states [9]. The total surface water storage capacity is about 250 km^3, and 53 km^3 is in Ganga River basin, 42.3 km^3 in Krishna, 30.2 km^3 in Godavari and 23.3 km^3 in Narmada and 16.3 km^3 in Indus River [10].

As water shortage in central and the peninsular appears, Indian government requested a team led by irrigation minister Dr. K.L. Rao to investigate inter-basin water diversion scheme, and as a result of which the Ganga-Cauvery Link Canal of length 2640 km was proposed in 1972 as shown in Fig. 1. This canal after completion was expected to draw water using pumps from Ganga River near Patna and pass through the basins of the Son, the Narmada, the Tapi, the Godavari, the Krishna and the Pennar rivers, and finally merge the diverted water into the Cauvery upstream of the Grand Anicut. Every year, in 150 days in a flood period these pumps could lift 1400 m^3/s of water to the elevation of 549 m for diversion to the Peninsular region which could leave an average environmental flow of 280 m^3/s for the Ganga River. The scheme had been thoroughly examined and found impractical due to the huge financial cost of about US$33 billion generated by using 5000–7000 MW of electricity [11].

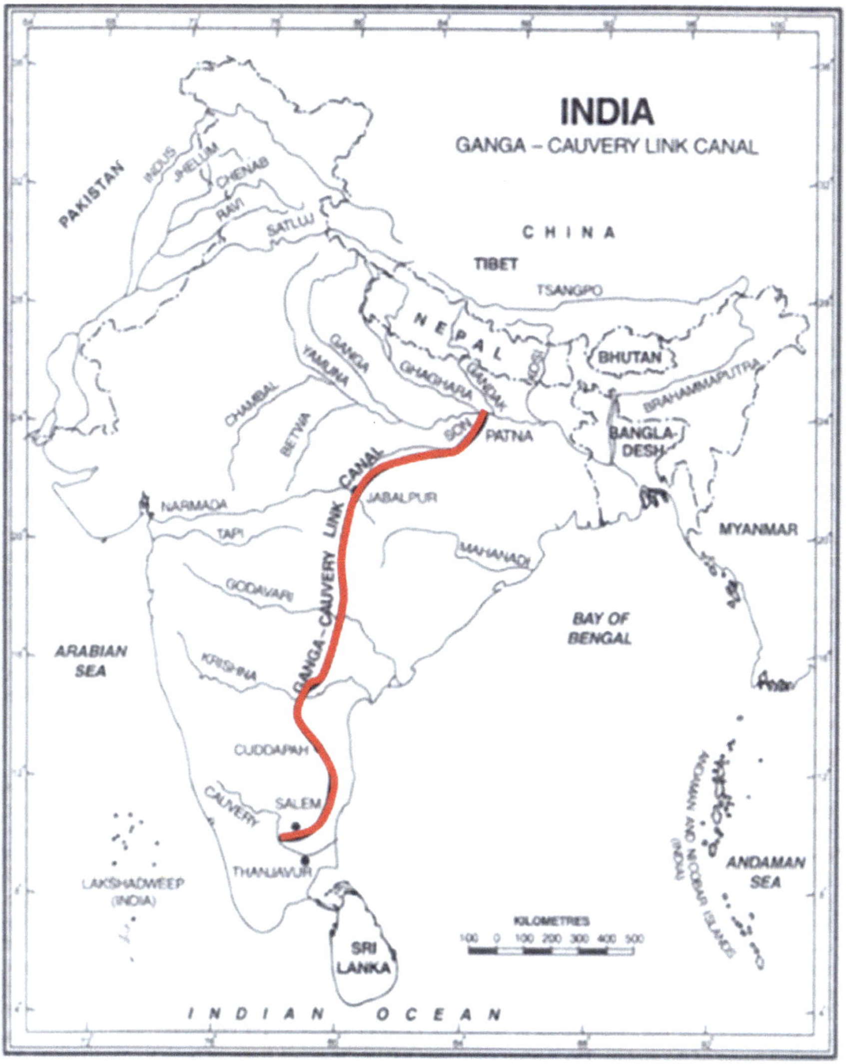

Fig. 1 Ganga–Cauvery link canal

So, the scheme was considered infeasible and thus limited to the idea only. In the 1980s, National Water Development Agency (NWDA) explored the feasibility of intra-state link [12] and came up with proposals under National Perspective Plan (NPP) with two components namely Himalayan Rivers Development and Peninsular Rivers Development. Later in 2005, an Intra state component was added to the plan.

Once completed, Himalayan Rivers have been expected to meet the water demand of 1300 km^3/year in 2050. However, these kinds of projects face major objections by Indian scholars and public as they have negative ecological, environmental and social impacts such as population displacement, displacement of aqua lives, increase in poverty due to decrease in income for river-based businesses such as fisheries and can create inter-state or regional disputes. As India shares 54 rivers with Bangladesh [13], and has ongoing disputes with several river dams, it has only been able to solve the dispute with Ganges River so far. Similarly, inter-state water dispute such as "Cauvery River Water Dispute" between the two states of Tamil Nadu and Karnataka which began in the 1970s still exists and has not yet been resolved.

Due to ongoing disputes on intra-link projects, its adverse impacts on the environment and the high costs associated, it is necessary to route for alternative solutions that are socio-economically feasible and can relieve from the water scarcity in future. For this reason, coastal reservoirs are being proposed at different coastal cities such as Mumbai, Tamil Nadu and Karnataka with existing water crisis.

3 Proposed Coastal Reservoirs

Due to temperate climates and flat fertile lands, coastal areas are densely populated and the population is increasing tremendously over the years. In India nearly 43% of its total population is currently living under the coastal states and out of which 13% population is living in coastal districts [14]. It is expected that population growth in coastal cities will be increased rapidly as experienced in other countries like China, Korea, and Australia. This increase in population in coastal districts and cities has led to an increase in demand for freshwater for both household and industrial sectors.

In general, Indian rivers are highly polluted due to industrial, agricultural and domestic wastes, especially faucal contamination, fertilizers, herbicides and pesticides. In India, even groundwater is also polluted by pathogenic bacteria and nutrients, chemicals and heavy metals. For example, more than 5 million people's drinking water is polluted by arsenic in west Bengal [15]. For this reason a safe and reliable water resource has become mandatory. It is also obvious from the initiation of desalinization plants that water crisis does exist in India and it is reasonable to propose coastal reservoirs. However, it is necessary to discuss first whether coastal reservoirs can effectively quench its thirst. The proposed coastal reservoirs for India is shown in Fig. 2. In outlines of these reservoirs, the redlines are coastal reservoirs, the white lines are clean water path, and black lines are polluted water path, short yellow lines are hydraulic gates.

Kolkata: is the capital of West Bengal state with city area of 1886 km^2 located on the east bank of the Hooghly river. The total population of 14.5 million including the population of suburbs as recorded in 2011 is expected to reach 19.1 million in 2030 which makes Kolkata the third most populous metropolitan area in India. The city's elevation is between 1.5 and 9 m and most of its land comes from sea reclamation and known as east Kolkata wetlands [16].

Rains by the south-west summer monsoon in Kolkata mainly occur between June and September. The highest monthly rainfall total occurs in July and August, and the averaged annual rainfall is about 1,800 mm/year. The Kolkata Municipal water

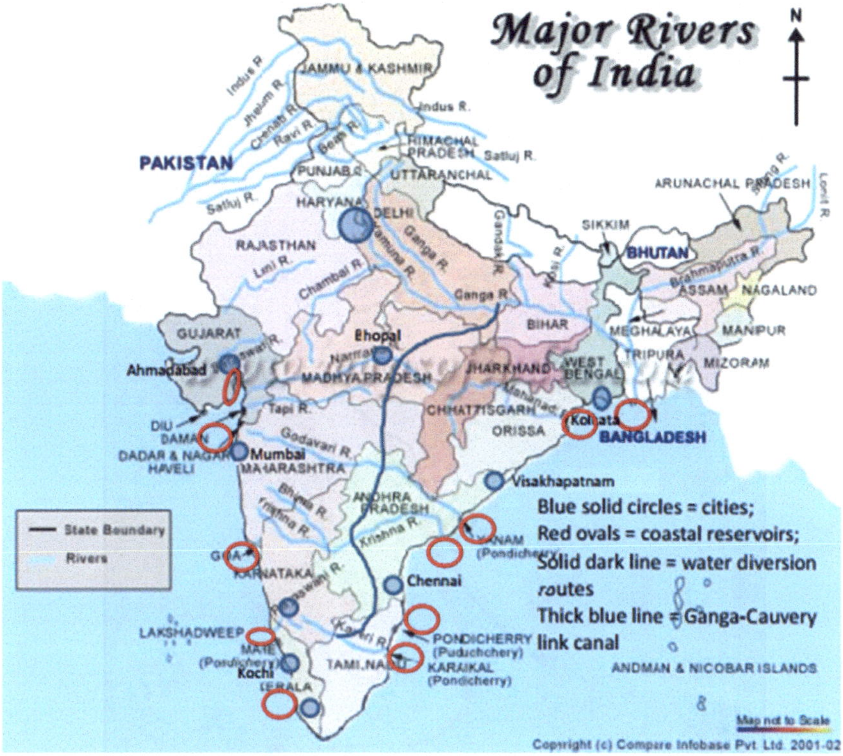

Fig. 2 Proposed coastal reservoirs for major cities in India

supply sources from the Hooghly River, in this city the water supply is free and it is one of the most water-rich cities of India. However, in 2016, a drought-like condition appeared in at least four districts of West Bengal and a district is declared "drought hit" only when more than 33% of its crop was destroyed. Kolkata is also one of the top 3 most flood-prone coastal cities, just next to Shanghai and Dhaka. A coastal reservoir is proposed for Kolkata to eliminate its flood disasters and water supply as shown in Fig 3.

Bhubaneswar: is the capital of the Indian state of Odisha. It is the largest city in Odisha and is a center of economic and religious importance in Eastern India. Odisha is one of the 29 states of India, it has 485 km of coastline along the Bay of Bengal. It is also the 3rd most populous state of India. Bhubaneswar is often referred to as a "Temple City of India". The city is located between the Daya River to the south and the Kuakhai River to the east. Both are branches of the Mahanadi River that is a major river in East Central India. Its catchment is about 141,600 km^2 as shown in Table 2 and the length of the total river course is about 858 km. Every year 66.9 km^3 of water flows to the sea through its mouths. The proposed coastal reservoir for Bhubaneswar's water supply is shown in Fig. 4. By developing runoff

Fig. 3 Proposed coastal reservoir for Kolkata

lost to the sea from Srikakulam River, city Visakhapatnam can get sufficient water for its development, only 100 km from its river mouth to the city.

Chennai and Other Southern Peninsular Cities. The recent development of Chennai's desalination plants such as Nemmeli and Minjur plants indicates water scarcity in the peninsular region. Since 1970s many water diversion schemes such as "Ganga-Cauvery link canal" as shown in Fig. 1 were proposed to provide additional water to the Peninsula where the mega cities like Chennai, Bengaluru, Madurai, Thiruvananthapuram, Kochi, Mangaluru are located. India's National River Linking Project (NRLP) also aims to divert water from Brahmaputra, Mahanadi and Godavari Rivers to peninsular rivers like Krishna, Pennar and Cauvery in the south. However, it is necessary to analyze whether the water shortage exists in the southern peninsula. In the southern peninsula, there are rivers like Godavari, Krishna, Pennar, Palar, Cauvery, Valgal, Vap, and the total runoff shown in Table 2 is more than 232 km^3/year. The distance from Chennai to Mangaluru is about 600 km whereas distance from Bangaluru to the southernmost tip is 550 km and the total area is about 165,000 km^2. Dr. Rao's diversion scheme could only divert 44 km^3/year of water at most to this

Fig. 4 Proposed coastal reservoir for Bhubaneswar

region at the cost of US$33 billion [11], and NRLP could only divert 70 km^3/year of water for this region at the cost of US$19.2 billion. Also, these projects would affect 1320 villages with displacement of 264,000 people. This indicates that the current solutions such as desalination plants and inter-basin water diversion schemes are unnecessary and both financially and socially infeasible as in fact there is no water shortage in this area, but there is the shortage of storage reservoirs. To quench the peninsular thirst, the coastal reservoirs at the river mouths of Godavari, Krishna, Pennar, Palar, Cauvery, Valgal, Vap are suggested and a canal is constructed to link these coastal reservoirs if needed in future. Thus, every year at least 100 km^3/year of additional water can be provided to the east side of the peninsula.

The distance from Godavari river mouth to Chennai is only 550 km, only 400 km from Krishna River mouth to Chennai. Chennai is the capital city of Tamil Nadu state. It is the fourth-largest city and fourth-most populous urban city in India with a population of 8.6 million in 2011. It is also the 36th-largest urban area by population in the world. This city is also nicknamed as "The Detroit of India" due to the presence of major automobile manufacturing industries. Chennai's average rainfall is 1298 mm/year [17]. Its water demand in 2015 was 438 GL/year, but its supply capacity was 360 GL/year. The gap is expected to increase significantly as the demand will increase to 766 GL/year in 2030.

Chennai's water crisis can be solved by building coastal reservoirs at Pennar and Palar river mouths. Pennar and Palar river have a catchment area of 55,200 km^2 and 18300 km^2 respectively with calculated annual runoffs of 6.3 km^3/year and 2 km^3/year. Both rivers are about 120 km and 67 km away from Chennai CBD. The Poondi reservoir (91GL) and Chembarambakkam Lake (103GL) in Palar are major sources of water supply to Chennai city. After commissioning of the Telugu Ganga project to supply nearly 365 GL/year of Krishna River water to the Chennai city, the dependence on Palar river water has reduced drastically.

Further, Pennar River and Palar river can generate water in a total of 8.3 km^3/year which is approximately 10 times more than the Chennai's water demand in 2030. This runoff can be collected at the river mouths by building coastal reservoirs which can easily resolve the water crisis of Chennai. Chennai will not only have sufficient water to sustain its developments but also further environmental impacts and financial burdens can be reduced by decommissioning of the desalination plants and other diversion schemes. The proposed coastal reservoirs for Chennai's water supply are shown in Figs. 5 and 6 where the shallow water in the river can be developed as wetlands to pre-treat the river water.

The above coastal reservoirs can also help Bangalore to solve its water shortage problems. Bangalore is the capital city of Karnataka state and is the fifth most populous city of India with a population of about 8.42 million in 2015. The population is expected to reach 9.43 million in 2025 and 25 million in 2050. Because of the presence of largest IT hub, this city is nicknamed as "Silicon Valley of India". It is

Fig. 5 Proposed coastal reseroir in Pennar River

Fig. 6 Proposed coastal reseroir in Palar River

the second fastest growing metropolitan city of India. Its elevation is over 900 m above sea level and recorded annual rainfall is about 1000 mm/year.

Currently, the river Cauvery (or Kaveri in Indian) provides around 80% of the total water supply (i.e., 292 GL/year to the city) with the remaining 20% being obtained from the Thippagondanahalli and Hesaraghatta reservoirs of the Arkavathi river. According to an estimate the demand for water in the city may rise to 930 GL/year by year 2036 and the city could face a shortage of 365 GL/year.

Cauvery is a large Indian river flowing in Karnataka and Tamil Nadu, damping into the Bay of Bengal through two principal mouths in Poompuhar, Tamil Nadu. Its catchment area is 81,155 km^2 from four states, 54% from Tamil Nadu, 42% from Karnataka, 3.5% from Kerala, and 0.2% from Puducherry. As a result of the historical conflict between these four states, Indian government established a tribunal in 1990 to resolve the conflict. This tribunal allocated a runoff of 11.8 km^3/year to Tamil Nadu and 7.6 km^3/year to Karnataka; 0.8 km^3/year for Kerala and 0.2 km^3/year to Puducherry out of total runoff of 21.4 km^3/year. Karnataka and Tamil Nadu being the major shareholders, Karnataka was ordered to release 5.4 km^3/year of water to Tamil Nadu in a normal year from June to May [18] as shown in Fig. 7.

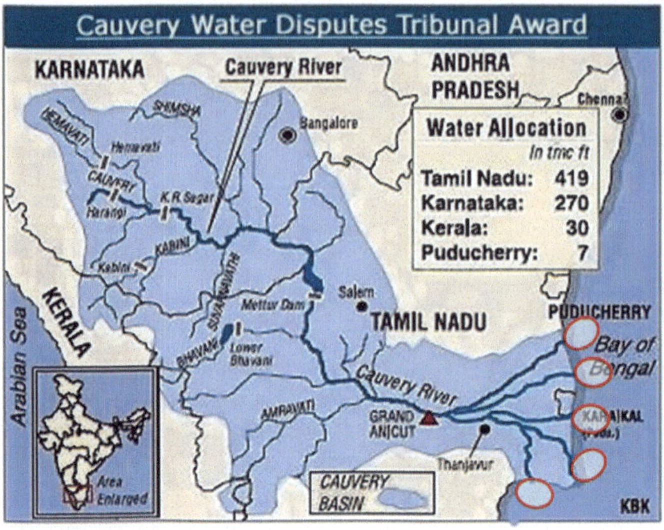

Fig. 7 Coastal reservoirs at Cauvery river mouth

The dispute however, did not end there. For example, in August 2016, Tamil Nadu asked Karnataka to release 1.4 km^3 Cauvery water by following the Cauvery Water Disputes Tribunal 2007. However, Karnataka declined to release water as the water in Karnataka was only enough for drinking purposes. In September 2016, Supreme Court asked Karnataka to release water to Tamil Nadu based on humanitarian grounds as a result of which a general strike was called in Karnataka and two people were killed in violence. Finally, the Supreme Court's order was not followed.

The water dispute and Bangalore's water crisis can be solved by coastal reservoirs at the Cauvery river mouths as shown in Fig. 7. Based on the report of Cauvery Fact-Finding Committee which included 38 years of flow data at Anicat on Coleroon, the estimated annual flow and exploitable water in this basin 21 km^3/year, and 19 km^3/year, respectively. However, the figures were less than the factual as the runoff from the 8000 km^2 delta was not included in their calculations.

The estimated extra runoff from excluded 8000 km^2 delta is 2.3 km^3/year, and the minimum flow from Anicat to the sea is 2 km^3/year (=21–19), so the total runoff to the sea should be more than 4.3 km^3/year. Except for environmental flow to the sea (1.3 km^3/year), the remaining 3 km^3/year of water can be collected and stored in the reservoirs at the river mouths. In this way, the irrigation water demand in the delta can be fully met and therefore the water demand in Tamil Nadu can be reduced. Therefore, the water supply to Karnataka state can be increased to 3 km^3/year indirectly once the downstream water supply is fully met. This could be more realistic if Karnataka state financially supports the construction of coastal reservoirs at the river mouths. Hence one can see that the coastal reservoirs can supply sufficient water to Tamil Nadu and Karnataka, and it also promotes the harmonious relationship between upstream and downstream water users. Therefore, Karnataka should increase its inland reservoir's storage capacity, and Tamil Nadu should concentrate on the construction of coastal reservoirs at its river mouths. If so, both states will have sufficient water to meet their demands and the water dispute can be settled down.

Kerala and Kochi: For the coastal state Kerala and city Kochi, the rainfall is very high up to 3500 mm/year [20]. Its runoff to the sea can be developed from the Vembanad Lake and the Kaithapuzha Lake from the Meenachil River and Muvathupuzha River. The Meenachil River is 78 km long and has a watershed area of 1208 km^2. The river has a total annual yield of 2,349 GL/year and an annual utilizable yield of 1110 GL/year [21]. Kollam is an old seaport and city of Kerala located on the banks of Ashtamudi Lake whose surface area is 61.4 km^2, and the catchment area is about 1,700 km^2 [19]. A coastal reservoir is proposed in Fig. 8.

Muvattupuzha River is one of the major perennial rivers in Central Kerala with a length of 121 km before it drains into the Vembanad Lake. This river constitutes a total catchment area of 1554 km^2 and average annual streamflow of this river is 3,560 GL. The proposed coastal reservoir for Kerala State is shown in Fig. 9.

Mangaluru is the chief port city of the Indian state of Karnataka. It is located about 352 km west of the state capital, Bengaluru. The average annual precipitation in Mangalore is 3,796 mm/year [22]. The highest rainfall recorded is 330 mm over 24 h on 22 June 2003. In the year 1994, Mangalore received very heavy annual rainfall of 5,018 mm/year. Water supply for Mangaluru (or Mangalore) city in Karnataka

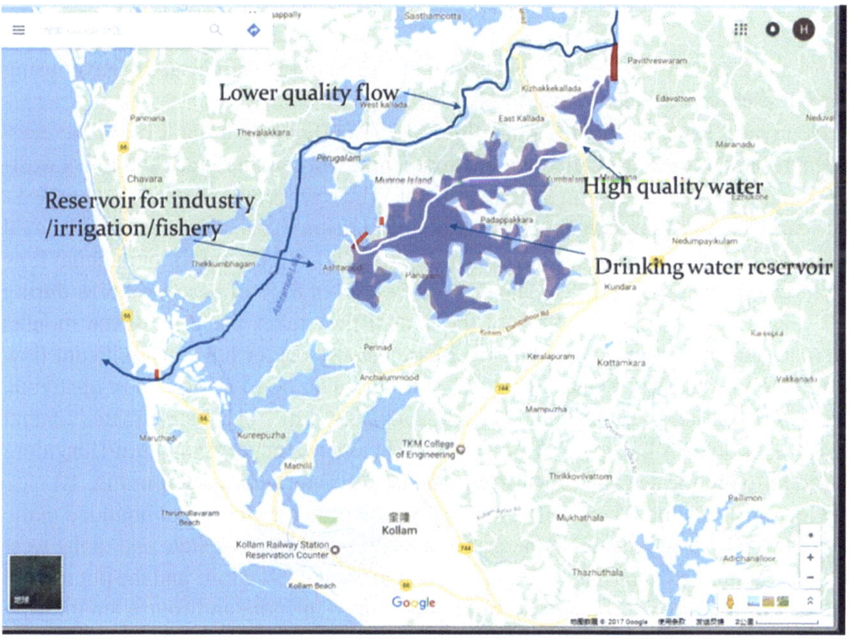

Fig. 8 Proposed coastal reservoir for Kollam

Fig. 9 Proposed coastal reservoir for Kerala and Kochi

State can be met by developing water from the Netravati River/Gurupura River using coastal reservoirs as shown in Fig. 10.

Netravathi River originates at an altitude of 1,000 m above sea level in the evergreen tropical rain forest called the Western Ghats (mountain range) and flows westward for about 103 km to join the Arabian Sea. The river drains an area of about 3657 km^2 carrying an average annual flow of about 11.5 km^3/year. The maximum annual flow was 16.7 km^3 (2.5 times the minimum) in 1980 and the minimum annual flow observed was 6.7 km^3 in 1987. The river carries about 85% of its annual flow during July and August, and about 98% of the annual flow takes place for seven months (June to December) in a year. The rivers are usually dry or have insignificant flow from February to May. Seawater intrudes into the river up to about 22 km upstream. The catchment area of Gurupur River is about 824 km^2, and its estimated annual runoff is about 2.6 km^3/year. To provide sufficient water to Mangaluru/Bangalore and to mitigate the seawater intrusion, a costal reservoir is proposed in Fig. 10.

Goa City: is situated on the western coast of Peninsular India and bounded in the north by Maharashtra State, in the east and south by Karnataka State and in the west by the Arabian Sea. Goa has a total geographical area of 3702 km^2 and the population is 1.46 million. Goa is the smallest of all the states in India and fourth smallest by population. Goa is mainly drained by Mandovi and Zuari rivers. Goa receives an overall annual rainfall of 3483 mm with a maximum of 5090 mm and a minimum of 2612 mm [6]. For Goa state, its capital city Panaji's water supply could be developed

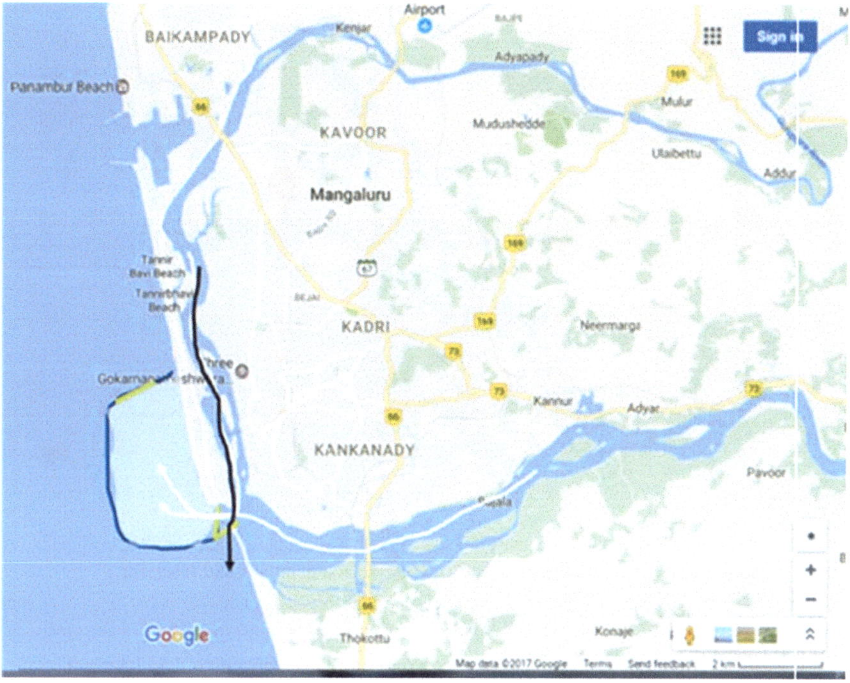

Fig. 10 Proposed coastal reservoir for Mengaluru/Bangalore

from a coastal reservoir inside Mandovi River, located in either northern or southern passage of Divar Island as shown in Fig. 11 below. This reservoir when built can meet the increasing water demand of the city in future. Due to readily available water resources, good annual average and small population, this city is less likely to suffer from water crisis as compared to the other coastal cities across India.

Mumbai: is the capital city of the Indian state of Maharashtra and it is also population-wise second largest city in India with population of 12.4 million in 1990, 20.7 in 2014 and expected to reach 28 million in 2030 [23]. Mumbai's metropolitan area covers 4355 km^2 whereas the city area is 603.4 km^2. It receives an average annual rainfall of about 2258 mm/year [24] which is mostly due to the southwest monsoon rains between June and September. The recorded maximum annual rainfall was 3,452 mm/year in 1954, and the highest daily rainfall was 944 mm on 26 July 2005. Mumbai's water supply mainly comes from two river basins, i.e., Ulhas and Vaitarna rivers. The basic information of these dams for water supply is shown in Table 3. In 2011, the water supply was 1423 GL/year, and it is estimated to be 1603 GL/year in 2021. The Ulhas basin drains an area of about 4,637 km^2 into the Arabian Sea and receives an average annual rainfall of 2,943 mm/year from southwest monsoon from June to October. The Ul has rises from Sahyadri hill ranges in the Raigad district of Maharashtra at an elevation of 600 m above mean sea level.

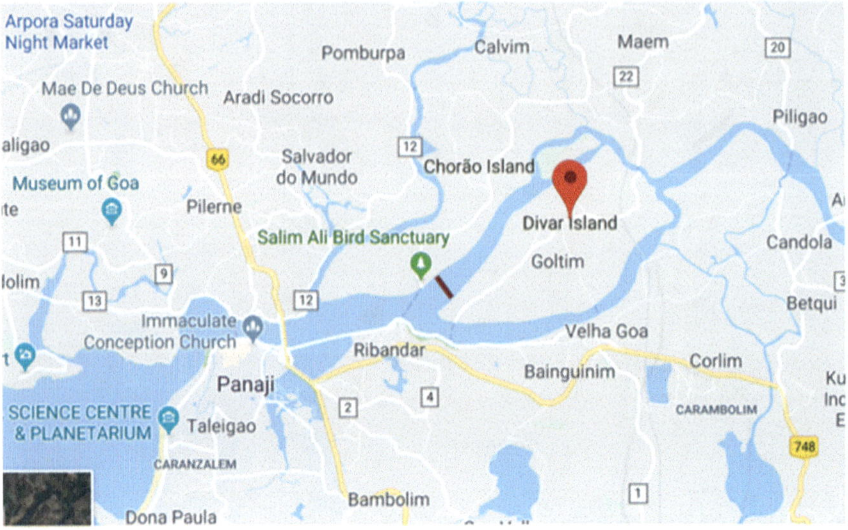

Fig. 11 Potential development of river water for City of Goa using coastal reservoir

Table 3 Dams and Lake near Mumbai

Dam	Year constructed	Capacity (GL)	Annual supply (GL)
Modak Sagar/(Upper Vaitarana)	1957	356	417
Tansa Lake	1892–1925	144	177
Vihar Lake	1860	27	40
Tulsi Lake	1879	8	6.6
Bhatsa	1983	710	620
Middle Vaitarna	2012	193	166

The total length of this west-flowing river from its origin to its outlet is 122 km and the annual runoff is estimated to be 6700 GL/year.

The river Vaitarna is located in the region North of Mumbai and South of the Tapi River which comes from the Sahyadri hill range at Trimbak and travels 120 km to the Arabian Sea. The Vaitarna basin is 2019 km^2, and its estimated annual runoff is about 3300 GL/year. The proposed coastal reservoirs for Mumbai's water supply are shown in Fig. 12, in which a barrage by rubber dam is constructed in Ulhas River to divert good quality water into the Thane Creek, a barrage is suggested to separate the sea. These two river basins drain more than 10,000 GL/year of water into the sea, which can supply 6 cities whose size is similar to Mumbai. Another barrage or

Fig. 12 Proposed coastal reservoirs for Mumbai

rubber dam is suggested to divert good quality water from Vaitarna River into the coastal reservoir which starts from Jhow Island and links with Amala Island.

Likewise, other coastal reservoirs can be constructed to develop the runoff from Daman Ganga River (2318 km^2), Par River (907 km^2), Auranga River (700 km^2), Ambika River (2715 km^2) and Purna River (18,929 km^2). The total catchment area is 25600 km^2. The annual rainfall in Surat is 1.2 m/year and 2.3 m/year in Mumbai, thus we can assume the average rainfall in these catchments is 1.7 m/year. If the runoff coefficient is 0.45, then the total runoff from these rivers is 20 km^3/year. This huge runoff can sustain west Indian future development.

Gujarat: is a relatively dry state in India, 5% of the country's population live in its 6.4% of the country's land with only 2% of the country's freshwater resources. Every year, the state generates 38 km^3/year of surface water. However, only 20 km^3/year can be developed due to the lack of suitable dam sites. The Gulf of Khambhat extends from north to south and its width varies from 25 km at the inner end to 150 km at the outer mouth, the total surface area is 17000 km^2. There are four rivers draining its annual flow to the sea. They are the Narmada River, Vishwamitri River, Mahi River and Sabarmati River. The total catchment area of these rivers is 160,900 km^2 and the total annual runoff is 60.8 km^3/year (see Table 2). State Gujarat has an area of 196,024 km^2 with a coastline of 1,600 km, and a population in excess of 60 million in 2011. Its largest city is Ahmedabad. There are 185 rivers in this state, only 8 of them are perennial in the southern part. To sustain Gujarat's development, from 1960s the government started to investigate the feasibility of barrage in the gulf of Khambhat for harvesting tidal power, land reclamation, freshwater development and train/road transport, etc.

In 1969, a 42.5 km barrage was proposed by the Gujarat State to prevent sea tide water ingress and store rain water. In 1975, Prof. E. M. Wilson of United Nation Mission suggested a dam between Ghogha and Dahej to the electricity authority for tidal power generation. In 1986, the Irrigation Department of Gujarat recommended a 46 km long dam to store 3.4 km^3 of water. In 1988, the Central Electricity Authority suggested two separate basins for tidal power and fresh water storage (see Fig. 13). In 2009, The Government of Gujarat accepted the Expert Advisory Group's recommendations to carry out more studies.

In 2016, as shown in Figs. 13 and 14, the project is mainly aimed for water resources development to create the world largest fresh water reservoir to meet demand for irrigation, domestic and industrial water supply. Associated components are use of top of the dam across the Gulf as a surface transport link, potential development of fisheries, reclamation of saline land around the fresh water reservoir. The

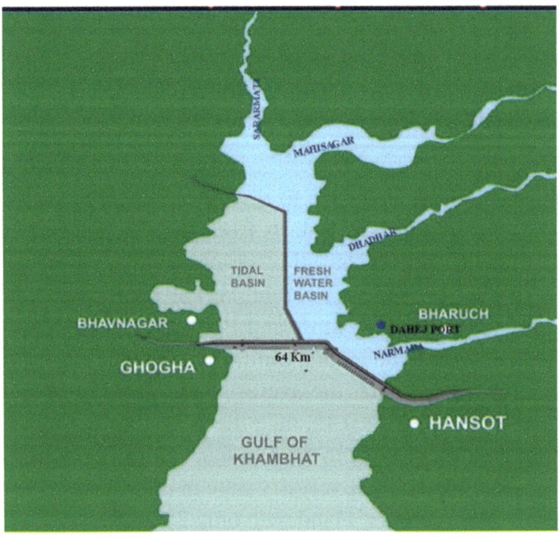

Fig. 13 Proposed coastal reservoir for Gujarat

dam length is 32 km long, the reservoir's storage is about 10 km^3 and its surface area is about 2000 km^2, the average depth is about 5 m.

As the Gujarat reservoir is the 1st generation coastal reservoir, this type of reservoirs often has water quality problems like those in Australia and South Korea. Special caution should be paid for its hydraulic detention time. The comparison of these three coastal reservoirs is shown in Table 4. The total runoff to the Gujarat reservoir is about half of the runoff (i.e., 30 km^3/year) from Narmada River (catchment area = 102 km^2, annual runoff = 45.6 km^3), Vishwamitri River (1185 km^2, 0.47 km^3), Mahi River (36 ^{k} km^2, 11 km^3), Sabamati River (21700 km^2, 3.8 km^3) if the existing dams were excluded. Table 4 clearly shows that Gurajat reservoir's detention time ranges between Alexandrina Lake and Sihwa Lake, same as the evaporation time. Therefore, one may conclude that it is possible that the reservoir's water quality is too poor to be used. In order to improve the water quality and also develop the runoff lost to the sea, the 2nd generation coastal reservoir should be applied, i.e., (Fig. 14,

Table 4 Comparison of Gujarat reservoir with other failed reservoirs

	Gujarat reservoir	Alexandrina lake	Sihwa lake
Country	India	Australia	South Korea
Annual flow (km^3)	30	11.6	0.15
Water storage (km^3)	10	1.7	0.32
Surface area (km^2)	2000	620	42.5
Storage/Annual evaporation (year)	10/2.3 = 4.3	1.7/0.8 = 2.1	0.32/0.043 = 7.4
Hydraulic detention time (year)	10/30 = 0.33	1.7/11.6 = 0.15	2.1
Problem		High salinity	Algal blooms

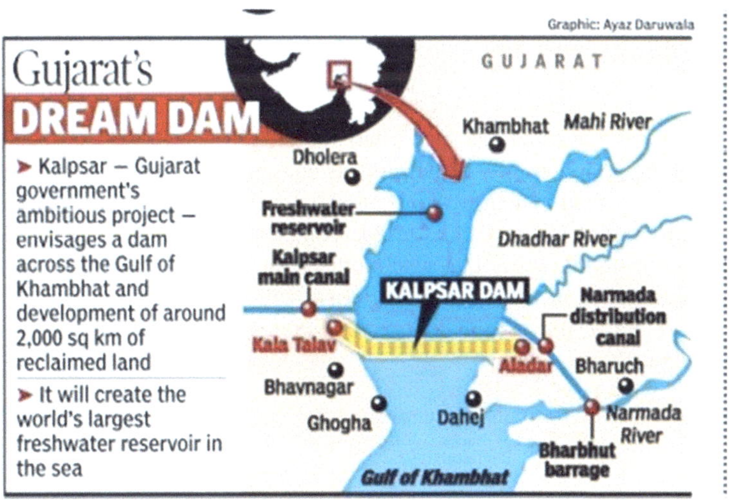

Fig. 14 Proposed coastal reservoirs for Kalpasar

the proposed 30km long seawall for the Kalpasar Coastal Reservoir)

(1) the shallow water should be excluded from the high-quality freshwater storage zone;
(2) the poor quality water should be able to be discharged into the sea without storage, or the by-pass channels should be included.

4 Conclusion

This chapter discusses possible coastal reservoirs in India, and the preliminary results show that coastal reservoirs could effectively mitigate flood disasters, and convert the disastrous floodwater into sweet water resources in dry seasons. Once local runoff is developed, downstream water demand is met and flood disasters are mitigated, the upstream water crisis could be solved, e.g., Chennai and Bangalore, thus India will have sufficient freshwater from the sea without desalination. Water pollution is another problem for India, the wetland pre-treatment can be developed to purify the river water prior to storage. The coastal reservoirs in its Arabian sea could develop at least 132 km^3/year of floodwater from rivers. These coastal reservoirs, if designed properly, could supply sufficient water to the southern and northwestern regions like cities Trivendrum, Mangaluru, Mumbai, etc. Kalpsasar coastal reservoir could be also useful to quench the thirst in Indian regions like Gujarat, Rajastan and Maharashtra and even its neighboring countries like UAE, MUSCAT and other Gulf countries to exchange their energy resources like oil/gas for water.

The analysis shows that the total runoff in India is about 1887 km^3/year (the average runoff coefficient is about 0.47), among it 1033 km^3/year of water could be developed by the inland strategy and at least 854 km^3/year of floodwater reaches the sea, for which coastal reservoirs can harvest in the sea. It is enough for India to quench its thirst, or India is not running out of the water, but water is running out of India.

It is beneficial that India has such a large stretch of coastline that provides the country with good opportunity to develop its coastal economic using its abundant water resources, cheap human resources and extensive coastal land. Among these three elements, water is critical. Coastal reservoirs would manage to capture and hold large amounts of usable water that would otherwise go to waste. As the population of India continues to increase, the water crisis will embellish, so as action needs be taken more drastically as time goes on, coastal reservoirs will more prominently become the most viable option.

5 Recommendations

Similar to any dam development, coastal reservoir's design, construction and management needs comprehensive investigation. The existing coastal reservoirs in the world tells that many do not achieve their designed purpose and some are failed. It is highly recommended that future coastal reservoir development should be guided/managed by experienced experts like IACRR members.

Like dam development, a coastal reservoir needs to consider rainfall, runoff, extreme whether conditions, but more than that, coastal reservoir development should consider seawater pollution, sea-level-rise and water quality. Its water quantity should include domestic, industrial and agricultural water use.

Coastal reservoir development needs experts like decision-makers, city-planners, engineers, researchers, suppliers, contractors and financial investors. IACRR is a pool of these experts.

References

1. Yang SQ (2004) Global and China's water crisis and its solutions in the 21st century. Tianjin University Press, ISBN 7-5618-2069-0 (in Chinese)
2. World Health Organization (2018) 2.1 billion people lack safe drinking water at home, more than twice as many lack safe sanitation. http://www.who.int/news-room/detail/12-07-2017-2-1-billion-people-lack-safe-drinking-water-at-home-more-than-twice-as-many-lack-safe-sanitation, Accessed 31 May 2018
3. The Times of India (2018) India set to become water scarce country by 2025: Report—Times of India. https://timesofindia.indiatimes.com/home/environment/developmental-issues/India-set-to-become-water-scarce-country-by-2025-Report/articleshow/47403894.cms, Accessed 31 May 2018
4. En.wikipedia.org (2018) List of countries by total renewable water resources. https://en.wikipedia.org/wiki/List_of_countries_by_total_renewable_water_resources, Accessed 31 May 2018
5. Publicationsdivision.nic.in. (2007) India 2007. http://www.publicationsdivision.nic.in/others/india_2007.pdf, Accessed 31 May 2018
6. Hussain M (n.d.) Geography of India. 5th edn. New Delhi: McGraw Hill Education (India) Pvt Ltd
7. Zade M, Ray SS, Dutta S, Pan S (2005) Analysis of runoff pattern for all major basins of India derived using remote sensing data. Curr Sci 88(8):1301–1305
8. Bhattacharya A, Reddy S, Ghosh M, Naika HR (2015) Water resources in India: its demand, degradation and management. [ebook] Int J Sci Res Publ 347. http://www.ijsrp.org/research-paper-1215/ijsrp-p4854.pdf, Accessed 31 May 2018
9. Cwc.gov.in. (2018) National Register of Large Dams. http://cwc.gov.in/main/downloads/NRLD_06042018.pdf, Accessed 31 May 2018
10. India-wris.nrsc.gov.in. (2011) Surface water resources. http://india-wris.nrsc.gov.in/wrpinfo/index.php?title=Surface_water_resources, Accessed 31 May 2018
11. Shah T, Amarasinghe U, McCornick P (2014) India's river linking project: the state of the debate. https://www.researchgate.net/publication/240637675_India's_River_Linking_Project_The_State_of_the_Debate1, Accessed 31 May 2018
12. Nwda.gov.in. (2018) Overview: national water development agency. http://nwda.gov.in/content/innerpage/overview.php, Accessed 31 May 2018

13. En.wikipedia.org. (2017) Sharing the water of the Ganges. https://en.wikipedia.org/wiki/Sharing_the_water_of_the_Ganges, Accessed 31 May 2018
14. Centre for Coastal Zone Management and Coastal Shelter Belt, G. (2017) Coastal states of India. Iomenvis.nic.in. http://iomenvis.nic.in/index2.aspx?slid=758&sublinkid=119&langid=1&mid=1, Accessed 31 May 2018
15. Mazumdar D, Dasgupta U (2011) Chronic arsenic toxicity: studies in West Bengal, India. Kaohsiung J Med Sci. https://ac.els-cdn.com/S1607551X11001124/1-s2.0-S1607551X11001124-main.pdf?_tid=b658872c-9904-4a57-b4e2-5901adde6c7d&acdnat=1527773972_7ecce57e7fbc9d1051a9b4e6a1c1541f, Accessed 31 May 2018
16. En.wikipedia.org. (2018) Kolkata. https://en.wikipedia.org/wiki/Kolkata, Accessed 31 May 2018
17. Tnenvis.nic.in. (n.d.). http://www.tnenvis.nic.in/WriteReadData/UserFiles/file/1_CHENNAI_RAINFALL.pdf, Accessed 31 May 2018
18. Thenewsminute.com. (2018) Cauvery dispute explained: what the cauvery water dispute between Karnataka and TN is all about. https://www.thenewsminute.com/article/explained-what-cauvery-water-dispute-between-karnataka-and-tn-all-about-76528, Accessed 31 May 2018
19. En.wikipedia.org. (2018) Ashtamudi Lake. https://en.wikipedia.org/wiki/Ashtamudi_Lake Accessed 31 May 2018
20. Shodhganga.inflibnet.ac.in. (2018) http://shodhganga.inflibnet.ac.in/bitstream/10603/3647/5/05_chapter%201.pdf, Accessed 31 May 2018
21. En.wikipedia.org. (2018) Meenachil river. https://en.wikipedia.org/wiki/Meenachil_River, Accessed 31 May 2018
22. Mangalore C, Mangalore T, Data M (2018) Climate Mangalore: Temperature, Climograph, Climate table for Mangalore—Climate-Data.org. En.climate-data.org. https://en.climate-data.org/location/2791/, Accessed 31 May 2018
23. Worldpopulationreview.com. (2018) http://worldpopulationreview.com/world-cities/mumbai-population/, Accessed 31 May 2018]
24. Chatterjee B (2018) 97% of annual rain received, Mumbai set to get excess rainfall. https://www.hindustantimes.com/, https://www.hindustantimes.com/mumbai-news/97-of-annual-rain-received-city-set-to-get-excess-rainfall/story-kzuSQST30B1FtGOltYpwdJ.html, Accessed 31 May 2018
25. Dubbudu R, Dubbudu R (2018) Per capita water availability down 70% in 60 years—FACTLY. FACTLY. https://factly.in/per-capita-water-availability-down-70-in-60-years/, Accessed 31 May 2018

Heavy Metal Accumulation in Dominant Edible Fish Species of River Kolab in Koraput District of Odisha, India

Kakoli Banerjee, Biswajita Mahapatra, and Gopal Raj Khemundu

Abstract The present study aims to assess the diversity of fin fishes along with the water quality and the major toxic heavy metals accumulated in the organs of two edible major carps: Labeo rohita and Catla catla collected from the two different sites of the River Kolab, i.e. Upper Kolab (site 1) and Lower Kolab stretches (site 2) in Koraput District of Odisha, India. The study recorded fourteen (14) least concerned followed by three (03) near threatened and two (02) Vulnerable species under IUCN category. Highest Shannon-Weiner species diversity index at site 1 revealed that site 1 has more fish diversity than site 2 and it is supported by the Index of Dominance whose value is just vice versa. This may be because of its proximity to the two major Industries: Hindustan Aeronautics Ltd. (HAL) and National Aluminium Company (NALCO). Temperature and turbidity of the water and heavy metal content were comparatively higher at the region of site 2, receiving effluents compared to river water, which received less effluent discharge (site 1). High metal accumulation in fishes at site 2 indicates the impairment of ambient stream due to discharge of industrial and municipal effluents into the river. The pattern of accumulation was in the order of Zn > Cu > Pb in both for muscle and intestine irrespective of species and sites, excepting in muscle of Labeo rohita in site 1 where Zn > Pb > Cu. It is to be noted that in case of intestine of Catla catla at site 2, Pb accumulation (1.299 ppm dry weight) is above the permissible level as per WHO/FAO (1984). This is an alarming signal for human community. Correlation coefficient computed between physico-chemical variables and accumulated metals revealed the regulatory effect of temperature and pH on metal accumulation. As river Kolab is the main source of drinking water and these fishes are the main fishery resources, consumed by many people in the study area, hence, proper detoxification of heavy metals in effluent and municipal waste is necessary before discharging them into the water body. It is also recommended that awareness programmes should be taken up among the people regarding the hazards of consumption of polluted water and fishes.

K. Banerjee (✉) · B. Mahapatra · G. R. Khemundu
Department of Biodiversity & Conservation of Natural Resources, Central University of Odisha, P.O. NAD, Sunabeda, Koraput, Odisha 763004, India

© Springer Nature Switzerland AG 2022

S. Yadav et al. (eds.), *Wastewater Assessment, Treatment, Reuse and Development in India*, Earth and Environmental Sciences Library,
https://doi.org/10.1007/978-3-030-95786-5_2

Keywords River Kolab; Shannon-Weiner diversity index · Index of dominance · Bioaccumulation · *Labeo rohita* · *Catla catla*

1 Introduction

Riverine ecosystem is the lifeline of any country sustaining multifarious floral and faunal diversity and play important role in providing certain ecological services [1, 2]. Rivers, watersheds and aquatic ecosystems are the biological engines of the planet. The anthropogenic pressure on freshwater resources is extreme, and has led to critical endangerment of freshwater biodiversity at both global and regional scales [3]. Rivers are connecting links between various habitats covering a large surface area [4]. Conserving diversity and abundance of freshwater fishes for their ecological function and their economic value is a significant part of river conservation, restoration and management.

India is one of the mega biodiversity countries in the world. It occupies the ninth position in terms of freshwater mega biodiversity [5]. India is bestowed with more than 3.0 million hectares of water spread area which harbours about 10,000 species accounting for about 11.7% in Indian waters. Out of 2,319 fish species found in India, 838 species are purely freshwater [6]. Odisha has a vast expanse of freshwater comprising of rivers, canals, lakes, reservoirs, swamps, ponds and tanks (Table 1). These water bodies spread over an area of about 6.66 lakh hectares and are inhabited by diverse group of fishes. These are Indian Major Carps, Mahseer, Minor Carps, Large and small Cat fishes, Air- breathing fishes, Clupeids, Weed fishes and other fishes [7].

Table 1 Odisha—river length and catchment areas in km^2 [7]

Name of rivers	Total catchment (Km^2)	C.A inside Odisha (Km^2)	Total length (Km)	Length inside Odisha (Km)
Mahanadi	1,41,134	65,628	851	494
Brahmani	39,116	22,516	765	461
Baitarani	14,218	13,482	440	360
Rushikulya	8,963	8,963	175	175
Vansadhara	11,377	8,960	239	176
Nagvali	9,275	4,500	217	125
Kolab	**20,427**	**10,300**	**343**	**270**
Indrāvati	41,700	7,400	530	167
Bahuda	1,118	890	96	74
Subarnarekha	19,277	2,983	446	81
Budhabalanga	4,838	4,838	199	199

**C.A* Indicates Catchment Area

Table 2 Orissa—industrial effluents discharge into rivers [59]

Sl. No.	Name of the river	Industrial effluent [kilolitre per day/KLD]
1	Mahanadi—Sambalpur	736
	– Cuttack	2,780
	– Paradeep	5,280
2	Brahmani	9,480
3	Kathajodi	750
4	Kuakhai	1,000
5	Daya	1,570
6	Rushikulya	2,180
7	Baitarani	5,360
8	Bansadhara	630
9	Bahuda	580
10	Subarnarekha	460
11	Budhabalanga	530
12	**Kolab**	**2,900**

Pollution of natural systems by toxic metals is a global environmental problem of increasing importance [8]. The study of toxic and trace metals in the environment is more important than other pollutants due to their non-biodegradable nature, accumulated properties, and their long biological half-life. Once they enter the environment, it is difficult to completely remove them from the environment. The effluent discharges of the rivers in Odisha are detailed in Table 2.

Heavy metals are known to have atomic weights varying from 22 to 92 in the Periodic Table from period 3 to 7 in contrast to trace metals which have <0.01% of mass of organism [9]. Environmental Protection Agency (EPA) has designated 8 metals (arsenic, cadmium, chromium, copper, lead, mercury, nickel and zinc) on the priority list of toxicity on human beings. It is being reported that lead in excess lead to brain and kidney disorders (otherwise known as fanconic syndrome), zinc causes anaemia and human cancer, copper causes depigmentation and cardio-vascular diseases etc. [10]. The degree of toxicity varies differently with respect to different metals in marine and freshwater ecosystems [11, 12]. Apart from human beings, fishes are also affected by metal toxicity and thereby act as good indicators for heavy metal [13].

Metal accumulation in fishes depends on the phenomenon of uptake and excretion [14]. Gills are the most active site of accumulation because of its increased surface area and binding efficiency on cell surfaces. However, this adsorption is dependent on the attraction of metal to specific ligands. Organs like liver and kidney secrete specific metal binding proteins where the metals get accumulated [15]. The three possible ways of metal uptake by fishes are: (i) the body surface i.e. through skin (ii) the gills, and (iii) the alimentary tract. Reports have proved that the mucus secretion in fish bodies prevent their skin from uptaking heavy metals [16]. According to Singh

[17], water parameters also play a pivotal role in accumulation pattern of heavy metals in fishes owing to the different aquatic ecosystems (rivers, lakes or oceans). However, no comprehensive work has been done in the fishes found in Kolab River in Koraput region of Odisha. The present research programme is an attempt to scan the finfish tissues and intestine of (*Labeo rohita* and *Catla catla*) in connection to bioaccumulation of heavy metals collected from upper Kolab and lower Kolab region of the river during pre monsoon season.

2 Materials and Methods

2.1 Study Area

The Koraput district lies at 17.4°–20.7° N latitude and 81.24°–84.2° E longitude. The district experiences minimum 12 °C and maximum 38 °C temperature, with three seasons are i.e. summer, winter and rainy. Summer occurs only for 1–2 months (preferably during April and may), rainy season lasts from June to October and winter is from November to March and winter season is generally longer than other parts of Odisha. The annual rainfall is 1,522 mm. The area has an undulating topography with hills, valleys, waterfalls and springs. The site selected for the study has the following characteristics (Fig. 1).

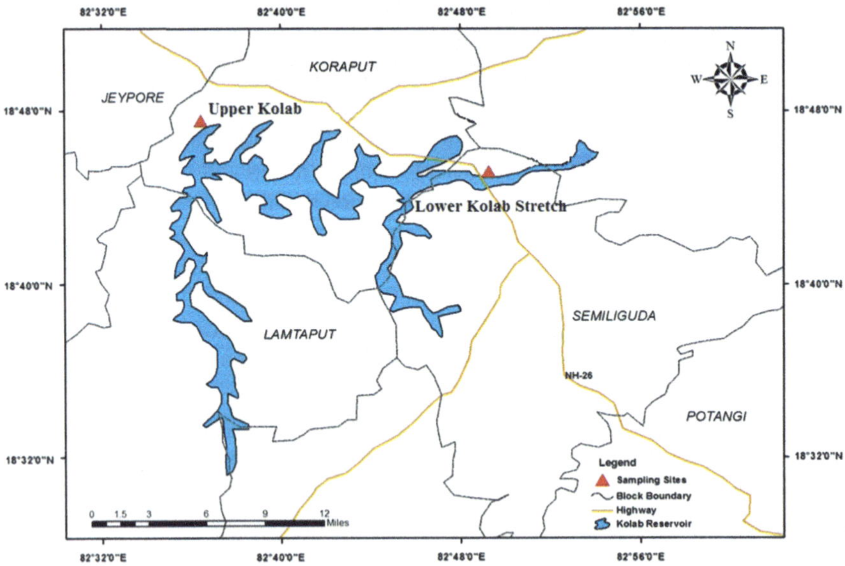

Fig. 1 Map of Kolab river and its tributaries showing the sampling sites

The river Kolab, known as "Sabari" in its lower reaches, forms a tributary of river Godavari. The river originates from the Eastern Ghats in Koraput district at an altitude of 1200 m and descends to 810 m from mean sea level at dam site near village Karanga traversing about 78 km with average bed slope of 6.25 m/km [18]. The catchment area intercepted at the dam site is 1630 km^2. This is bounded by latitude 18° 23′ and 18° 47′ N and longitude 82° 32′ N and 83° 32′ E. The Kolab reservoir is situated in the Kolab River in Koraput district of Odisha. Koraput is one of the most vulnerable districts of the country depending on delicate ecology and tribal socio-economy. The topography of the region is hilly, undulating and rolling. The degraded soils with exposed rocks resulted from severe erosion is the common landscape.

2.2 *Data Collection for Finfish Diversity Study*

Monthly fish data was collected during April–May, 2015 with the help of local fishermen of the surrounding villages belonging to Jagannath Sagar Primary Fisherman Co-operative Society (PFCS) registered under Government of Odisha located at Jeypore, Koraput and Maa Mangala PFCS, located at Sunabeda, Koraput registered under Government of Odisha. In order to record fishes available in the dam, direct physical count method was used. Shannon-Weiner species diversity index and Index of Dominance was calculated for both the study sites as per standard formulae:

$$\bar{\mathrm{H}} = -\sum_{i=1}^{s}\left(\frac{ni}{N} \times \ln \frac{ni}{N}\right)$$

$$\mathrm{Id} = \sum [\mathrm{ni}(\mathrm{ni}-1)/\mathrm{N}(\mathrm{N}-1)]$$

where,

$\bar{\mathrm{H}}$ Shannon-Weiner species diversity index
Id Index of Dominance
n_i Importance value of each species
N Total no. of importance value

Water samples were collected at the same time from the respective sites and in-situ analysis of water temperature, pH and turbidity was analyzed by standard methods. The identification of the fishes was done using standard literatures [7, 19] and IUCN categories were documented through website (http://www.iucnredlist.org). The selected fishes (*Labeo rohita and Catla catla*) were subjected to heavy metal analysis, brought to the laboratory were preserved in ice box. The dissected muscles and intestines were dried at 110 °C to remove moisture content and kept ready for heavy metal analysis.

2.3 Heavy Metal Analysis

Selected fish species viz. *Labeo rohita* and *Catla catla* (about 8 in number) were brought to the laboratory and dissected for muscles and intestine with the help of clean stainless-steel blade [20]. The samples were placed in petridish and oven dried at 110 °C for 48 h. 0.5 g of each dry tissue was weighed and put in a separate test tube. Then nitric acid (HNO_3) and perchloric acid ($HClO_4$) in 2:1 ratio was added to each sample and left over-night at room temperature. Samples were digested in a water bath set to 100 °C water boiling temperature and the content was warmed for about 2 h until all the tissues dissolved. The digested sample was cooled to room temperature followed by the addition of 5 ml. of distilled water. The solution was transferred to 25 ml. volumetric flask and made up to mark with 1% HNO_3. Heavy metal concentration (in μg/g dry-weight) were determined using Atomic Absorption Spectrophotometer (AAS). The cathode lamps used were of copper (wavelength nm 324.7, lamp current mA 3.0), lead (wavelength nm 217.0, lamp current mA 5.0) and zinc (wavelength nm 213.9, lamp current mA 5.0).

2.4 Statistical Analysis

The data collected were expressed as Mean ± Standard deviation. In order to find out the inter relationship between the physico-chemical parameters and the heavy metal content in fish tissues (muscle) and intestine, Spearman's correlation coefficient was computed using SPSS 13.1.

3 Results

3.1 Water Quality

The temperature of Kolab river water samples of upstream (site 1) and downstream (site 2) during study period varied from 29.7 ± 0.48 °C at site 1 to 31.2 ± 0.03 °C at site 2 respectively. The downstream water sample has slightly higher temperature than the upstream water samples during the study period. The pH of Kolab river water samples of upstream (site 1) and downstream (site 2) ranges from 7.5 ± 0.17 to 7.3 ± 0.13 respectively and the turbidity values of Kolab river water samples during the study period range from 6.2 ± 0.03 NTU at site 1 to 8.1 ± 0.1 NTU at site 2 respectively (Table 6).

Statistical analysis computed with the data sets at sites 1 and 2 revealed that temperature has a negative correlation with turbidity at 1% level of significance ($r = -0.9998$, $p < 0.01$). pH and temperature showed a positive correlation (r =

0.9596, $p < 0.01$) which is a significant characteristic feature of water during pre-monsoon season. This proves that with an increase in temperature, pH of the water also increases and vice versa. The negative relationship between temperature and turbidity shows that in pre-monsoon season, there is no run-off from the adjacent land masses; hence the water is clean and clear with respect to other seasons. The decreased turbidity level allows a better growth of phytoplankton which is a live feed for the fishes. This typical characteristic of water has been witnessed both at sites 1 and 2 respectively.

3.2 *Fish Diversity*

The study revealed the occurrence of nineteen (19) freshwater fish species belonging to five (5) orders, eight (8) families in two tributaries of Kolab River in Koraput district of Odisha i.e. Upper Kolab and Lower Kolab stretches. At site 1 (Upper Kolab stretch), fish diversity was represented under five (5) Orders i.e. Order Cypriniformes, Order Osteoglossiformes, Order Siluriformes, and Order Perciformes. Order Cypriniformes was dominant with ten (10) fish species under a single Family Cyprinidae (*Labeo rohita, Labeo bata, Cirrhinus mrigala, Cirrhinus reba, Puntius terio, Ctenopharyngodon idella, Cyprinus carpio, Esomus danricus, Catla catla, Amblypharyngodon mola*). Order Perciformes represented two (2) families; each Family represented by one (1) species each i.e. Family Anabantidae (*Anabas testudineus*) and Family Gobiidae (*Glossogobius giuris*). Order Siluriformes represented three (3) species under three (3) families, each family represented by one (1) species each i.e. Schilbeidae (*Ailia coila*), Siluridae (*Ompok bimaculatus*) and Heteropneustidae (*Heteropneustes fossilis*) (Fig. 2).

At site 2 (Lower Kolab stretch) fish diversity was represented under five (5) Orders i.e. Order Cypriniformes, Order Channiformes, Order Siluriformes, Order Osteoglossiformes, Order Perciformes. Order Cypriniformes consisted seven (7)

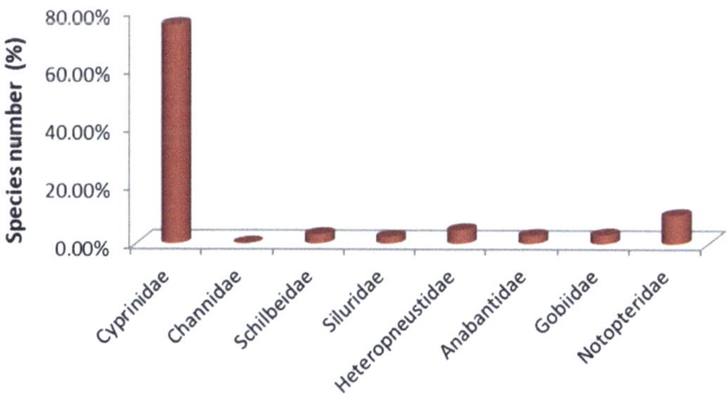

Fig. 2 Family wise fish species abundance in site 1 of Kolab river

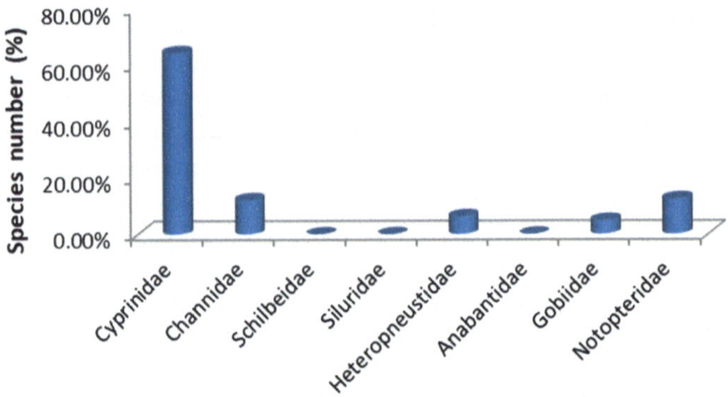

Fig. 3 Family wise fish species abundance in site 2 of Kolab river

species under family Cyprinidae represented by seven (7) species (*Labeo rohita, Catla catla, Puntius terio, Hypophthalmichthys molitrix, Esomus danricus, Labeo calbasu, Amblypharyngodon mola*). Order Channiformes represented two (2) species under single Family Channidae (*Channa punctatus, Channa orientalis*). Order Siluriformes representing one (1) species under Family Heteropneustidae (*Heteropneustes fossilis*). Order Perciformes represented one (1) species under the Family Anabantidae (*Anabas testudineus*). Order Osteoglossiformes representing one (1) species under the Family Notopteridae (*Notopterus notopterus*) (Fig. 3).

The list of fish species recorded under present IUCN category is shown in Table 3. Almost all the fishes recorded are edible excepting two (2) species which are ornamental (*Esomus danricus, Glossogobius giuris*). Moreover, it was recorded that fourteen (14) species i.e. *Labeo rohita, Labeo bata, Labeo calbasu, Catla catla, Cirrhinus mrigala, Cirrhinus reba, Puntius terio, Esomus danricus, Amblypharyngodon mola, Channa punctata, Heteropneustes fossilis, Anabas testudineus, Glossogobius giuris, Notopterus notopterus* comes under Least Concern category followed by three (03) species i.e. *Hypophthalmichthys molitrix, Ailia coila* and *Ompok bimaculatus* comes under Near Threatened category and only two (02) species i.e. *Cyprinus carpio* and *Channa orientalis* are coming under Vulnerable category respectively.

The catch statistics per day per catch at both the sites is represented in Tables 4 and 5 respectively. The diversity index was found to be higher (mean $\bar{H} = 2.28$ at site 1 and mean $\bar{H} = 1.96$ in site 2) in the months of April at both the sites. The relative abundance calculated at both the sites revealed maximum diversity of *Labeo rohita* (16.9%) and (21.32%) at sites 1 and 2 respectively, followed by *Catla catla* (13.3%) and (15.47%) at sites 1 and 2 respectively (Figs. 2 and 3). This is also supported by the Index of Dominance calculated for April and May at sites 1 and 2, where Id is higher at site 2 (mean Id = 0.16) than site 1 (mean Id = 0.14) respectively (Tables 4 and 5). The family wise representation of the species dominance is shown in Figs. 2 and 3.

Table 3 List of Fishes with their IUCN category of Kolab River

Sl. No.	Zoological name	Edible/ ornamental fish	IUCN status	Site 1	Site 2
1	*Labeo rohita*	Edible	LC	+	+
2	*Labeo bata*	Edible	LC	+	−
3	*Labeo calbasu*	Edible	LC	−	+
4	*Catla catla*	Edible	LC	+	+
5	*Cirrhinus mrigala*	Edible	LC	+	−
6	*Cirrhinus reba*	Edible	LC	+	−
7	*Puntius terio*	Edible	LC	+	+
8	*Cyprinus carpio*	Edible	VU	+	−
9	*Hypophthalmichthys molitrix*	Edible	NT	+	+
10	*Esomus danricus*	Ornamental	LC	+	+
11	*Amblypharyngodon mola*	Edible	LC	+	−
12	*Channa punctata*	Edible	LC	−	+
13	*Channa orientalis*	Edible	VU	−	+
14	*Ailia coila*	Edible	NT	+	−
15	*Ompok bimaculatus*	Edible	NT	+	−
16	*Heteropneustes fossilis*	Edible	LC	+	+
17	*Anabas testudineus*	Edible	LC	+	−
18	*Glossogobius giuris*	Ornamental	LC	+	+
19	*Notopterus notopterus*	Edible	LC	+	+

VU Vulnerable, *NT*-Near threatened, *LC* Least Concern
(+) Sign indicates the presence of the species in the above represented particular area and (-) sign indicates the absence of the species in the above represented particular area

3.3 Heavy Metal Accumulation in Fishes

Of the three metals studied in the present programme, zinc and copper are essential elements while lead is a non-essential element for most of the living organisms. Irrespective of species the pattern of concentration of different metals was comparable in the present programme. Highest concentration of zinc was observed in both the species followed by copper and lead excepting muscle of *Labeo rohita* at site 1 where Zn (5.973 ± 0.366 ppm dry weight) > Pb (0.949 ± 0.216 ppm dry weight) > Cu (0.649 ± 0.016 ppm dry weight) (Figs. 4, 5, 6, 7, 8, 9, 10 and 11). The pattern of accumulation was in the order of Zn > Cu > Pb both for muscle and intestine in case of both the species. Zinc, copper and lead accumulated as per the order *Catla catla* > *Labeo rohita* at site 1 in muscle, whereas it is vice versa in case of intestine. In case of site 2, zinc accumulated as per the order *Labeo rohita* > *Catla catla* both in case of muscle and intestine. In case of copper, the accumulation pattern was of the

Table 4 Catch statistics of fin fishes (per catch per day) during April–May 2015 at site 1

No of days	1	2	3	4	5	6	7	8	9	10	11	12	13	14	15
Labeo rohita	32	31	30	33	34	32	31	35	33	30	28	26	24	22	23
Catla catla	27	26	28	25	23	27	25	30	28	26	25	24	21	19	20
Cirrhinus mrigala	04	03	05	00	01	01	03	06	02	08	06	04	00	00	03
Cirrhinus reba	15	13	11	12	14	16	18	17	20	19	19	16	15	11	13
Puntius terio	14	12	10	11	13	15	17	16	18	20	17	15	13	10	11
Cyprinus carpio	12	10	10	09	11	13	15	12	14	13	11	10	09	08	11
Hypophthalmichthys molitrix	13	11	11	10	12	14	15	13	12	14	12	11	10	09	12
Esomus danricus	08	06	06	07	08	07	06	04	00	03	02	00	03	04	05
Amblypharyngodon mola	15	13	14	17	16	18	20	19	17	16	14	12	13	11	10
Labeo bata	07	05	06	04	03	06	05	02	03	01	00	00	04	00	02
Ailia coila	09	08	06	05	07	04	03	05	04	02	00	01	00	02	03
Ompok bimaculatus	07	00	00	03	04	02	00	01	03	04	02	00	00	01	04
Heteropneustes fossilis	09	08	06	05	07	04	06	08	00	07	08	05	03	04	02
Anabas testudineus	08	07	05	04	03	03	01	02	01	01	00	00	03	01	03
Glossogobius gluris	06	05	03	04	02	03	03	01	02	01	01	02	02	03	05
Notopterus notopterus	17	16	19	20	15	14	14	12	16	14	13	11	15	17	16
Mean $\overline{H}_{30}$ of April = 2.37 Mean Id_{30} of April = 0.138															
No of days	16	17	18	19	20	21	22	23	24	25	26	27	28	29	30
Labeo rohita	21	19	17	16	14	12	10	08	07	05	06	03	02	01	01
Catla catla	17	15	13	14	11	09	07	05	06	04	02	03	01	01	01

(continued)

Table 4 (continued)

No of days	16	17	18	19	20	21	22	23	24	25	26	27	28	29	30
Cirrhinus mrigala	05	02	02	04	07	00	01	03	00	03	01	02	02	00	01
Cirrhinus reba	12	14	09	07	05	08	06	07	04	05	05	03	00	01	00
Puntius terio	10	12	08	06	06	07	05	04	05	03	03	02	00	01	01
Cyprinus carpio	07	06	04	04	02	03	03	01	01	00	00	01	01	00	00
Hypophthalmichthys molitrix	08	07	05	06	03	04	04	02	01	01	00	00	01	00	01
Esomus danricus	01	00	03	02	00	00	03	01	01	00	02	00	01	01	00
Amblypharyngodon mola	11	12	13	10	14	12	09	07	04	05	04	02	01	00	01
Labeo bata	02	03	01	00	02	00	01	02	01	00	01	00	00	00	01
Ailia coila	00	04	00	00	01	00	03	00	02	01	00	00	01	00	00
Ompok bimaculatus	03	04	02	01	00	00	01	02	02	00	00	01	00	01	00
Heteropneustes fossilis	02	06	05	02	03	01	02	01	01	02	02	01	01	00	01
Anabas testudineus	05	04	03	02	02	00	00	01	02	00	03	00	01	00	00
Glossogobius gluris	04	03	02	04	00	01	00	03	00	02	00	01	00	00	01
Notopterus notopterus	13	12	13	10	11	08	06	07	06	03	04	02	00	01	00
No of days	1	2	3	4	5	6	7	8	9	10	11	12	13	14	15
Labeo rohita	01	03	05	07	06	04	06	05	05	08	09	07	07	06	08
Catla catla	02	03	04	02	05	03	06	04	03	02	05	07	09	08	07
Cirrhinus mrigala	01	02	00	00	01	01	00	00	01	00	01	00	00	01	00
Cirrhinus reba	00	01	00	01	00	00	01	00	00	01	00	01	00	01	00
Puntius terio	01	02	04	03	02	01	01	03	00	02	00	04	05	06	08

(continued)

Table 4 (continued)

No of days	1	2	3	4	5	6	7	8	9	10	11	12	13	14	15
Cyprinus carpio	00	01	00	00	01	01	00	01	00	00	01	00	01	00	01
Hypophthalmichthys molitrix	01	00	01	00	01	01	01	00	00	01	01	00	00	01	00
Esomus danricus	00	00	01	01	00	00	01	00	00	02	00	01	00	00	01
Amblypharyngodon mola	02	02	01	02	01	01	02	00	00	01	01	02	01	00	01
Labeo bata	01	00	01	01	00	00	01	00	00	00	00	02	01	01	00
Ailia coila	00	00	01	01	00	00	01	01	00	00	00	01	02	01	00
Ompok bimaculatus	00	01	00	00	01	02	00	00	00	01	00	00	01	01	02
Heteropneustes fossilis	01	00	01	00	01	01	02	01	02	00	01	02	00	01	02
Anabas testudineus	00	01	00	00	01	00	00	00	02	00	01	03	00	02	00
Glossogobius gluris	01	01	00	00	00	00	00	01	00	00	02	00	01	01	01
Notopterus notopterus	05	03	04	00	01	02	02	01	02	01	00	00	01	00	01

Mean $\overline{H}_{30}$ of May = 2.19
Mean Id_{30} of May = 0.142

No of days	16	17	18	19	20	21	22	23	24	25	26	27	28	29	30	31
Labeo rohita	10	09	07	10	13	12	11	13	12	14	16	18	17	19	20	18
Catla catla	09	10	08	09	11	13	14	13	10	11	10	12	13	15	18	20
Cirrhinus mrigala	00	00	01	02	02	03	06	05	07	05	08	07	09	08	09	11
Cirrhinus reba	02	01	02	03	02	05	07	09	07	08	10	09	10	11	13	15
Puntius terio	07	05	04	06	07	08	06	07	07	09	10	09	12	10	13	14
Cyprinus carpio	01	00	02	02	01	04	05	07	06	07	06	08	10	12	14	13
Hypophthalmichthys molitrix	00	01	02	02	04	03	07	06	05	09	08	10	12	13	14	13

(continued)

Table 4 (continued)

No of days	16	17	18	19	20	21	22	23	24	25	26	27	28	29	30	31
Esomus danricus	03	02	00	00	01	02	04	06	00	07	05	04	06	07	09	11
Amblypharyngodon mola	02	01	02	04	03	05	08	07	06	08	10	12	11	15	17	16
Labeo bata	00	02	01	00	00	02	03	00	04	03	00	02	05	03	04	09
Ailia coila	00	01	00	01	02	00	00	04	00	00	05	07	03	04	06	08
Ompok bimaculatus	00	00	00	01	04	00	00	05	00	00	03	00	00	06	04	07
Heteropneustes fossilis	00	00	01	01	00	02	03	04	01	05	04	06	08	07	06	07
Anabas testudineus	00	01	00	02	00	03	00	04	00	00	04	06	00	00	05	06
Glossogobius gluris	02	00	00	00	01	01	02	03	00	00	05	00	04	00	04	05
Notopterus notopterus	02	01	01	02	03	05	08	07	10	09	11	12	10	13	12	14

Table 5 Catch statistics of fin fishes (per catch per day) during April–May 2015 at site 2

No of days	1	2	3	4	5	6	7	8	9	10	11	12	13	14	15
Labeo rohita	30	29	28	32	33	31	30	32	31	27	26	25	24	23	21
Catla catla	25	24	23	22	22	20	21	26	27	25	22	21	24	19	17
Puntius terio	13	11	10	11	12	14	16	15	17	20	16	15	11	10	13
Hypophthalmichthys molitrix	12	10	11	09	11	13	14	12	11	13	11	10	10	08	09
Esomus danricus	08	07	06	06	07	08	05	04	00	03	02	01	03	00	05
Labeo calbasu	04	02	04	00	01	00	02	05	02	00	06	03	00	01	03
Channa punctata	08	07	09	05	04	06	08	10	12	08	05	06	04	04	05
Heteropneustes fossilis	07	10	09	07	08	04	06	03	05	07	03	05	06	04	02
Glossogobius giuris	07	05	03	04	01	03	02	01	03	01	04	00	01	03	03
Notopterus notopterus	17	16	18	16	15	13	14	11	15	12	13	10	14	13	15
Channa orientalis	08	05	06	04	03	01	00	01	02	01	03	00	01	04	03

Mean $\overline{H}_{30}$ of April = 1.98
Mean Id_{30} of April = 0.162

No of days	17	18	19	20	21	22	23	24	25	26	27	28	29	30
Labeo rohita	17	15	14	15	12	10	08	07	05	05	03	04	02	01
Catla catla	16	12	13	10	10	07	05	06	04	02	01	02	00	01
Puntius terio	13	08	06	05	07	04	04	03	02	00	01	00	00	01
Hypophthalmichthys molitrix	07	04	06	02	04	00	02	01	01	01	00	00	01	01
Esomus danricus	00	00	02	00	01	03	00	01	00	02	00	00	01	01
Labeo calbasu	02	00	04	00	00	01	03	00	03	01	00	02	00	01
Channa punctata	03	01	00	02	03	00	04	00	03	00	01	00	01	00

(continued)

Table 5 (continued)

No of days	17	18	19	20	21	22	23	24	25	26	27	28	29	30
Heteropneustes fossilis	01	05	02	02	01	02	03	01	04	02	00	00	02	01
Glossogobius giuris	05	02	02	00	01	00	00	03	01	00	02	00	01	00
Notopterus notopterus	11	13	10	09	08	05	06	04	03	02	01	00	01	01
Channa orientalis	05	00	02	01	01	00	00	03	02	00	03	01	02	01

No of days	1	2	3	4	5	6	7	8	9	10	11	12	13	14	15
Labeo rohita	01	03	02	05	04	03	06	05	04	07	07	06	05	06	08
Catla catla	02	01	03	02	04	02	06	04	01	02	03	05	08	07	07
Puntius terio	01	02	03	03	01	01	00	02	00	01	00	03	04	00	05
Hypophthalmichthys molitrix	01	00	01	02	01	00	01	01	00	00	01	00	01	01	00
Esomus danricus	01	00	00	01	00	01	00	00	02	02	01	00	00	02	03
Labeo calbasu	00	02	01	00	00	01	01	01	02	01	00	00	00	01	03
Channa punctata	01	00	01	02	02	01	00	00	02	03	00	01	03	00	01
Heteropneustes fossilis	02	01	01	03	00	02	00	01	00	01	01	00	00	03	00
Glossogobius giuris	00	01	00	01	00	00	01	01	00	02	02	00	00	01	00
Notopterus notopterus	01	03	02	03	04	03	01	01	00	01	02	00	01	02	01
Channa orientalis	01	02	00	01	01	00	00	01	01	03	02	01	00	00	02

Mean $\overline{H}_{30}$ of May = 1.95
Mean Id_{30} of May = 0.166

No of days	16	17	18	19	20	21	22	23	24	25	26	27	28	29	30	31
Labeo rohita	09	09	07	10	12	10	11	13	10	13	15	18	16	17	18	19
Catla catla	08	09	03	10	11	12	14	12	10	09	11	12	12	14	17	18

(continued)

Table 5 (continued)

No of days	16	17	18	19	20	21	22	23	24	25	26	27	28	29	30	31
Puntius terio	06	04	04	03	02	05	06	07	07	08	10	09	11	09	12	11
Hypophthalmichthys molitrix	01	01	01	03	03	02	04	05	07	06	07	09	12	11	13	11
Esomus danricus	03	02	01	00	04	02	05	06	04	02	05	06	08	07	09	10
Labeo calbasu	00	00	01	04	03	00	06	00	02	01	00	03	05	01	03	04
Channa punctata	01	03	00	00	04	03	05	06	02	03	05	07	09	08	08	09
Heteropneustes fossilis	00	01	01	02	01	03	03	04	02	04	06	06	07	09	06	08
Glossogobius giuris	02	00	01	00	02	03	04	03	02	01	05	06	04	02	03	06
Notopterus notopterus	00	01	03	02	04	07	06	07	11	15	14	13	11	15	13	14
Channa orientalis	02	01	00	03	02	03	05	04	03	01	05	06	05	04	05	07

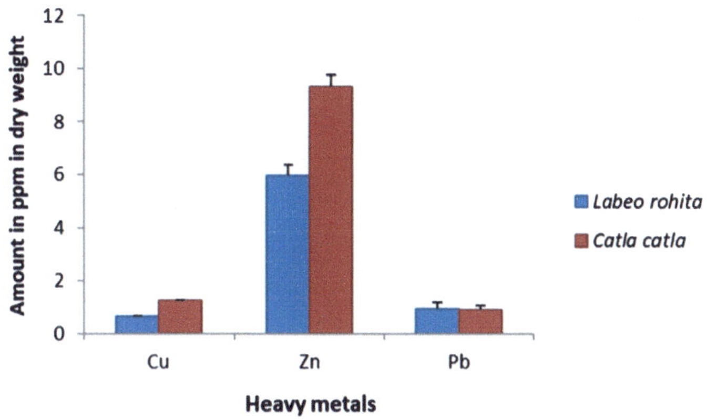

Fig. 4 Heavy metal accumulation in muscle cells of *Labeo rohita* and *Catla catla* in site 1

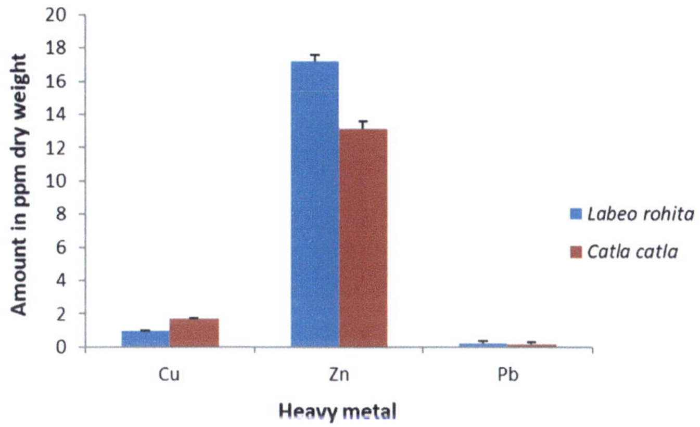

Fig. 5 Heavy metal accumulation in intestine cells of *Labeo rohita* and *Catla catla* in site 1

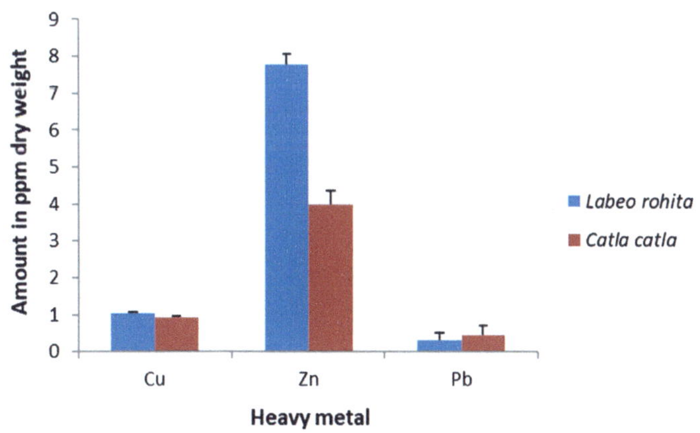

Fig. 6 Heavy metal accumulation in muscle cells of *Labeo rohita* and *Catla catla* in site 2

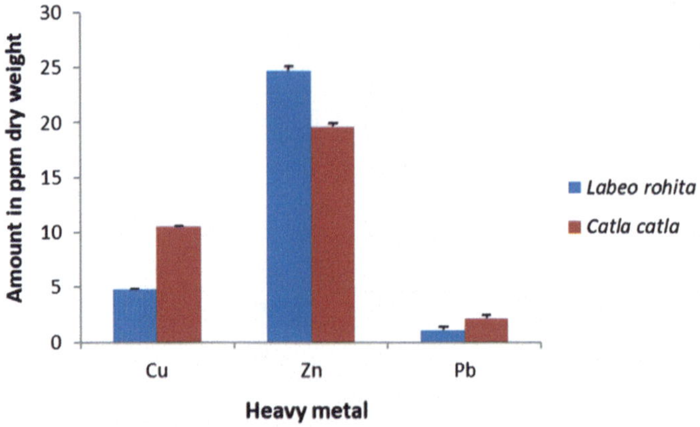

Fig. 7 Heavy metal accumulation in intestine cells of *Labeo rohita* and *Catla catla* at site 2

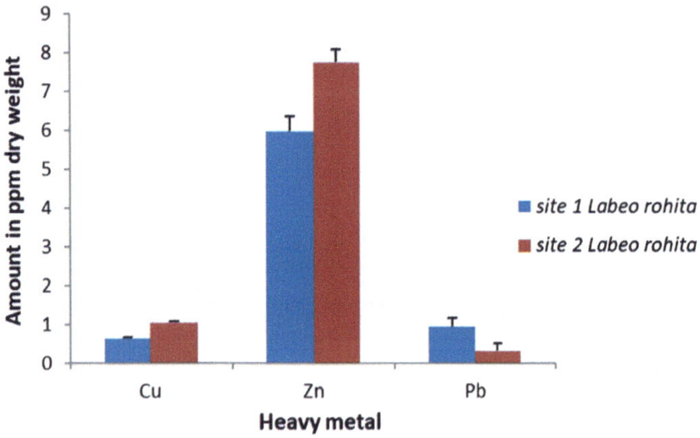

Fig. 8 Heavy metal accumulation in muscle cells of *Labeo rohita* in site 1 and 2

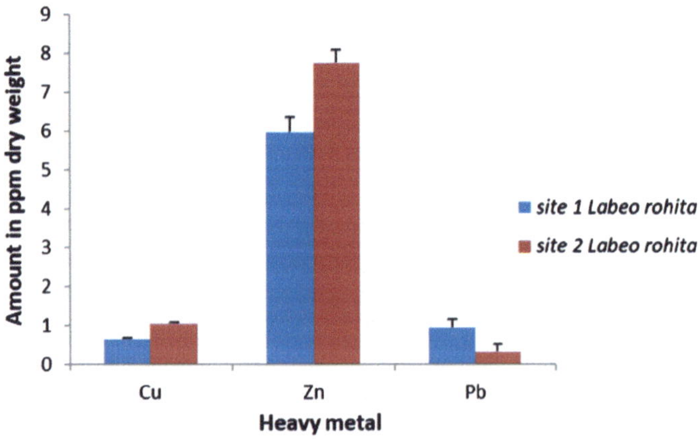

Fig. 9 Heavy metal accumulation in intestine cells of *Labeo rohita* in site 1 and 2

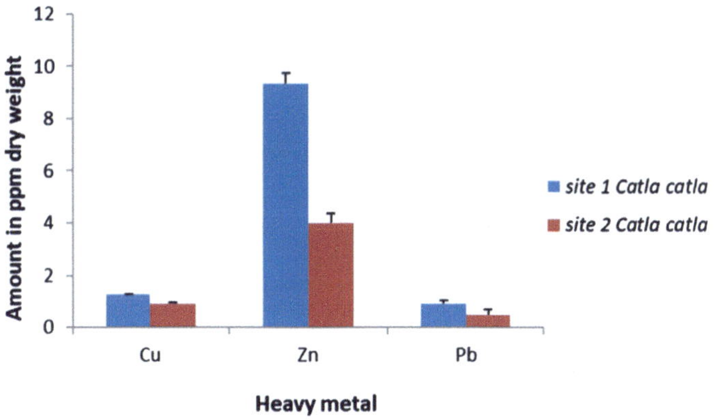

Fig. 10 Heavy metal accumulation in muscle cells of *Catla catla* in site 1 and 2

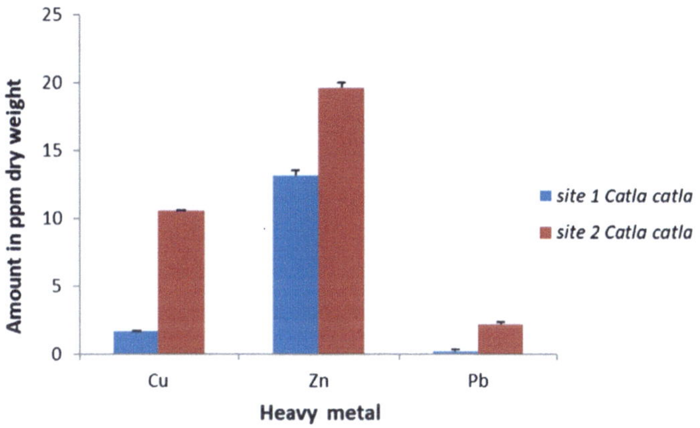

Fig. 11 Heavy metal accumulation in intestine cells of *Catla catla* in site 1 and 2

order *Catla catla* > *Labeo rohita* at site 1 in both muscle and intestine and in site 2 of intestine except copper in muscle at site 2 (Figs. 4, 5, 6, 7, 8, 9, 10 and 11). Lead accumulated as per the order *Labeo rohita* > *Catla catla* in case of muscle and intestine at site 1 whereas it is just the reverse in case of site 2 (Figs. 4, 5, 6, 7, 8, 9, 10 and 11).

The accumulation of zinc, copper and lead in muscle and intestine of both the fishes reveals a significant negative correlation ($p < 0.01$) with temperature and pH at sites 1 and 2 respectively excepting muscle of *Labeo rohita* at site 1. With respect to turbidity, the zinc, copper and lead accumulation showed a positive correlation ($p < 0.01$) at sites 1 and 2 in case of both the species excepting muscle in *Labeo rohita* at site 1 (Tables 6, 7a, 7b, 8a, 8b, 9a, 9b, 10a, 10b). Considering the accumulation pattern of zinc, copper and lead in muscles and intestine of *Labeo rohita* and *Catla catla*, it has been witnessed that all the metals in fish muscle and intestine are within

Table 6 Values of physico-chemical variables measured at sites 1 and 2 (Mean ± SD) of Kolab River water

Sl No	Parameters	Site 1	Site 2	Acceptable limit (as per BIS Specification for drinking water)
1	Temperature	29.7 ± 0.48 °C	31.2 ± 0.03 °C	5 °C
2	pH	7.5 ± 0.17	7.3 ± 0.13	7.0
3	Turbidity	6.2 ± 0.03 NTU	8.1 ± 0.1 NTU	6.5–8.5 NTU

Table 7a Inter-relationship between turbidity, temperature, pH with heavy metals (Cu, Zn and Pb) in muscle of *Labeo rohita* at site 1

	Turbidity	Temperature	pH	Cu	Zn	Pb
Turbidity	1					
Temperature	−0.99983	1				
pH	−0.95423	0.959634	1			
Cu	−0.97812	0.981822	0.995572	1		
Zn	0.61859	−0.60385	−0.35529	−0.44158	1	
Pb	−0.99942	0.998618	0.943529	0.970492	−0.6449	1

Table 7b Inter-relationship between turbidity, temperature, pH with heavy metals (Cu, Zn and Pb) in intestine of *Labeo rohita* at site 1

	Turbidity	Temperature	pH	Cu	Zn	Pb
Turbidity	1					
Temperature	−0.95544	1				
pH	−0.83385	0.959634	1			
Cu	0.929786	−0.99701	−0.97849	1		
Zn	0.911293	−0.99223	−0.98717	0.998876	1	
Pb	0.960769	−0.99983	−0.95423	0.995402	0.989743	1

Table 8a Inter-relationship between turbidity, temperature, pH with heavy metals (Cu, Zn and Pb) in Muscle of *Catla catla* at site 1

	Turbidity	Temperature	pH	Cu	Zn	Pb
Turbidity	1					
Temperature	−0.95544	1				
pH	−0.83385	0.959634	1			
Cu	0.93472	−0.99797	−0.97558	1		
Zn	0.846154	−0.96577	− 0.99974	0.980316	1	
Pb	0.980316	−0.99491	−.92642	0.986486	0.93472	1

Table 8b Inter-relationship between turbidity, temperature, pH with heavy metals (Cu, Zn and Pb) in Intestine of *Catla catla* at site 1

	Turbidity	Temperature	pH	Cu	Zn	Pb
Turbidity	1					
Temperature	−0.95544	1				
pH	−0.83385	0.959634	1			
Cu	0.922613	−0.99536	−0.98224	1		
Zn	0.891042	−0.98533	−0.99355	0.997176	1	
Pb	0.980316	−0.99491	−0.92642	0.980609	0.963123	1

Table 9a Inter-relationship between turbidity, temperature, pH with heavy metals (Cu, Zn and Pb) in Muscle of *Labeo rohita* at site 2

	Turbidity	Temperature	pH	Cu	Zn	Pb
Turbidity	1					
Temperature	−0.95544	1				
pH	−0.83385	0.959634	1			
Cu	0.986241	−0.99109	−0.91363	1		
Zn	0.922613	−0.99536	−0.98224	0.973684	1	
Pb	0.846154	−0.96577	−0.99974	0.922613	0.986241	1

Table 9b Inter-relationship between turbidity, temperature, pH with heavy metals (Cu, Zn and Pb) in Intestine of *Labeo rohita* at site 2

	Turbidity	Temperature	pH	Cu	Zn	Pb
Turbidity	1					
Temperature	−0.95544	1				
pH	−0.83385	0.959634	1			
Cu	0.986241	−0.99109	−0.91363	1		
Zn	0.891042	−0.98533	−0.99355	0.953821	1	
Pb	0.960769	−0.99983	−0.95423	0.993399	0.981981	1

Table 10a Inter-relationship between turbidity, temperature, pH with heavy metals (Cu, Zn and Pb) in Muscle of *Catla catla* at site 2

	Turbidity	Temperature	pH	Cu	Zn	Pb
Turbidity	1					
Temperature	−0.95544	1				
pH	−0.83385	0.959634	1			
Cu	0.986241	−0.99109	−0.91363	1		
Zn	0.980316	−0.99491	−0.92642	0.999466	1	
Pb	0.960769	−0.99983	−0.95423	0.993399	0.996616	1

Table 10b Inter-relationship between turbidity, temperature, pH with heavy metals (Cu, Zn and Pb) in Intestine of *Catla catla* at site 2

	Turbidity	Temperature	pH	Cu	Zn	Pb
Turbidity	1					
Temperature	−0.95544	1				
pH	−0.83385	0.959634	1			
Cu	0.995871	−0.97829	−0.88052	1		
Zn	0.922613	−0.99536	−0.98224	0.953821	1	
Pb	0.986241	−0.99109	−0.91363	0.997176	0.973684	1

the permissible limits of FAO/WHO [21]. However, there is a signal of caution in case of lead accumulation in intestine of *Catla Catla,* where it is more than permissible level. However, long term studies are required to understand the seasonal variation in accumulation and its effect on human health.

4 Discussion

4.1 Water Quality

The developmental activities, pollution and overfishing are responsible for degradation of water quality [22]. Rivers play an important role in linking habitats of plants and animals in the extent that they cover. The liaison work of rivers between the upstream and downstream areas including the banks of the river renders a chance to inhabit species among migrants from other regions. Such a combination demands a diversity measurement objective and a proper understanding of populations for ecosystem studies. Analyzing the role of fishery resource use on community livelihoods is important for an effective management framework. The river Kolab, the only source of drinking water in and around Koraput mine area of Odisha, is located in the vicinity of HAL and NALCO, two major industries of the Government of India. The physico-chemical variables like temperature, pH and turbidity are primary factors for sustenance of fish diversity of river. Hence, attention has been given for monitoring the selected physico-chemical parameters on which the metal concentration in aquatic phase also varies. The fly-ash generation, industrial coolant discharge, pipe line corrosion, municipal sewage, dumping ores from mines and paper and pulp board mills are some of the major sources of zinc, copper and lead in the present study area which may be attributed to the presence of NALCO, HAL and BILT paper mill located in and around the river stretch. This is the major reason for availability of zinc, copper and lead in the muscle and intestine of selected fish species: *Labeo rohita* and *Catla catla.*

Table 11 Comparison of heavy metal concentrations in ppm dry weight in fin fishes collected from different parts of the world

Location	Species	Organs	Zn	Cu	Pb	References
River Nile, Egypt	*Clarias gariepinus*	Muscle	12.6–38.42	39.79–10.3	2.54–0.432	[60]
River Ravi, Pakistan	*Catla catla*	Muscle	–	4.03 ± 1.04	–	[33]
River Kabul, Pakistan	*Cirrhinus mrigala*	Muscle	–	–	41.0 ± 23.8	[51]
	Cyprinus carpio	Muscle	–	–	34.2 ± 23.5	
River Karnaphuli, Bangladesh	*Apocryptes bato*	Muscle	–	15.29 ± 3.82	15.22 ± 1.32	[50]
Taihu Lake, China	*Pelteobagrus fluvidraco*	Muscle	–	0.034 ± 0.002	0.052 ± 0.002	[53]
	Cyprinus carpio	Muscle	–	0.037 ± 0.002	0.087 ± 0.001	
River Sultej, Punjab	*Labeo rohita*	Muscle	36.29 ± 3.17	2.87 ± 0.65	4,32 ± 0.16	[55]
River Yamuna, Uttar Pradesh	*Labeo rohita*	Muscle	–	3.350–3.569	1.112–1.861	[36]
River Ganga, Uttar Pradesh	*Channa punctata*	Muscle	6.42–18.84	0.42–0.96	1.86–2.89	[54]
	Aorichthys aor	Muscle	8.62–14.86	0.80–1.20	2.46–3.89	
	Labeo rohita	Muscle	35.24 ± 12.35	6.27 ± 1,34	1.03 ± 0.23	[56]
	Puntius ticto	Muscle	12.35 ± 4.96	1.65 ± 0.44	1.73 ± 0.84	
	Labeo rohita	Muscle	25.36 ± 2.04	3.88 ± 0.15	1.12 ± 0.03	[57]
	Catla catla	Muscle	7.87 ± 2.58	15.24 ± 2.04	2.03 ± 0.11	
River Kallada, Kerala	*Catla catla*	Muscle	4.32	3.89	89.86	[52]
	Labeo rohita	Muscle	4.86	3.97	69.12	
Sukinda Valley, Jajpur, Odisha	*Guduia sp.*	Intestine	–	–	39.43–43.24	[49]
		Muscle	–	–	10.00–11.15	
	Channa punctata	Muscle	–	–	0.60–0.66	
		Intestine	–	–	10.03–11.33	
Hirakud Reservoir, Bargarh, Odisha	*Puntius sp.*	Muscle	23.78 ± 2.49	1.61 ± 0.62	44.3 ± 5.76	[58]

(continued)

Table 11 (continued)

Location	Species	Organs	Zn	Cu	Pb	References
Kolab River, Koraput, Odisha	*Puntius sp.*	Muscle	54.99 ± 5.22	4.17 ± 1,49	6.89 ± 2.14	[58]
	Labeo rohita	Muscle	7.77–5.97	0.649–1.053	0.317–0.949	This study
		Intestine	0.97–4.81	17.17–13.16	0.228–1.142	
	Catla catla	Muscle	4.02–9.32	0.932–1.255	0.468–0.895	
		Intestine	1.71–10.57	19.62–24.74	0.166–2.199	

4.2 Fish Diversity

The study of freshwater fishes in the Indian subcontinent has been limited to scattered works on commercial fisheries and even to some major river systems such as the Ganges and Yamuna. Of the 2500 species of fish identified in the Indian subcontinent, 930 are classified as freshwater species. Temporal and spatial accumulation of fish in a river is very dynamic. The fisheries diversity study in the Kolab river has thrown light on existing fisheries resource and the catch statistics over a period of 60 days (April–May 2015). The results on fish diversity in present research programme clearly shows the assemblage of higher diversity in the upstream region (site 1) owing to the fact that this area is comparatively a more pristine zone than the lower stretch of river (site 2), which goes hand-in-hand with the studies of Alam et al. [4] and Hossain et al. [23]. The IUCN categories mentioned in (Table 4) shows that the ecology of Kolab River is comparatively much better than the Mahanadi River of Odisha which has been studied by Kumar et al. [24]. Qualitative research in fish diversity of Kolab River and specifically classifying them into respective families including their edible and ornamental quality has been done in the present research programme as shown in Table 3. Such type of studies has also been done by Tamboli and Jha [25] and Kumar et al. [24]. The first ever study on fishes of Odisha was put forth by Day [26], which included both marine and freshwater fishes. However, no detail investigation has been so far taken up on the freshwater fish diversity of Ganjam district of Odisha.

A fish diversity index was used to describe the diversity of community samples by a number [27]. The idea of "species diversity" depends on richness and evenness of species in a community [28]. Diversity index increases when the number and variety of the species increases. In a given area, the value of the diversity index is maximum when all species are equally dominant [22]. Site 1 found the highest Shannon-Weiner diversity index, where Site 2 saw the lowest. Similar observations were also found by Nabi et al. [29]. The present study shows the relative abundance of *Labeo rohita* and *Catla catla* at both the sites, which was also confirmed by Hossain et al. [23]. A significant increase in Shannon Weiner index values was also found at site 2. This may also be due to the even distribution of all species at site 1, proving it to be a more congenial environment. The present study on fin fish diversity recorded sixteen (16) species at site 1 and eleven (11) species at site 2 which shows the diversity of fishes of Kolab River (Tables 4 and 5). Rivers with their tributaries and distributaries are sites for globally vulnerable species (Table 3). The more sensitive the species are, the greater the need for fish protection.

4.3 Heavy Metal Accumulation in Fishes

Toxic metal pollution are one of the major concerns of mining areas around the world [30]. They are the most dangerous as they have a longer life in the environment which directly or indirectly affects living organisms in the route of food chain [31].

Increased level of heavy metals as a result of heavy mining, especially during the rainy season has adverse effects on the environment which contaminates surface water and groundwater bodies [32]. The metal levels of biological samples depend on the ambient water chemistry. Therefore, the common physico-chemical parameters like pH, temperature and turbidity were also analyzed. The correlation coefficient (Tables 7a, 7b, 8a, 8b, 9a, 9b, 10a, 10b) computed with the metal concentration in the fish muscle during the study period has shown negative correlation with pH ($p < 0.01$) and temperature ($p < 0.01$), but positive correlation with turbidity ($p < 0.01$) excepting in muscles of *Labeo rohita* which clearly speaks that with the decrease in pH, the metal ions get transformed into dissolved state and hence accumulation of heavy metals in the fishes is more. The significant positive correlation ($p < 0.01$) between turbidity and accumulation of metals shows that more is the suspended particulate matter in the ambient media, more is the accumulation. This has also been supported by Rauf et al. [33]. In the river, fish are often at the top of the food chain and has tendency to concentrate heavy metals from water [34]. Dural et al. [35] also observed highest levels of zinc, copper and lead in the liver and gills of three freshwater species which are (*Sparus aurata, Dicentrachus labrax* and *Mugil cephalus*). Kumar et al. [36] while working with *Labeo rohita* also reported that the amount of heavy metals being ingested from contaminated water has resulted in zinc, copper, lead accumulation in fish which possess adverse effect on human health. The concentration of zinc, copper and lead in *Labeo rohita* and its effect on human health has also been studied by Kumar et al. [36].

Fish is an important component in the diet of people of India s and is a significant item in their diet. It contains 25% animal protein [37] covering 75% (105.6 million tonnes) of estimated world fish production used for direct human consumption [38]. It is said that fish consumption in developing countries increased by 57% in 2020 [39]. Fish protein contains two types of omega-3 polyunsaturated fatty acids; Eicosapentenoic acid (EPA) and docosahexenoic acid (DHA). Omega-3 (n-3) fatty acids helps in reduction of cholesterol levels and reduces the incidence of heart disease, stroke and preterm delivery. Fish also contains vitamins and minerals that is required as a part of human diet, however the intake of polluted fish can be the main route of exposure to heavy metals [40].

In the present study, the accumulation level of zinc, copper, and lead as per FAO/WHO [21], in muscle and intestine of *Labeo rohita* and *Catla catla* (Figs. 4, 5, 6, 7, 8, 9, 10 and 11) shows that the accumulation is far below the permissible levels at both the sites. It is observed that site 2 is a more stressful zone than site 1, where accumulation of lead was found more in case of the intestine of *Catla catla* which is an alarming signal for proper monitoring of lead discharge units from time to time. Accumulation of heavy metal in body tissues of fishes are dependent on its exposure period and the concentration of metal it is exposed to. High levels of heavy metals were found in the case of the intestine in comparison to muscle in the present study (Table 11). Although fish intestines are seldom consumed, it usually accumulated heavy metals which might represent good biomonitors of metals present in the surrounding environment [41]. This result also goes at-par with previous observations made by Deb and Fukushima [42] who recorded metals to be in high concentrations in the gills, intestine and digestive glands. These organs have a relatively high potential for metal accumulation. Heavy metal contaminations have devastating effects on the ecological balance of the recipient environment and the diversity of aquatic organisms [43, 44].

Heavy metals enter in humans through bioaccumulation and biomagnification and may cause serious health hazards in two ways: (i) they accumulate and thereby disrupt function in vital organs and glands such as heart, kidneys, bone, liver, etc.; (ii) they displace the vital nutritional minerals from their original place, thereby, hindering their biological function. Toxic heavy metals may knock down immune, reproductive, nervous and endocrine systems in animals and these effects can be at organ, tissue and cell level [45]. Even essential metals like Cu, Co, Zn, Fe, and Mn also become toxic at high concentration [46]. Fishes are widely used to evaluate the health of aquatic ecosystems because pollutants build up in the food chain and are responsible for the adverse effect and death in aquatic systems [47].

Destruction and degradation of aquatic health are a major challenge to the fish diversity [48]. Previous literature data and the identification of point and nonpoint sources of contamination will help us in resource conservation and maintenance. This study has not only helped in indicating the metal accumulation status in the two dominant fishes i.e. *Labeo rohita* and *Catla catla.* However, it has also generated a signal of caution for their consumption at large by the common people in and around Koraput, although a long term study may justify the degree of deterioration of the fish tissue.

5 Conclusion

The entire study advocates the following points:

- The statistical panorama suggests the negative impact of pH and temperature on the magnitude of bioaccumulation. However, turbidity exhibits a positive correlation which may be due to adsorption of the metal species on suspended particulate matter.
- Site 1 is more congenial compared to site 2 as confirmed by high Shannon-Weiner index value and low Index of Dominance value. The fish diversity study of river Kolab exhibits the presence of nineteen (19) finfish species.
- The population of *Labeo rohita* exhibits pronounced dominancy over *Catla catla* in the present study area.
- Pb accumulation occurred in the intestine of *Catla catla* by crossing the barrier of safe limit as prescribed by WHO [22].
- The present study area needs constant monitoring of bioaccumulation patterns in consumable aquatic fauna because of the presence of major industries in and around its vicinity.

6 Recommendations

The management policy advocates liming of the effluents released from the industries to the river to trigger the alkaline condition of the water body so that the heavy metal species get compartmented in the sediment bed instead of entering in the body tissues of aquatic organisms. The dose of liming has to be optimized depending on the pH of the wastewater generated from these industries. Conservation of natural resources in the river Kolab needs to be taken up through awareness programs involving the industrialists, policymakers, fishermen, and people of ranks of the society of Koraput in the loop.

7 Conflicts of Interest

The authors declare no conflict of interest.

Acknowledgements Authors acknowledge their deep sense of gratitude towards the authorities of the Central University of Odisha and University Grants Commission, Delhi, India for their moral and financial support towards the work respectively. They also remain thankful to Dr. B.B Nayak and Dr. N.K Dhal of CSIR-IMMT, Bhubaneswar for providing their laboratory support for smooth conductance of the present work.

References

1. Li L, Binghui Z, Lusan L (2010) Biomonitoring and bioindicators used for river ecosystems: definitions, approaches and trends. Procedia Environ Sci 2:1510–1524
2. Moog O, Schmutz S, Schwarzinger I (2018) Biomonitoring and bioassessment. In: Schmutz S, Sendzimir J (eds) Riverine ecosystem management. Aquatic Ecol Ser 8. Springer, Cham. https://doi.org/10.1007/978-3-319-73250-3_19
3. Ritcher BD, Braun DP, Mendelson MA, Master LL (2003) Threats to imperilled freshwater fauna. Conserv Biol 11:1081–1093
4. Arle J, Mohaupt V, Kirst I (2016) Water monitoring of surface waters in Germany under the water framework directive—a review of approaches. Methods Results 8:217. https://doi.org/10.3390/w8060217
5. Agarwal T, Danai P (2017) Biodiversity Evaluation. Int J Curr Res Acad Rev 5(3):86–92
6. Lakra WS, Sarkar UK (2010) NBFGR-Marching ahead in cataloguing and conserving fish genetic resources of India. Fishing Chimes 30:102–107
7. Pati S, Pati DK (2008) Freshwater fishes of Orissa. Fishes and shell fishes of Orissa, pp 32–35
8. Malina G (2004) Ecotoxicological and environmental problems associated with the former chemical plant in Tarnowskie Gory, Poland. Toxicology 205:157–172
9. Vodyanitskii YN (2012) Standards for the contents of heavy metals and metalloids in soils. Eurasian Soil Sci 45(3):321–328
10. Moore JW, Ramamoorthy S (1984) Heavy metals in natural waters—applied monitoring and impact assessment. Springer Series on environmental management. Springer. New York, 789 p. https://doi.org/10.1007/978-1-4612-5210-8

11. Jaishankar M, Tseten T, Anbalagan N, Mathew BB (2014) Toxicity mechanism effects of some heavy metals. Interdiscip Toxicol 7(2):60–72
12. Phillips DJH (1980) Quantitative aquatic biological indicators. Their use to monitor trace metal and organochlorine pollution. Applied Science Publisher Ltd., London, pp 412–455.
13. Salmeron FP, Melendez CME, Martinez TL (1990) Hepatotoxic and nephrotoxic effect of lead on the tilapia Sarotherodonaureus. J Anales de la Escuela Nacional de Ciencias Biologicas, Mexico 33:1–4
14. Olsson PE, Kling P, Hogstrand C (1998) Mechanism of heavy metals accumulation and toxicity in fish. In: Langston WJ et al (Eds) Metal metabolism in aquatic environment. Published by Chapman and Hall, London. Chapter 10, pp 321–350
15. Stagg RM, Shuttleworth TJ (1982) The accumulation of copper in Platichthys flesus L. and its effects on plasma electrolyte concentrations. Elsa lgl 20(4): 491–500
16. Dallinger RF, Prosi HS, Back H (1987) Contaminated food and uptake of heavy metals by fish. Oceologia (Berl.) 73(1):91–98
17. Singh P (2014). Studies on seasonal variations in physico-chemical parameters of the river Gomti (U.P.) India. Int J Adv Res 2(2):82–86
18. Anonymous (2004) India's initial national communication (NATCOM) to the United Nations framework convention on climate change. Govt. of India
19. Talwar PK, Jhingran AG (1991) Inland fishes of India. Oxford and IBH Publishing Co-Pvt Ltd New Delhi, p 1158
20. Dybem B (1983) Field sampling and preparation subsamples of aquatic organism for analysis of metals and organochlorides. FAO Fisher Tech 212:1–13
21. FAO, (1984) List of maximum levels recommended for contaminants by the joint FAO/WHO codex Alimentarius commission. Second Ser CAC/FAL Rome 3:1–8
22. Bassem SM (2020) Water pollution and aquatic biodiversity. Biodivers Intl J 4(1):10–16
23. Hossain MS, Das NG, Sarkar S, Rahman MZ (2013) Fish diversity and habitat relationship with environmental variables at Meghna river estuary, Bangladesh. Egypt J Aqu Resear 13:22–45
24. Kumar ST, Charan GB, Kumar SS (2013) Review of the research on the fish diversity in the river mahanadi and identifying the probable potential ornamental fishes among them with reference to threats and conservation measures. Res J Anim Vet Fish Sci 1(3):16–24
25. Tamboli RK, Jha YN (2012) Status of cat fish diversity of River Kelo and Mand in Raigarh District, CG. India. ISCA J Bio Sci 1(1):71–73
26. Day F (1869) On the fishes of Orissa. Proc Zool Soc London 369–387. https://doi.org/10.1111/j.1469-7998.1869.tb07340.x
27. Magurran AE (1998) Ecological diversity and its measurement. Princeton University Press, Princeton, pp 230–450
28. Chowdhury MSN, Hossain MS, Das NG, Barua P (2010) Environmental variables and fisheries diversity of the Naff River Estuary. Bangladesh. J Coastal Conser 15(1):163–180
29. Nabi MRU, Mamun MAA, Ullah MH, Mustafa MG (2011) Temporal and Spatial distribution of fish and shrimp assemblage in the Bakkhali river estuary of Bangladesh in relation to some water quality parameters. Mari Biol Res 7:436–452
30. Briffa J, Sinagra E, Blundell R (2020) Heavy metal pollution in the environment and their toxicological effects on humans. Heliyon 6(9):e04691
31. Ali H, Khan E, Ilahi I (2019) Environmental chemistry and ecotoxicology of hazardous heavy metals: environmental persistence, toxicity and bioaccumulation. J Chem 2019:1–14
32. Obasi PN, Akudinobi BB (2020) Potential health risks and levels of heavy metals in water resources of lead-zinc mining communities of Abakaliki southeast Nigeria. Appl Water Sci 10(184). https://doi.org/10.1007/s13201-020-01233-z
33. Rauf A, Javed M, Jabeen G, Javid A, Hussain SM (2015) Concentration of heavy metals in the water and fish tissues of the river Ravi. Pakistan J Bio & Env Sci 6(2):141–147
34. Rajeshkumar S, Li X (2018) Bioaccumulation of heavy metals in fish species from the Meiliang Bay. Taihu Lake, China Toxicol Reports 5:288–295
35. Dural M, Goksu MZL, Ozak AA (2007) Investigation of heavy metals in economically important fish species captured from the Tuzla Lagoon. Food Chem 102:415–421

36. Kumar R, Tripathi RM, Gupta AK (2014) Seasonal variation of heavy metal concentration in *Labeo rohita* and *Edaphodon kawai* fishes of River Yamuna, Allahabad. Uttar Pradesh. IOSR-JESTFT 8(6):27–34
37. Bahansawy M, Khidr A, Dheina N (2009) Seasonal variations of heavy metals concentrations in mullet, *mugil cephalus* and *Liza ramanda* (Mugilidae) from Lake Manzala. Egypt J Appl Sci Res 5(7):845–852
38. FAO (2006) the state of world fisheries and aquaculture, food and agriculture organization of United Nations, Rome, p 162
39. Retnam A, Zakaria M (2010) Hydrocarbons and heavy metals pollutants in aquaculture. Proceed Post Grad Quolloquium Semest 320
40. Bogut I (1997) Water pollution by heavy metals and their impact on fish and human health. Hrvatske vode 5:223–229
41. Zhang Z, He L, Li J, Wu Z (2007) Analysis of heavy metals of muscle and intestine tissue in fish-in Banan section of Chonqing from three gorges reservoir, china. Polish J Environ Stud 16(6):949–958
42. Fukushima T, Deb SS (1999) Metal in aquatic ecosystems: mechanisms of uptake, accumulation and release. Int J Environ Stud 56(3):385
43. Anandhan R, Kavitha V (2019) Effect of aluminium chloride toxicity against histopatholgy of gill of fresh water fish Labeo rohita (Ham). Int J Curr Adv Res 8(4A):18052–18054
44. Forstner U, Wittmann GTW (1981) Metal pollution on the aquatic environment. Spring-verlag Berlin Heidelberg New York, p 486
45. Geeraerts C, Belpaire C (2009) The effects of contaminants in European eel: a review. Ecotoxicoly 19:239–266
46. Alloway BJ, Ayres DC (1993) Chemical principles of environmental pollution. Blackie Academic U K, pp 140–149
47. Javed M, Usmani N (2011) Accumulation of heavy metals in fishes: a human health concern. Int J Environ Sci 2(2):659–669
48. Lin HY, Bush A, Linke S, Possingham HP, Briwn CJ (2017) Climate change decouples marine and freshwater habitats of a threatened migratory fish. Diver Distri 23:751–760
49. Dutta K, Ghosh AR (2013) Heavy metals in liver and kidney of Channa spp. in Sukinda valley. Jajpur, Odisha. The Ecoscan 3:191–197
50. Ahmed ASS, Sultana S, Habib A, Ullah H, Musa N, Hossain MB, Rahman MM, Sarker MSI (2019) Bioaccumulation of heavy metals in some commercially important fishes from a tropical river estuary suggests higher potential health risk in children than adults. PLoS ONE 14(10):e0219336
51. Ali H, Khan E (2018) Assessment of potentially toxic heavy metals and health risk in water, sediments, and different fish species of River Kabul, Pakistan. Hum Ecol Risk Assess. https://doi.org/10.1080/10807039.2018.1438175
52. Emmanuel J, Lalsha SR, Baijulal B, Jyothi J (2014) Biomonitoring of heavy metal accumulation in selected inland fishes of Kerala. J Aqua Biol Fisher 2:241–246
53. Rajeshkumar S, Li X (2018) Bioaccumulation of heavy metals in fish species from the Meiliang Bay, Taihu Lake, China. Toxicol Rep 5:288–295
54. Gupta A, Rai DK, Pandey RS, Sharma B (2009) Analysis of some heavy metals in the riverine water, sediments and fish from river Ganges at Allahabad. Environ Monit Assess 157:449–458
55. Kaur S, Khera KS, Kondal JK (2018) Heavy metal induced histopathological alterations in liver, muscle and kidney of freshwater cyprinid, Labeo rohita (Hamilton). J Entomol Zool Stud 6(2):2137–2144
56. Maurya PK, Malik DS (2018) Bioaccumulation of heavy metals in tissues of selected fish species from Ganga River, India, and risk assessment for human health. Huma Ecol Risk Assess. https://doi.org/10.1080/10807039.2018.1456897
57. Maurya PK, Malika DS, Yadav KK, Kumar A, Kumar D, Kamyab H (2019) Bioaccumulation and potential sources of heavy metal contamination in fish species in River Ganga basin: possible human health risks evaluation. Toxicol Rep 6:472–481

58. Panda BP, Panda B, Parida SP, Mahapatra B, Dash AK, Pradhan A (2019) Heavy metal accumulation in some fishes preferred for consumption by Egrets in Odisha. India. Nat Env Pollut Technol 18(3):975–979
59. Orissa Department of Water Resources, ODWR (2012) Contribution to river basin study. Orissa, India, pp 34–78
60. Ibrahim AT, Omar HM (2013) Seasonal variation of heavy metals accumulation in muscles of the African Catfish Clarias gariepinus and in River Nile water and sediments at Assiut Governorate, Eqypt. J Biol Earth Sci 3(2):B236–B248

Cake and Membrane Filtration in Mitigating Global Water Demand

N. K. Brahma, A. Mahapatra, S. N. Roy, and Sarajit Basu

Abstract The paradigm shifts of Cake and Membrane filtration designs are rapidly changing to compensate the demands of drinking water (DW), potable water (PW) and Industrial water (IW) of exponentially growing population 9.5 billion in 2030 and for recycling of wastewaters (WW) to prevent pollution and global infection WBD (Water born Diseases). The fundamentals of cake and gel formation in membranes and the designs essentially involved in separation and purification of highly polluted toxic and colloidal wastewater through jute pulp cake filtration and are need to be modified, for cost-effective and potential application. To review and to validate the above concept, the previous developed and reported technologies of the authors have been considered. Few of those designs are reproduced and modified. In this paper we have once again emphasised, that the technologies which are being reported would be essential and appropriate to apply for WW purifications and to compensate the demands of DW, PW and IW water in future.

Keywords Membrane · WBD (Water Born Diseases) · DW (Drinking Water) · PW (Potable Water) · IW (Industrial Waste) · CMF (Cake and Membrane filtration) · Arsenic and bacteria

N. K. Brahma
Department of Chemical Engineering, IIT-Kharagpur, Kharagpur, West Bengal, India

N. K. Brahma · S. Basu
Chemical Engineering Division, WBSC/IEI, Kolkata, India

N. K. Brahma · A. Mahapatra · S. Basu
Institute of Genetic Engineering, Badu, Kolkata, West Bengal, India

S. N. Roy (✉)
Ceramic Membrane Division, CSIR-CGCRI, Kolkata, India

Adamas University, Kolkata 700126, India

S. Basu
Department of Chemical Engineering, IIT-Bombay, Bombay, India

Jadavpur University, Kolkata, India

© Springer Nature Switzerland AG 2022

S. Yadav et al. (eds.), *Wastewater Assessment, Treatment, Reuse and Development in India*, Earth and Environmental Sciences Library,
https://doi.org/10.1007/978-3-030-95786-5_3

1 Introduction

The authors attempt to present the essentialities of water purification (EWP) methods and their technological innovations. Design conditions of MBST (Membrane Based Separation Technology) will depend upon the increasing flux (Jx), the reductions of CP (Concentration Polarization). Reuses of membrane and the energy involved to increase and decrease the pressure gradient (ΔP), corresponding to Flux, $Jx = F\Delta P$ (F is the permeability constant of Membrane), the MBST has been considered to be as efficient water purification system. Jute pulp cake filtration is an alternative innovation, where the cake is formed on the surface of jute pulp, attached to clarifier basin, rotating around the sediment basin to separate the flowing toxic colloidal WW (wastewater). The said cake is used to separate pollutants at 2 microns (μm) cake pores. Cake filtration and MBST are considered as integrated process, where MF (Micro filtration), UF (Ultrafiltration), RO (Reverse Osmosis) and NF (Nanofiltration) membranes are used as per the demand of separations for DW. CFt is cost-effective, eco-friendly and appropriate to maintain water in primary treatments as large quantities of tertiary and industrial effluents flowing to canals and wet lands. Water treating in primary and secondary clarifiers can be made by various chemical and biological systems. Jute pulp cake filtration can be made before primary and secondary treatment to remove sediment particles and slurry from water to purify. MBST is used also for desalination and for removals of heavy metals including As, Cr and Phosphate. The selection of membranes and to design different modules like plate, wickle, hollow fibres, cylinders, etc. are considered to be used in medical sciences, AK (Artificial Kidney), RO in desalinations and plate/cylinder modules in food industries. Microbes (Bacteria, fungi and Algae) and water borne dieses (WBD) are to be removed from the drinking water. Algae took important role in clarifications of waste water, reserved in basin to capture CO_2. Algal biomass is used for bio-fuel, food, fertilizer and medicine to mitigate energy and pollution. Heavy metals like Arsenic (arsenic As^{3+}/As^{5+}), Chromium (Cr^{6+}), Copper (Cu^{2+}), Cd (Cadmium) and lead (Pb) originated from industrial and domestic effluents can also be treated by microbes (***Pseudomonas putida***, ***Aspegillus nieger***, ***Candida species***) and by designing water basins with ***Water Hyansinth*** culture and WLC (Wet land Cultivation) as bio-augmenting and bio-remediation for healthy socio-economic environment.

Water (H_2O) has another name is life. Without water there is no life. National Aeronautics and Space Administration (NASA) and Indian Space Research Organization (ISRO) are jointly venturing and searching the presence of water in planetary system. Their objective is to search water in planetary system and to transfer life and technology to a new discovered planet for an alternative solution to optimize the increasing pressure on global population and pollution. Presently they are expanding their research towards the existence of life in a planet. The presence of water is therefore considered to be an indispensable component after solar rays (i.e., Sun). Global population is 9.5 billion and 23 trillion tones of solid wastes, and is increasing every year. The discovery along with water resource could be one such solution to shift human civilization to another planet.

However, for sustaining the management of global water resource, so long the planetary system is not being stable and justified, membrane and cake filtration would be essential, as the fundamental demand of the life processes on mother earth. Its purification, recycling and reuse are essential. One sustainable resource management would be Zero Discharge, where 80% of used water would be recycled and is used mostly for various purposes of industrial activities. In this case origin and used water would be is essential. We have vast ocean and truly to say 3/4 of the earth is covered by water. This water is basically non-PW, non DW and saline (SW). Middle East, UAE has used vast desalined plants to mitigate their water. RO (Reverse Osmosis) 10,000 m^3 desalining water-plants are already under use [1, 2]. MBST (Membrane Based Separation Technology) is applied to purify the said used water for drinking purpose as major breakthrough industrial application in history of civilization. As per record, 120–150 L water day^{-1} $person^{-1}$ is to be essential for living purpose, which would be reduced to 60–65 L day^{-1} $person^{-1}$ in 2020, if all water is recycled, purified, and if river water, wastewater, and used water are combined together and if wastage of water is prevented and if rain water harvesting is properly done, then this management would be fruitful. Along with increasing water demands, the infection level is also increase (i.e., in general people use water from specific places, where population and pollution are dense. This community interactions support may also support to generate water born diseases). World population in 2020 will be 9.5 Billion and ¾ of the population will be in India and China, in which India, Bangladesh, Sri-Lanka, and Pakistan, will suffer unpredictable level of infections, because of technological deficiencies and political corruptions. 0.01% of the total 7.5% of available global water will be used as surface and available water, i.e. 1 trillion liters of global water reserve, only 0.01 trillion liters of water is used for living purposes. 9.5 billion × 150 lit = 2525 billion liters water would be required (1–3 × 10^9 lit) for survivals, including 2–5 lit × 9.5 = 22–30 billion lit of DQW (Drinkable Quality Water) without WBD (Water Born Diseases) with required minerals for mankind [1–7] as shown in Fig. 1 where it describes flowingly that 0.1% of the total global water is used for domestic and industrial purposes.

While the first arsenic patients were seen in West Bengal, India, in 1983, it was only considered and confirmed in 1993. Currently, the pollution has affected 59 of Bangladesh's 64 districts, with arsenic levels exceeding the national (50 ppb) approved cap.. It is estimated that 35–77 million people in Bangladesh, out of a total population of 125 million, are at risk of drinking contaminated water. Pink and red colors in Fig. 2 reflect possible infected areas in Bangladesh, where the presence of arsenic is acute and removal of arsenic is urgently and ultimately necessary, which is critical for the safety of human habitats.

The authors attempt to present the said problem through technological innovation specifically through MBST (Membrane Based Separation Technology) and jute pulp cake filtration [1–5]. Cake filtration and MBST as integrated phenomena would be cost-effective, eco-friendly and appropriate. Both Cake and MBST require efficient energy to develop gradient, so it is essential that parallel to the development of

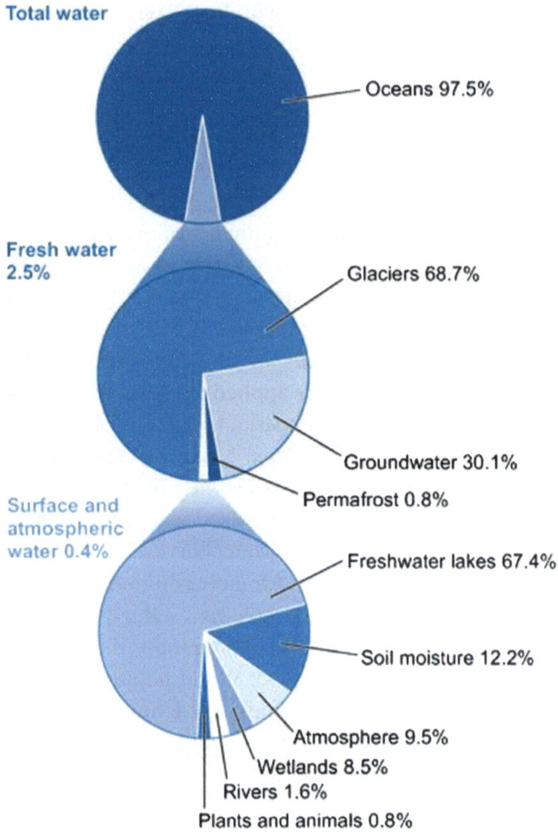

Fig. 1 Schematic diagram shows the distribution of fresh water globally is reproduced

membrane modules (i.e., membrane properties, selections of polymers and their structures) and cake filtration models also to be changed. MF (Microfiltration), UF (Ultrafiltration), NF (Nanofiltration) and RO (Reverse Osmosis) are used according to the decreasing pore sizes till 0.001 μm. These are the essential steps in separation and optimizing DWQ, covering the treatments of WW (waste water) (Brahma et al. 1993) through Cake filtration, Slurry treatments, Solid waste management to remove pollutants from soil and water and to develop a healthy socio-economic environment and to provide arsenic and bacteria free water would be the ultimate demand.

In light of the above discussion, a Hand Pump Membrane Module (HPM) using RO membrane was created and published in the newspaper "Times of India" on June 25, 2013 under the title "Prof's low-cost response to city arsenic challenge", as shown in Fig. 3. It is manually operated in order to obtain water that is free of arsenic and microbes. Water will be sold for 60 paisa per liter. Manual and nonconventional solar, wind, and biogas dependent HPM will be sufficient in remote parts of developing countries, such as West Bengal and other Indian states, where arsenic and microbe-free drinking water is in high demand and there is no electricity. The demand for

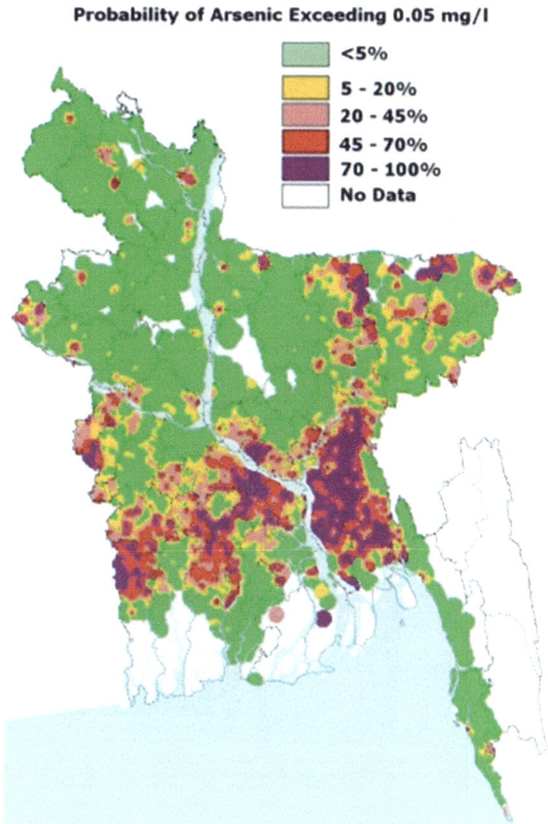

Fig. 2 Show distribution of arsenic concentration in tube well water samples in Bangladesh

HPM has increased in rural areas of West Bengal, including Birbhum, Murshidabad, Bankura, Malda, and Nadia. This HPM has maintained a standard of 0.5 ppm, as recommended by the WHO. The developed low-cost HPM model could be useful and necessary for removing arsenic from polluted ground water.

2 History of DW and Involvement of Arsenic

The Ancient Civilization, 4000 BC was developed around water sources (i.e. Egypt Neil River). However, they were not aware about water quality and quantity. They were aware only about tests and salinity. Water Treatment apparently signified the improvements of quality and tests of water. In ancient Sanskrit and Greek civilization, the treatment of water was made by charcoal and sand filtrations followed by sunlight, boiling and sedimentation the contamination of drinking water was removed. During 1800s the meaning contamination of water was first identified by the presence of microbes and heavy metals. Arsenic contamination as "inheritance dust" was also a

TOI, 25.06.2013

Prof's low-cost answer to city arsenic threat

Krishnendu Bandyopadhyay | TNN

Kolkata: A recent Unicef report has revealed 89% of people in the state has access to water, but the quality is questionable. In a bid to prevent millions from being affected by water-related diseases, a former professor of IIT-Mumbai, Sarajit Basu, has invented a low-cost hand-operated machine that can purify water.

"The poor quality of drinking water often plays havoc with human life, crippling the rural economy," said Basu, a Humboldt Fellow who has been working on the low-cost technology for quite some time now.

A hand-operated reverse osmosis water purifier system, the machine invented by Basu, can filter 2 to 3 litres of polluted or brackish water in a minute. "This technology (for which he got a patent) can purify water with high arsenic content and that with a salinity of 1500 to 5000 PPM. Since 30% of our population still lives without electricity, I have incorporated a hand-pump in the machine. It can serve the need of 60 families or 300 people — an average size of a small village," the former IIT professor said. He also has an electric-motor variant for villages with electricity.

Explaining the mechanism behind the machine, Basu said, "When two liquids of different concentration are separated by a semi-permeable membrane, the fluid has a tendency to move from low to high solute concentrations for chemical potential equilibrium. The reverse osmosis is the process of forcing a solvent from a region of high solute concentration through a semi-permeable membrane to a region of low solute concentration by applying a pressure in excess of the osmotic pressure. This machine is doing exactly the same."

In the process, seawater and brackish water or arsenic water is pressed against one surface of the membrane and causes transport of salt/arsenic depleted water across it. This results in generation of potable drinking water from the low-pressure side.

While the potable water is collected from one outlet, the polluted water can be re-transferred to the machine for further purification. If the machine is operated by each family for eight-12 minutes, their requirement (3 litre per capita/day) will be met. The only requisite for installing the machine is it should be set up at the centre of a village, near a tubewell.

Basu has also worked on the financial feasibility of the machine. A litre of potable water costs 5 paise, while the cost of each machine is Rs 50,000. In a year or so, a community that buys the machine can recover the cost and will be in a position to install another machine. But the operator must be trained. The hand-pump operator can also be paid on a monthly basis, which will lead to generation of employment.

Environmentalist Somendranath Ghosh said, "This can well be included in Nirmal Gram Yojana."

Fig. 3 Shows newspaper cutting where a hand driven reverse osmosis water purifier machine called Hand Pump Membrane module (HPM) invented by one of the author late Prof. (Dr.) Sarajit Basu, which can filter 2–3 L of polluted or brackish water in a minute

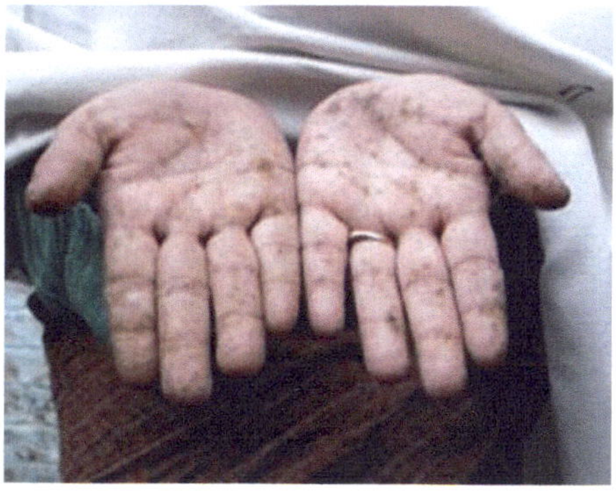

Fig. 4 Signs of arsenicosis: spots on the hands

threat parallel to WBD. In 55 AD, Nero poisoned Britannica with arsenic to secure the Roman throne. During 15th / sixteenth centuries, Italian Borgi was used arsenic for political assassinations. Napoleon may have been poisoned by arsenic contaminated tainted wine. As^{3+} is more toxic than As^{5+}. Both are present in ground water at contaminated sites. Although arsenic was used in medicine for over 2500 years long and is still using to prevent many microbial and parasitic infections. Arsenic contamination can cause cancer, skin lesions. Men and children are more affected than women.

2.1 *Toxic Effects of Arsenic to Human Health*

Arsenic is toxic substance to human health and toxicity depends on the amount of arsenic intake, which is classified into acute, sub-acute and chronic toxicity respectively. It is a silent killer. It is 4 times as poisonous as mercury and its lethal dose (LD) for human is 125 mg. Drinking water contamination causes the last variety of toxicity. Undetectable in its early stages, arsenic poisoning takes between 8 and 14 years to impact on health, depending on the amount of arsenic ingested, nutritional status, and immune response of the individual. Figure 4 shows problems of the human body which is generally affected by arsenicosis.

3 Possible Water Clarification and Desalination

The demand for water can also be optimized by using RO in the desalination process, as shown in Fig. 5, where it maintains pressure gradient and energy, as well as membrane quality and fouling for the recirculation of 100% desalination water, directly recovered from the sea was sent 40% to RO and 60% to pressure exchange pump. The retentate was combined with 60% of the bifurcated original sea water and returned to a 40% flow to the RO factory. To preserve the pressure gradient and

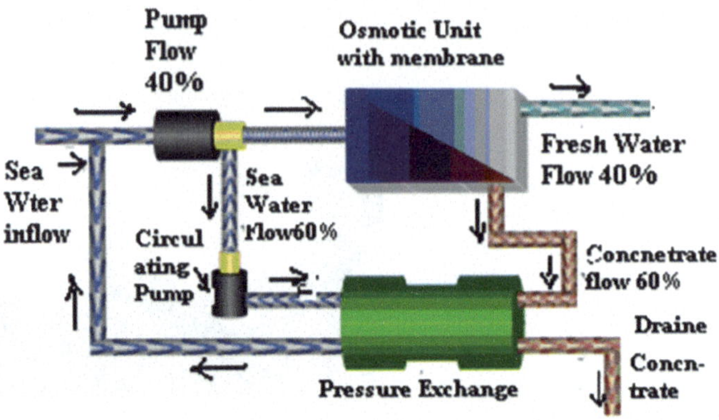

Fig. 5 Hybrid membrane application of RO in desalination process

avoid membrane fouling, 100% saline water flow was first distributed to 40% and 60%. The key goal of RO in desalination is to combine the use of UF and MF hybrid membranes.

A Jute pulp cake filtration model was proposed by these authors as shown in Fig. 6 which can be used in primary treatment. Jute fibers are used to make filter beds. Jute pulp cake filter beds move clockwise to shift filter beds, are connected to the flurry water entry to the sediment basin, and are connected to a pressure gradient displaying device (i.e., when the cake fails to filter and the pressure (ΔP) is increased, the filter bed will automatically turn on to the next filter bed for use). Basu et al. [8]

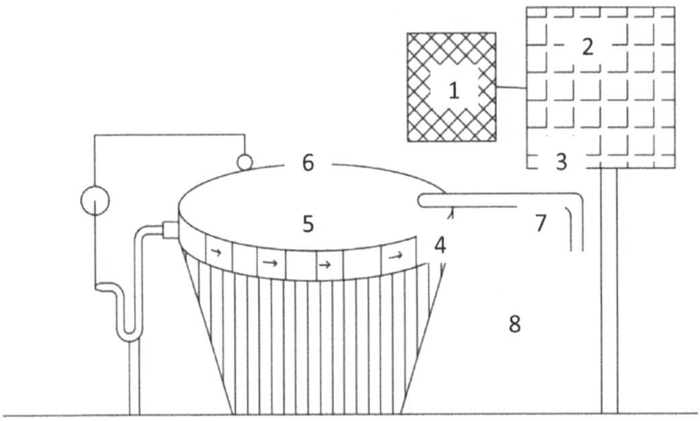

Fig. 6 Proposed a jute pulp cake filtration model. Legends: 1. Iron greed, 2. The iron frame to set the greed with jute pulp, 3. The perforated greed with jute pulp, 4. Attaching clamp to the belt, 5. Clarifier basin, 6. Moving belt, 7. Sediment slurry channel, 8. Relay for controlling cake formations, pressure gradient and for switching to the next filter bed, serially connected surrounding the sediment basin

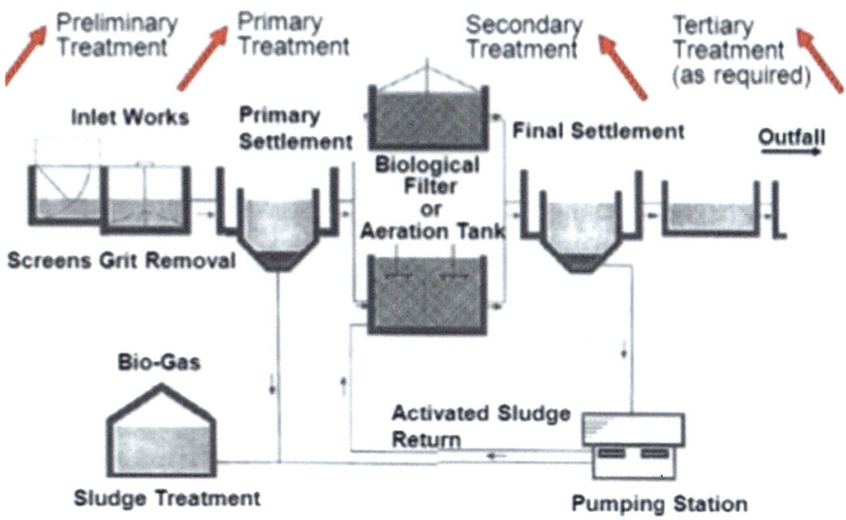

Fig. 7 A schematic diagram of conventional waste water treatment plant (WWTP)

and Brahma et al. [9] proposed that arsenic treatments can be made in secondary and tertiary clarifiers.

4 Conventional Wastewater Treatment

It consists of a combination of physical, chemical and biological processes and operations to remove solids, organic matter and, sometimes, nutrients from wastewater and are presented in Fig. 7.

5 Wetland for Wastewater Treatment

An engineered wetland is an ecological wastewater treatment system that mimics and enhances the efficacy of naturally occurring wetlands' purification processes (Fig. 8). Long roots are useful for such treatments as heavy metal absorption, through Bio-augmentation, and Bio-remediation. The process maintains biodiversity while also removing heavy metals and preventing surface water pollution. To treat wastewater, constructed wetlands are planned and built similarly to natural wetlands. A shallow depression in the ground with a flat bottom characterizes them. In built wetlands, the flow is regulated so that the water is uniformly distributed among the wetland plants.

The overall theme is that the plants, microorganisms, and substrates all work together to filter and purify the water. To begin, water is slowed as it reaches the

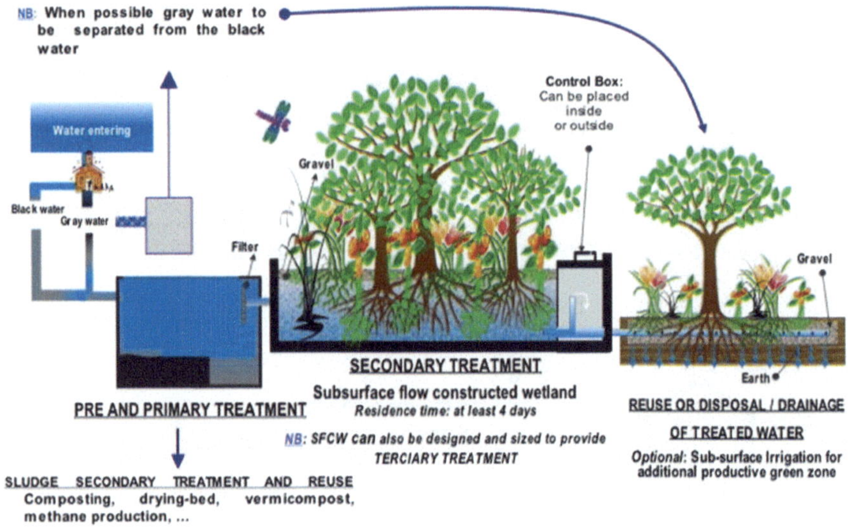

Fig. 8 A schematic diagram of waste water treatment by WLC (Wet Land Culture) using plant roots, wet & agriculture land

wetland, allowing solids to settle. Plant roots and the substrate expel the larger particles in the wastewater during the water flow to the built wetland. Pollutants and nutrients in wastewater are naturally broken down and taken up by bacteria and plants (Fig. 9) [10, 11], allowing them to be removed from the water. The pathogens present in wastewater will be killed by the retention period in the wetland, which varies depending on the design and desired quality level, as well as plant secretion of antibiotics. After treatment in a constructed wetland, water can be safely released into surface waters or used various purposes.

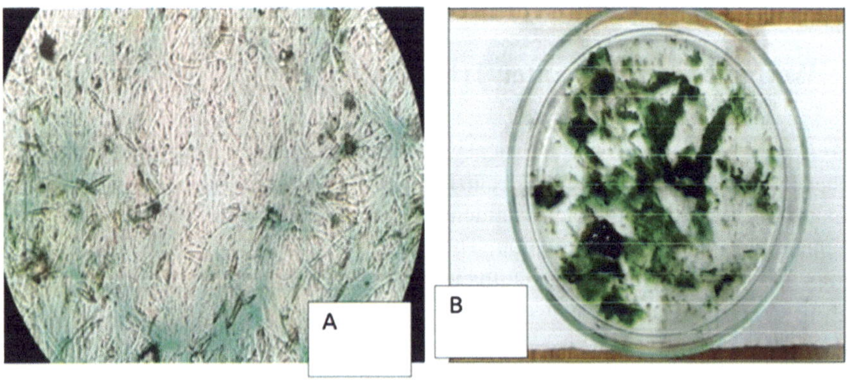

Fig. 9 Algae observed in Microscope at X400 magnification **a**, and **b** Algae is grown on Petri-dishes

6 Conclusions

Present status of global water purification and to prevent global pollution and infections are challenging. According to exponentially growing population and pollution, it is expected that within 2030, there would be increasing demand of water. Perhaps, we would face World War-III for water. There would be increasing draught areas, disaster for increasing infection and biodiversities. There would be ice melt and due to increasing global warming, there would be flood and increasing water pollution due to flood due to increasing sea level. The demands of increasing need technology compared to greed technology are to be considered. Human relations would be in threat, due to questions of political power and for preventions of corruptions. Food scarcities' and environmental pollution will be increased due to population explosions. The sustainability of our mother earth is under threat. JPCF, MBST would be the necessary technology, which would be integrated with WWT, primary, secondary and tertiary treatments.

Recommendations

(i) Improve Zero Effluent Discharge (ZED) technology.
(ii) Increase Need technology (like MBST, Vaccination, Algal based Food and bio-fuel) compared to greed technology (like luxurious automobile, Flats, TV, IT based instruments).
(iii) Essential population control as per the demand of water according to WHO recommendations.
(iv) Metering system of domestic and industrial water supply.
(v) (v) Optimize PPP-C (Population, Pollution and Poverty as control by corruptions for the application of EAT (Enforced Appropriate Technology), unbiased and only on human factor.
(vi) (vi) Increase the concept of Universal Vaccination (ie. application of Plasmid Genetic Engineering and RNA Technology.
(vii) (vii) Understanding of MBST and Cake concepts.

Acknowledgements We are grateful to our beloved Late Prof. (Dr.) S. Basu, a pioneer scientist and teacher in the field of MBST (Membrane Based Separation Technology), inspired us to write this article, describing various MBST activities viz. MF, UF, RO membrane applications and his ideas to review this paper and to provide documents of his achievements on MBST.

References

1. Brahma NK (2008) Zero discharge and global warming: role of MBST, O3, UV and ME (water recycling), chemical business (ISSN: 0970-3136, 22(6):18–24
2. Brahma NK (2005) Cake filtration in waste water treatments. ICE, 47(4):254–261, cited in the paper in Indian science abstract (2007), ISSN: 0019-6339, 43(3), in Chemical technology division

3. Brahma NK, Nag A (1993) Continuous measurement of BOD in waste water. J Inst Eng, IEE-EN, UDC 628(3123):72–76
4. Porter MC (2005) Handbook of industrial membrane technology, Crest Publishing House, New Delhi-110002
5. Wijmans JG, Nakao S, Van der berg JWA, Telstra FR, Smolders CA (1985) Hydrodynamic resistance of concentration polarization boundary layers in ultra filtration. J Membr Sci 22
6. Van der Berg GB, Smolders CA (1990) Flux declining ultra filtration process. Desalination 77:101
7. Van der berg GB, Racz IG, Smolders CA (1989) Mass transfer coefficients in cross flow ultra filtration. J Membr Sci 47:25–51
8. Basu S, Brahma NK (2017) Desalination through Reverse Osmosis (RO), a sustainable and cost-effective future technology for energy and water management. In: Indian engineering congress, IEI, Kolkata
9. Brahma NK, Basu S (2008) O_3, UV and MBST based industrial wastewater treatments and sequestration of CO_2 for minimizing global warming as part of zero-effluent discharge. Chem Bus 22(8):18–26
10. Brahma NK, Mahapatra A (2019) Global Warming (GW) and microbes: understanding life/GW through biological evolution, molecular and bioprocess engineering ODS, IEI/WBSC, 12th January
11. Brahma NK (2001) Proposed bio-methanation of petrochemical crude waste, Chap. (4). Microbial control of environmental pollution (Chap. 14) and design of measurements in continuous process (Chap. 21) published in a text book entitled "environmental pollution and management of wastewaters by microbial techniques". In: Pathade, Goel YC (eds) College of Science, Department of Pollution studies and Post Graduate Department of Microbiology. ABD publishers, Oxford Book Company, Jaipur, India

Wastewater Treatment and Reuse

Potential for Use of Treated Waste Water for Industrial Reuse in India

Rishi Rana and Rajiv Ganguly

Abstract The rate at which the population growth has increased has led to the steep increase in generation of wastewater in overall world. The major challenge faced by urban India remains in terms of the availability of fresh and clean water along with appropriate sanitation structure. Out of total water supplied for domestic use around 70–80% gets generated as wastewater. Many of the recent studies have also concluded that the supply is approximately equals the demand for India explaining the acute shortages in many parts. To meet the severity of water crises, industrial and agriculture water demand, wastewater resource, its collection, treatment and reuse is very important alternative for fresh water. The wastewater generated is generally discharged without treatment in open leading to insanitary environment finally causing the pollution to environment. Treatment of wastewater is a part of public health and sanitation. Wastewater treatment, whether on-site or off-site are the part of the full circle which helps in prevention to environmental pollution and safeguarding health issues. The wastewater after the treatment procedures plays a very crucial role for industrial as well as agricultural water demands. It has some economic advantages and act as a source of revenue for the urban local bodies. Treated wastewater although is economically viable, industrial reuse is limited by the availability of industrial clusters in the vicinity of the treatment plant. If this treated wastewater is reused by industries and agriculture then it frees up water which could be used to meet city's water demand. It is the need of the hour to properly treat the wastewater especially near the source of generation and reuse it so as to protect the environment and a reliable source of water supply can be provided. The major drawback remains the non-availability of the required data. With the drastic change in climatic conditions along with rising population, the main concern for a country like India must

R. Rana
Department of Civil Engineering, Jaypee University of Information Technology, Waknaghat, District Solan, Himachal Pradesh 173234, India
e-mail: rishi.rana@juit.ac.in

R. Ganguly (✉)
Department of Civil Engineering, Harcourt Butler Technical University, Kanpur, Uttar Pradesh 208002, India
e-mail: rganguly@hbtu.ac.in

© Springer Nature Switzerland AG 2022

S. Yadav et al. (eds.), *Wastewater Assessment, Treatment, Reuse and Development in India*, Earth and Environmental Sciences Library,
https://doi.org/10.1007/978-3-030-95786-5_4

be increased usage of the recycled and treated wastewater which can in longer run combat the scarcity of available water resources.

Keywords Wastewater treatment · Recycling · Reuse · Scarcity · India · Water resources

1 Introduction

Environmental issues have become more and more critical aspects for sustainability on earth in which we live. The major challenge faced all over the world these days remains the provision of safe drinking water and essential sanitation systems to all. There is plenty of water available on earth in order to sustain life but keeping in mind over-utilization and increase in pollution levels, we must do everything which can help in preservation and protection of supply of water [1, 2]. We also must purify and reuse the water we presently waste (i.e., wastewater). It was reported, that the major population of people still don't have access to improved waters sources and proper sanitation [3]. The diseases caused due to contaminated water are the major cause of mortality and morbidity worldwide [4]. The pathogens arising from contaminated water infects more than 250 million people every year resulting in about 10–20 million deaths and out of it major contribution is due to infection caused by poor sanitation practices. Now day's major concern worldwide revolves around over exploitation of the natural resources available to us along with the extreme pollution of environment. Fig. 1 shows water systems for the future.

The rapid urbanization along with migration of population to the urban centers has led to a wide decline in storage capacity and excessive over burdening of our natural environment due to the harmful emanations and high waste generation has reached to a very perilous situation. The declining quality of water and the related diseases are causing a major threat to human health as well as environment in evolving countries.

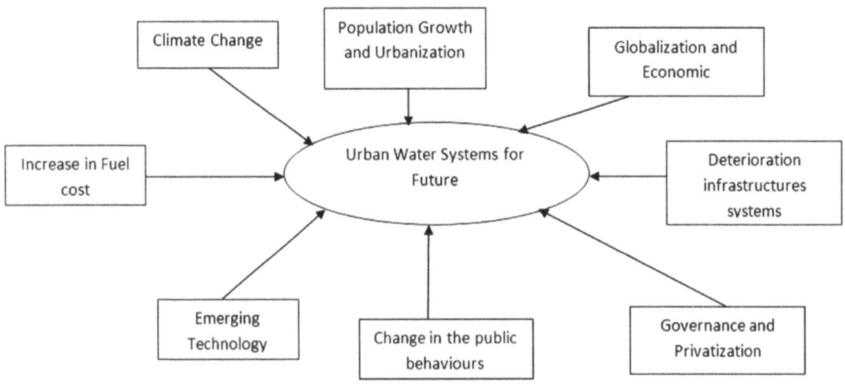

Fig. 1 Water supply systems for the future

Most of the people in this developing world are without the proper access to non-toxic and suitable sources of water [5, 6]. It is estimated that around 94% of the diseases in the world are caused owing to the contaminated water, scarce sanitation systems and declined hygienic environments [7]. Therefore, it is necessary to provide proper and essential water supply systems to achieve improved health conditions leading to increase in individual's productivity. It has also been proved in many studies that there is a direct relation among the improved water sources system and sanitation with increase in health and economies. With the improvement in both the water supply systems and sanitation measurements will led to health benefits including the improvement in quality of life and reduction in medical expenditure as water which is contaminated act as a vector which carries pathogens which further causes various diseases [8].

Terms like "waster crisis" and "shortage of water" are common in usage in today's scenario worldwide. Majority of our population is living in industrialized countries which still holds abundant amount of water but a certain amount of population still lives in developing countries which are regarded as "water poor". The global statics stated that around 1.1 billion people are deprived of suitable water sources and approximately 2.7 billion people are lacking appropriate sanitation systems. It was one of the missions of the development goals to reduce at least half the number of people who are deprived of proper sanitation and water supply facilities by the end of 2015. The countries which are still developing most of the investments are made in water and sanitation has emphasized water supply plants, wastewater treatment, and mains construction infrastructure external to the household. As such, these installations have created little beneficial impact on the lives of poor urban residents [4, 9].

The major factors that make changes in the delivery service and continuous supply of water supply schemes is the water quality and scheme is the water quality and schemes provided to the people [10]. For providing suitable supplies of water along with the sanitation facilities to the developing nations will ultimately lead to over utilization of ground water sources and the construction planning's at the sites. Due to lack of assimilative methodology which involves the people who have scientific knowledge in the field of water engineering, public health and sanitation engineering and also without any appropriate designing methodology creates more ruckuses giving rise to contamination of ground water by various micro-organisms and pathogens. These unhygienic conditions will lead to put general public on the risk of diseases thereby reducing the health aids. More and more general public especially the one living in rural areas have been involved in concern regarding the development of infrastructures for water supply and sanitation programs [2, 3].

One of the major challenges being faced by urban India has to be in terms of proper and suitable water supply and sanitation systems. The availability of a suitable water supply system has become a major concern in most Indian cities and towns. It is observed that many times it is often not sufficient to meet the ever increasing demands. The sharp increase in population has led to extreme decline in availability of water from 1,545 cubic meters in the year 2011 to 1,128 cubic meters in year 2016. The urban population of India is expected to be around 50% of the total population in the coming decades [9, 11]. This rapid increase in population is many times co-linked with the industrial growth along with more energy demand. The demand of water sources increases due to rise from industries, residential or households and industries especially the power plants which in turn add to the water woes leading to urban water stress. The cities like Delhi, Bangalore and Chennai have already faced the brunt of acute water shortage which has led to increase in the cost of fresh water production and industrial tariff [11]. The rapid and persistent growth of population coupled with industrialization and expanding and intensifying food production are all putting pressure on water resource which causes a significant increase of wastewater [12].

Continuing population growth, rapid industrialization, the uncontrolled disposal of the municipal, industrial and agricultural waste constitutes one of the most serious threats to the sustainability by contaminating the water, land and air pollution [1, 10]. There are many challenges those developing countries facing i.e., lack of necessary infrastructure such as electricity, roads and water supply, etc. On the list of priorities, wastewater drainage, sanitation and treatment tend to rank higher [4, 13]. With the swapping or reuse of the treated wastewater, the industries and agriculture frees up water which could be helpful in fulfilling the water demand of a country or cities [5, 14]. While it can be argued that it can provide economic benefits, it also puts utilities under pressure to supply potable quality drinking water to the population. This will also lead to change the ability of utilities to service the demand of commercial and industrial units. In addition to the challenges in meeting the demand, this may also lead to further increase in water tariffs in order to subsidize residential water supply.

Every community produces both liquid and solid waste along with air emissions. The liquid waste or wastewater is the supply of water for the community after it has been through various applications. When the organic matter decomposes in the untreated wastewater, many odorous gases are released leading to nuisance conditions [15]. Untreated wastewater contains many pathogenic microorganisms which are potentially mutagenic and carcinogenic as well as nutrients which help in stimulating the growth of aquatic plants. Wastewater collection, treatment and reuse of treated wastewater or reclaimed water are one such alternative that is gaining revenue. There is a long history of engineering for improving the quality of water and is named as wastewater treatment. There are physical, chemical and biological processes. Urban wastewater management has become a challenge in many developing countries like one in India because infrastructure development and regulations has not been keeping pace with the tremendous growth in population and urbanization.

2 Wastewater—A Growing Resource

Water is the most important natural resource and is required to sustain all kinds of life forms. Availability of water is one of the most critical components for sustaining all economic activities in a country. The country like India holds very vast resources of water due to the presence of rich rivers and sedimentary sinks to clench the ground water. The various demands for water include domestic, irrigation, industry and energy demands [11]. The amount of waste water produced from the class 1 cities (which holds around 60% of the overall population) holds about 17,000 MLD. Among all this metro cities generate approximately 60% of the total wastewater [4, 5, 13]. Subsequently, the quantity of water which is available for supply generally is not sufficient to meet the demands of the population, water conservation measures are employed. One of the most common methods of controlling water demand is the use of intermittent supplies, usually by necessity rather than design. This is where the water is physically cut-off for most of the day and hence limiting the consumers' ability to collect the water. The water quantity available for supply wastewater is termed as the stream of utilized water from residential areas, commercial and institutional sectors and various industrial activities. It can be defined as the combination of domestic effluent containing the excreta, fecal sludge and urine and is called as black water [1, 2, 4, 9]. It is also called as grey water when it contains the kitchen and bathing wastewater. Wastewater also includes the water from commercial establishments as well as from institutions, industrial effluents, storm water and other urban run-off, agricultural and horticulture effluents [16]. The industrial chemical effluents are not considered as they are potentially harmful and toxic and must be treated separately [9]. The challenge majorly lies in choosing various technologies for treatment of wastewater. Non-availability of data related to the various discharges all in further renders it difficult for estimating the volume and quality of wastewater generated. Data given by CPCB [5], Economic Survey of Delhi [15], Global Water Intelligence [14] explained that around 900 cities from both class-1 and II classifications displayed more collected wastewater than the treated [10]. It is also observed that about 60–70% wastewater is generated which may also varies among cities. Although government has taken various steps to reduce the water pollution but all the efforts goes in vain as maximum part of the untreated wastewater returns back to approximately 80% of the surface waters.

For an effective planning and construction of economical methods to curb wastewater, the characterization of raw wastewater is necessary. The designing of these treatment units must be done keeping in consideration the diverse nature of waste and thereby the characteristics of effluents [1, 2, 12]. The characteristic properties of wastewater varies from place to place which in turn depends upon the various factors like the food habits of the residing people, climatic conditions, water supply schemes and economic status of the community. With time there has been an increase in research on determining the characteristics of wastewater so as to examine the constituents present and have a detailed scan of their health effects along with environmental hazards. The expansion and progress in terms of the technologies used for treatment of wastewater is still not been able to keep balance with the emerging contaminants detection potential [17]. The hazards pertaining to the quality of water arise as a huge quantity of wastewater generated is disposed of openly in various water sources which can be further used as water supply systems. The safest and cleansed available sources of groundwater and surface water should be protected, used, and maintained for potable water supply purposes. Numerous parameters are used to determine the suitability of water and the health significance of contaminants that may be found in untreated and treated water.

Most of the wastewater contains organic, ammonical and many other harmful reactive compounds which are a substantial originator of high amount of biological oxygen demand. Discharging the untreated wastewater from both industrial and domestic sources interferes with the quality of water. this leads to reduction in the oxygen levels creating anaerobic conditions causing increased rate of mortality among aquatic animals, unbalancing the ecosystem and other nuisances [11]. Few industries generate wastewater which is very toxic and can be disastrous cause of health hazards to the species as well as environment. The discharge of such harmful chemicals in the aquatic environment leads to killing of aquatic life and thereby causing prolonged effects on the ecosystem as well. In many cases, the heavy metals which are present get accumulated in the food chain causing bio-magnification. The presence of a nutrients like nitrogen and phosphorous in huge amounts in the wastewater from industries and domestic sectors causes eutrophication of water bodies.

A typical wastewater treatment plants consist of various treatment procedures including unit operations and processes. Depending upon the degree of treatment, the processes can be subdivided into three types of treatment systems. They can be classified as primary, secondary and tertiary treatment processes [14]. Primary treatment processes include removal of suspended organics. Secondary treatment processes include the chemical unit processes. Tertiary treatment is engaged in removal of colloidal materials and nitrogen compounds as well. The major need for including tertiary processes in the treatment processes is to clarify wastewater which have chemicals and organics which are tend to persist and are problematic to remove. Processes like adsorption, ion exchange, activated sludge process and reverse osmosis [16].

The wastewater generated can be used in agriculture as well as industrial sectors irrespective of whether it is treated, untreated or sometimes partially treated especially in developing countries. As per the reports [13], by the end of the year 2015, there will be 88% growth in overall global population and majorly in developing countries. With the increase in urban water supply schemes there is equal surge in wastewater generation rate as only 10–20% of the wastewater generated from both the domestic and industrial sources is recycled [18]. Due to lack of financial schemes the growing volumes of wastewater are rendered to cheap alternatives like disposing them directly to irrigation systems. Applying appropriate treatment methodologies increases a tremendous prospective for the recycle and reuse of wastewater in numerous areas.

In developed countries, the domestic water is often reused but is way out of world health organization (WHO) standards suggested for per capita utilization as it contributes around 75% of wastewater generated [19]. Simplest way for reducing the scarcity of water and its pollution can be to increase the recycling and reuse of water and wastewater directly at the source of generation [3]. Developing country like India already falls under the category of water deficient and suffers from enormous demand of water due to the growing population, which can be handled by effectively recycling and reutilizing the wastewater.

Urban India is growing rapidly and this leads to number of problems pertaining to urban infrastructure and various services including water supply, urban sanitation systems, waste management and sewer systems. Around 87% of the urban population of country has been provided with the proper sanitation systems, the major cause of concern remains the collection and accurate disposal of the wastewater generated from such households. Solution to this severity is lies in reusing the treated wastewater and for this purpose many wastewater treatment plants are already been in use in many states.

3 Wastewater Reclamation and Reuse

Continued population growth along with the contamination of the both surface and ground water, uneven distribution of the water resources followed by the periodic draughts have led to the stricter laws formation by the water agencies and also simultaneously look out for new sources for water supply. Major environmental problem in India arises from the excessive contamination of surface and sub-surface sources of water. The concept of wastewater treatment in Indian scenario is not a new concept. Rapid industrialization has played a major role in the area of wastewater treatment and its reuse. With the concern and understanding for sustainability towards environment many treatment methodologies are being practiced today [4, 13]. In many Indian cities and towns, the available clean water supply systems have become scarce in order to fulfill the rising water demands. Keeping in mind the current critical situation, the used water which is collected from various communities should not be considered as waste but as a resource which can be recycled and reused [20]. The method and application of reutilization of water sources is becoming more suitable in many areas of the country due to acute shortage. Application of the treatment plants for wastewaters is part of the sustainable development as with the help of them the wastewater gets treated and further can be used among the same community [21]. The generalized approach remains the same i.e. minimize the utilization of freshwater and entry of pollutants and maximize the reuse of water. The introduction of the concept of reutilization and reusing of wastewater clears the picture regarding the water stress in overall world [22]. The major reuse of wastewater includes irrigation, industrial use, replenishment of surface water and ground water recharge.

The main purpose of the wastewater treatment systems is the provision of clean drinking water, therefore the selection of treatment technologies should only be the one best available which meets the regional as well as national quality standards prescribed. The principal goal of wastewater treatment plants from past many centuries have been primarily the same i.e. the generation of water which is free from all the chemicals and appeals to the people consuming them. Figure 2 shows the major division of wastewater treatment if treatment process is for making water drinkable.

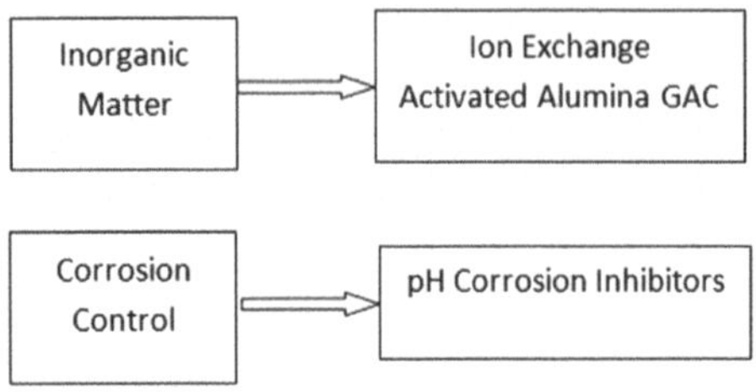

Fig. 2 Process for treatment of wastewater [11]

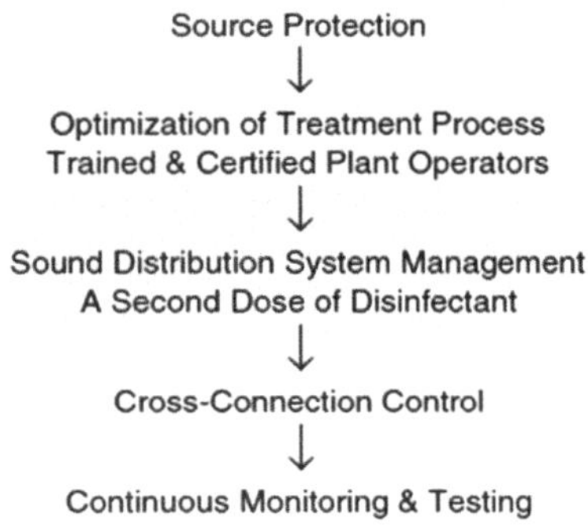

Fig. 3 Multiple barrier approach

The act for safe drinking water (SDWA) along with environmental protection act (EPA) has set some standards for drinking water and this was named as multiple barrier approach [23]. This multiple barrier approach has certain fundamentals like assessment along with and protection of drinking water sources, rationalizing the treatment processes, certifying the reliability of distribution systems and continuous monitoring and testing of the water before it reaches the public domain. Figure 3 summarizes the multiple barrier approach.

If the wastewater after certain level of treatment is still not fitted reuse for reuse, so the decentralized wastewater treatment systems may be engaged for reducing the amount contamination so to bring the water to a level which is safe for reuse. India's growth story has been impressive so far with the industry and service sectors being the growth drivers. The wastewater treatment and its reuse if considered as business can reduce the expenditure required in treatment processes thereby making it economical [24]. Reusing wastewater especially in the industrial sector helps in reduction of expenses incurred in water supply systems and wastewater treatment processes thereby leading to less stress on water sources. For many local governments, the wastewater collection systems are one of the most valuable assets they own [25]. The wastewater collection systems also protect public health and environment in tremendous ways. Due to the cost of treatment, health issues and safety concerns, the wastewater reuse applications have been limited primarily to non-usable uses like agriculture, irrigational and industrial systems [11]. At some dwellings where there exist no possibilities for freshwater supplies, such communities are investigating the prospects of direct and indirect reuse of wastewater.

Table 1 Categories of wastewater reuse and potential issues [5]

S. No.	Wastewater reuse category	Issues
1	Recycling, reclamation and Reuse of industrial wastewater; water for cooling operations; feed for boiler; water for construction	The constituents which are present in reclaimed water are associated with scaling, corrosion, formation of biological growth; human health concerns like transmission of contaminants in water
2	Agricultural irrigation; crop irrigation	If not managed properly can lead to surface as well as ground water contamination
3	Ground water recharge; replenishment	Possible contamination of ground water aquifer used as source of potable water source
4	Potable reuse; mixing properly in water sources	Elements in recycled water have few small amount of organic contaminants which can cause toxicological hazards.

The degree of treatment given to the wastewater is required for protection of human health as well as environment will depend upon the level of consistency of treatment systems. Few such categories are listed in Table 1.

The treated wastewater used in industrial units is mainly for the cooling processes. The cooling water plays a predominant role after reuse in industrial sectors as they can be further used for cooling processes in ponds or towers. These hold a large amount of the demand of water in industries. The demand in industrial sector highly varies so as to provide sufficient supply of water which hold a quality that adheres to the standards. In India, Chennai is one such state which leads the way in wastewater reuse technologies [26]. The reclaimed wastewater can also be used for landscape irrigation like irrigation of parks, grounds, golf courses, areas around the commercial sites and residences.

Recycling methodologies holds a place depending upon the conditions like the accessibility to ample supply of water sources, economic conditions, probability of the further use of reused water and the methodologies and policies for discharge of wastewater [1, 2, 12]. Figure 4 gives on overview of the scheme of reuse of wastewater by industries.

There are many hazards linked with the use of the wastewater which is partially treated or that remains untreated. These include pollution of ground water, soil pollution leading to undesirable consequences on the consumers as well as farmers who consume and use this untreated wastewater and related products. Despite of such adverse effects, wastewater is still being used as a source of livelihood. In developing countries, due to scarcity of water, farmers have no option left but to use wastewater in the form available [1, 2, 4, 9]. It is a common practice in developing countries to use the treated, untreated and partially treated wastewater in field of agriculture [4, 13]. Table 2 shows the environment protection act (EPA) guidelines for reuse of wastewater.

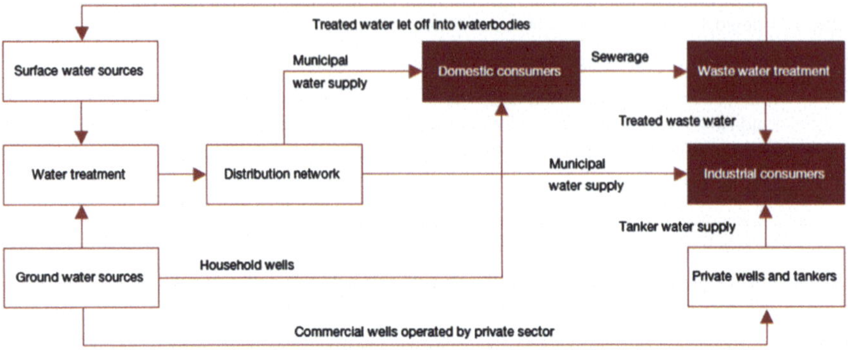

Fig. 4 Schematic diagram of reuse of wastewater by industries [4]

Table 2 EPA suggested guidelines for wastewater reuse [6]

Level of treatment	Reuse	Treated quality of water	Monitoring of water	Setback distances
Tertiary	For urban reuse Food crop irrigation Recreational activities	pH = 6–9 BOD_5 < 10 mg/l Turbidity < 2NTU E. coli = none	pH = every week BOD_5 = every week TSS = every day E. coli = every day	15 m (50 ft) to potable water supply
Secondary	Restricted Access area irrigation Food crop irrigation Nonfood crop irrigation	pH = 6–9 BOD_5 = 30 mg/l TSS = 30 mg/l E. coli = 200/100 ml	pH = every week BOD_5 = every week TSS = every day E. coli = every day	30 m (100 ft) to areas accessible to public and 90 m (300 ft) to potable water supply

The reuse of wastewater is a regular practice in developing countries like India [5]. Various techniques have been designed by the developing countries in regard to reuse of the wastewater which can be utilized by the developing countries. Due to the tremendous cost of disposal of the untreated wastewater not only financially but also causing the unstable ecosystem makes it a necessity to look forward to an effective water resource planning and reviewing recycling projects. Approximately 80% of the wastewater generated in developing countries (like China and India) is mainly used for purpose of irrigation [5, 14, 15].

Mostly, the wastewater generated in Indian cities and states is being reutilized in agricultural fields but the major drawback faced in this regard is the lack of policy framework and generation of standards [18]. The water and wastewater treatment sector accounts for a major financial part in both public and private sectors. Keeping this in view, alternative techniques for both treatment and recycling of wastewater in Indian states and cities must be designed. In India, major part of wastewater is untreated which is directly used in agricultural sector which is significantly lowers the risks as compared to its usage in industries or households. Major cities which are involved in such activities are Mumbai, Hyderabad, New Delhi and Ahmadabad. Out of 16,662.5 ml per day of discharge only about 23% is given treatment before it can be released and the rest is disposed of without any prior treatment [17].

There are both pros and cons to the wastewater reuse applications. The pros include the rise in rate of generation of employment, providing food security to the farmers, clean and suitable irrigation water along with recycled nutrients in wastewater. Due to all year along availability of wastewater, this assures employment opportunities to the poor farmers and labors. It can be observed from many reported literatures [26], that almost 43% of the paddy is grown in treated wastewater and is continuously used by urban areas. The high level of nutrients present in the wastewater is not only helpful to farmers as they act as an substitute of fertilizers which can be a reliable source of increasing crop intensity [9].

Alternatively, the negative impacts of untreated wastewater are a big threat to all mankind as it generates long term effects. Some of the chronic hazards include contamination of soil by increase in its salinity levels, rise in level of heavy metal accumulation in soil as well as ground water etc. which further restraint the crop choice leading to decline in crop productivity. Many researchers Goyal and Sharma, Phillippe et al. [2], Deccan [3] and Ghosh et al. [11] have concluded in their studies that there has been many instances where ground water have high slat concentrations due to interference of untreated wastewater. Untreated wastewater also has shown the presence of many bacteria's, viruses and pathogens, causing several diseases. Wastewater also tends to increase eutrophication process thereby creating unbalanced ecosystem and hampering sustainable development.

4 Conclusion

Wastewater use in agriculture has been a common phenomenon in a number of water scarce developing countries for more than a century now. It has been and is still supporting the livelihood of a number of urban and peri-urban farmers. However, with the growing population the volumes of urban wastewater have dramatically increased. The problem is further complicated with increased contamination of wastewater with new chemicals, with changing lifestyles of people and the addition of industrial effluents. The environmental and health related problems of the use of untreated wastewater has become prominent. There is an urgent need to address these problems before this untreated wastewater completely pollutes all the rivers/natural water bodies. Most of the developed countries have been able tackle this problem by appropriate treatment of wastewater and safe disposal with minimum environmental and health impacts. Time and again, developing countries have tried to adopt similar water treatment technologies from the western world and have failed. There are both social and economic reasons for this failure. It is very important to understand the social and economic context of a society/community/city before a technology is implemented. The different social economic aspects to be considered are—perceptions of people regarding water, education levels, awareness towards the environment and the willingness and ability to pay to protect their environment. In addition to this, the political will and institutional support are essential to make wastewater a safe asset for people in developing countries. With issues of climate change, increases in urban population and increased demand for water from competing sectors, wastewater recycling is becoming an important strategy to complement the existing water resources for both developing and developed countries and there are lessons, experiences, data and technology which can be shared for mutual benefit.

References

1. Babette N, Katharina S (2018) Comparing urban wastewater systems in India and Brazil: options for energy efficiency and wastewater reuse. Water Policy 1129–1111
2. Philippe R, Rohit C, Lukas U (2020) Governance arrangements for the scaling up small scale wastewater treatment and reuse systems—lessons from India. J Front Environ Sci 8. https://doi.org/10.3389/fenvs.2020.00072. Article 72
3. Deccan H (2019) NGT Order: 50 Karnataka SPTPs need to be upgraded. https://www.deccanherald.com/city/ngt-order-50-karnataka-stps-need-to-be-upgraded-732388.html. Accessed 27 June 2019
4. Goyal K, Kumar A (2020) Multi-criteria analysis to identify the factors affecting water reuse scheme in India. Water Environ J.https://doi.org/10.1111/wej.12642
5. CPCB (2019) About us. http://cpcb.nic.in/Introduction/. Accessed 27 June 2019
6. CPHEEO (2019) About us. http://mohua.gov.in/cms/cpheeo.php. Accessed 28 June 2019
7. Gupta PK, Sharma D (2018) Assessments of hydrological and hydro-chemical vulnerability of groundwater in semi-arid regions of Rajasthan, India. Sustain Water Resour Manag 1–15.https://doi.org/10.1007/s40899-018-0260-6
8. Gupta PK, Yadav BK (2019) Subsurface processes controlling reuse potential of treated wastewater under climate change conditions. In: Singh R, Kolok A, Bartelt-Hunt S (eds) Water conservation, recycling and reuse: issues and challenges. Springer, Singapore. https://doi.org/10.1007/978-981-13-3179-4_8
9. Singh A, Sawant M, Kamble SJ, Mihir H, Markus S, Enrique A, Absar K (2019) Performance evaluation of a decentralized wastewater treatment system in India. Environ Sci Pollut Res 26:21172–21188. https://doi.org/10.1007/s11356-019-05444-z
10. Chertow M, Gordon M, Hirsch P, Ramaswami A (2019) Industrial symbiosis potential and urban infrastructure capacity in Mysuru, India. Environ Res Lett 14(7.14):075003
11. Ghosh R, Kansal A, Venkatesh G (2019) Urban water security assessment using an integrated metabolism approach—case study of the National Capital Territory of Delhi in India. Resources 8(2):62
12. Angelakis A, Asano T, Bahri A, Jimenez B, Tchobanoglous G (2018) Water reuse: from ancient to modern times and the future. Front Environ Sci 6:26
13. Mustapha IH, Gupta PK, Yadav BK, VanBruggen JJA, Lens PNL (2018) Performance evaluation of duplex constructed wetlands for the treatment of diesel contaminated wastewater. Chemosphere. https://doi.org/10.1016/j.chemosphere.2018.04.036
14. Global Water Intelligence (2019) Effluent regulations threaten Indian sewage project chaos, vol 20. Indian Projects. https://www.globalwaterintel.com/global-water-intelligence-magazine/20/5/general/effluent-regulations-threaten-indian-sewage-project-chaos. Accessed 2 July 2019
15. Economic Survey of Delhi 2018–2019 (2019). http://delhiplanning.nic.in/sites/default/files/13%29%20Water%20Supply.pdf
16. Ranjan S, Gupta PK, Yadav BK (2018) Application of nano-materials in subsurface remediation techniques—challenges and future prospects. Chapter 6, In Recent advances in environmental

management, vol 1. CRC Press Taylor & Francis Group. ISBN: 9780815383147. Google Scholar
17. Singh A, Kamble SJ, Sawant M, Chakravarthy Y, Kazmi A, Aymerich E, Philip L (2018) Technical, hygiene, economic, and life cycle assessment of full-scale moving bed biofilm reactors for wastewater treatment in India. Environ Sci Pollut Res 25(3):2552–2569
18. Ranjan P, Vikram G, Manish KM, MdNasim S, Shubham K, Ranjan P (2020) Wastewater treatment in urban area: a review. Int J Environ Plan Dev 6(1). https://doi.org/10.37628/jepd.v6i1.574
19. Eshetu M, Todt D, Heistad A (2018) Treatment of source-separated blackwater: a decentralized strategy for nutrient recovery towards a circular economy. Water 10(4):463
20. Gupta PK, Ranjan S, Kumar D (2018a) Groundwater pollution by emerging industrial pollutants and its remediation techniques. In: Recent advances in environmental management, Chapter 2, vol 1. CRC Press Taylor & Francis Group, Google Scholar. ISBN 9780815383147
21. Gupta PK, Abhishek YBK (2018b) Impact of hydrocarbon pollutants on partially saturated soil media in batch system: morphological analysis using SEM techniques. In: Water quality management, Chapter 5; Water Sci Technol, vol 79, Springer Google Scholar. ISBN: 978-981-10-5794-6
22. Ricart S, Rico AM, Ribas A (2019) Risk-Yuck factor nexus in reclaimed wastewater for irrigation: comparing farmers' attitudes and public perception. Water 11:187. https://doi.org/10.3390/w11020187
23. Sathaiah M, Chandrasekaran M (2020) A bio-physical and socio-economic impact analysis of using industrial treated wastewater in agriculture in Tamil Nadu, India. Agric Water Manag 241:106394. https://doi.org/10.1016/j.agwat.2020.106394. ISSN: 0378-3774
24. Kirti G, Arun K (2020) A modelling approach to assess wastewater reuse potential for Delhi city. Water Supply 1 20(5):1716–1725. https://doi.org/10.2166/ws.2020.080
25. Lekshmi B, Sharma S, Sutar RS, Parikh YJ, Ranade DR, Asolekar SR (2020) Circular economy approach to women empowerment through reusing treated rural wastewater using constructed wetlands. In: Ghosh S (eds) Waste management as economic industry towards circular economy. Springer, Singapore. https://doi.org/10.1007/978-981-15-1620-7_1
26. Baawain MS, Al-Mamun A, Omidvarborna H, Amal-Al S, Choudri BS (2020) Public perceptions of reusing treated wastewater for urban and industrial applications: challenges and opportunities. Environ Dev Sustain 22:1859–1871.https://doi.org/10.1007/s10668-018-0266-0

Understanding Sustainable Arsenic Mitigation Technology Application in the Indian Subcontinent

Nadim Reza Khandaker and Mohammad Moshiur Rahman

Abstract Groundwater is the source of potable water in many regions of the Indian Subcontinent. Unfortunately, in the recent decades it has come to light that the groundwater in many aquifers has high concentration of arsenic. The arsenic contamination public health crisis in Bengal India and Bangladesh is in Biblical proportions, tens of millions of people are affected. A viable mitigation effort is the deployment of sustainable arsenic removal technologies for arsenic safe water supply to the affected regions. This paper introduces the available arsenic removal technologies, their mechanism of operation, and discusses the performance efficacy of the technologies in the groundwater matrix of the affected regions. The technologies range from specific arsenic adsorption media-based filters, coagulation co-precipitation arsenic removal technologies, membrane technologies and in situ remedial technologies. The removal technologies were sensitive to groundwater chemistry with respect to removal efficiency and system life. There has to be in place in fracture in the affected areas to monitor the performance of the arsenic removal technologies with respect to whether they are producing arsenic safe water and once system performance limit is reached remedial measures needs to be taken for sustainable mitigation. Many of the technologies generate arsenic waste that has to be dealt with and the communities need to develop handling and ultimate disposal facilities for sustainable mitigation.

Keywords Arsenic removal · Technology · Groundwater chemistry · Performance · Sustainability

---

N. R. Khandaker · M. M. Rahman (✉)
North South University, Plot: 15, Block: B, Bashundhara R/A, Dhaka 1229, Bangladesh
e-mail: mohammad.rahman@northsouth.edu

N. R. Khandaker
e-mail: nadim.khandaker@northsouth.edu

© Springer Nature Switzerland AG 2022

S. Yadav et al. (eds.), *Wastewater Assessment, Treatment, Reuse and Development in India*, Earth and Environmental Sciences Library,
https://doi.org/10.1007/978-3-030-95786-5_5

1 Introduction

Bangladesh and India are two countries of the same heritage sharing the same constraints of providing economic growth and prosperity to the majority of its population living with the constrains of poverty. In this resource challenged context both countries have the challenge to provide safe drinking water to its population. Historically, both courtiers had unacceptable infant mortality rates due to water borne diseases due to outbreaks of Cholera, Typhoid, Hepatitis, and other diarrheal diseases. In this context public health organizations in search of safe source of drinking water went for the utilization of groundwater. It is a public health norm that groundwater at a safe depth is pathogen free and can be consumed with some certainty without the risk of becoming ill from water borne diseases. This was heavily promoted in rural south Asia and majority of the rural population of the Subcontinent is reliant on groundwater for their safe source of potable water.

They say that sometimes road to hell is paved with good intentions and this is the case of the genesis of the arsenic in groundwater problem. Trying to save the population from water borne diseases and promoting groundwater as a safe source without testing for the presence of arsenic is a grave unheralded tragedy of this century. This if unabated is the single largest mass poisoning of population in the world. Communities in the six major continents of the world are affected. The map of the world below indicates the geographical localities where groundwater contamination with arsenic is a public health problem (see Fig. 1).

Of which the crisis is of an epidemic proportion in the greater Ganges Brahmaputra deltas of India, Bangladesh, and the plain lands of Nepal. The critically affected areas are the alluvial aquifers of West Bengal of India and Bangladesh (Fig. 2).

Tens of millions of people are affected and suffering from arsenic related diseases with chronic consumption of arsenic laced water. The challenge for water professionals is how to provide arsenic safe drinking and cooking water to the affected

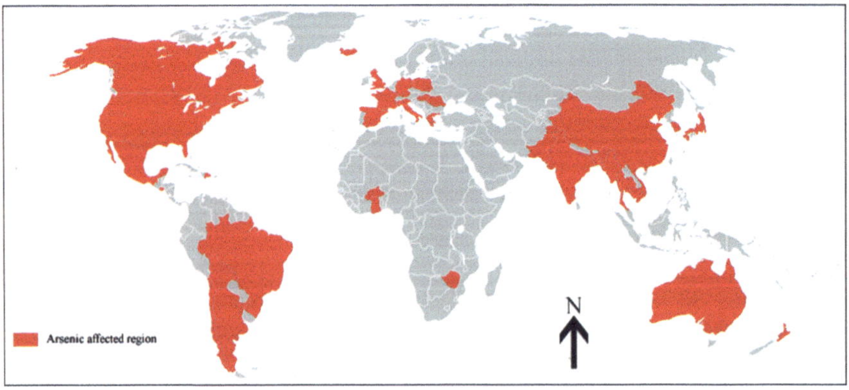

Fig. 1 Arsenic affected regions in the world [1–6]

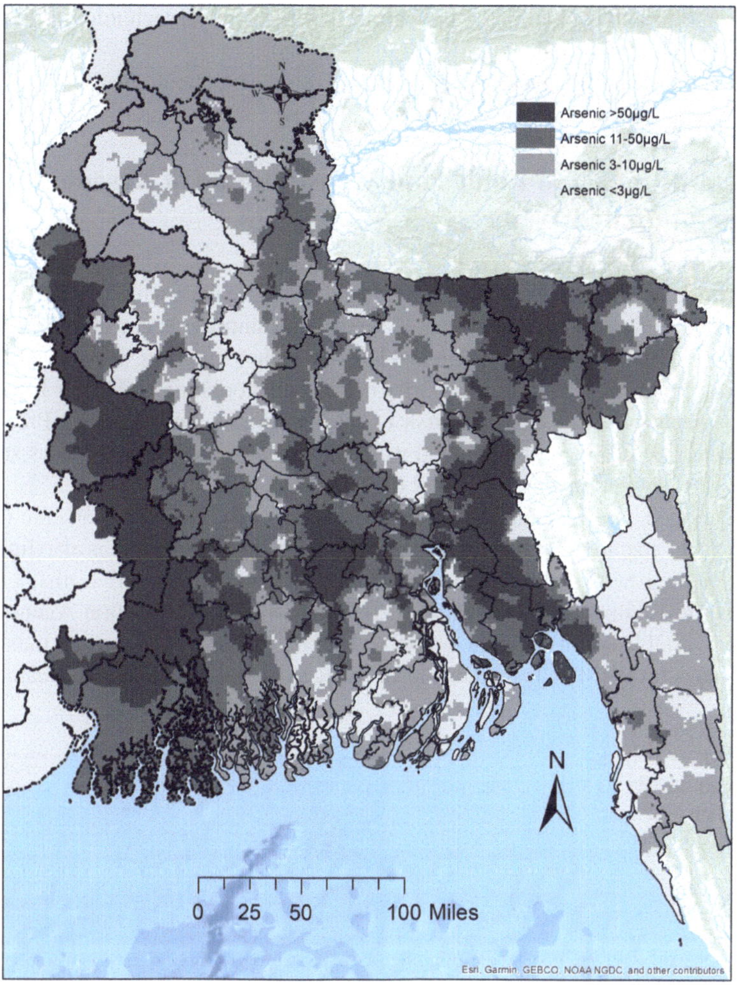

Fig. 2 Arsenic affected areas of greater Bengal Basin

communities. One of the solutions in the overall management portfolio is the effective introduction and sustainability of arsenic removal technologies. This paper will introduce germane issues to the overall application of arsenic removal technologies to the greater Bengal basin of India and Bangladesh. We will address the following.

- Toxicology of arsenic.
- Aquatic chemistry of arsenic.
- Fundamental mechanisms of arsenic removal technologies.
- Groundwater chemistry and its effect on performance of arsenic removal technologies.

- Description and evaluation of prevailing arsenic removal technologies in India, Bangladesh, and Nepal.
- Safe waste disposal generated from arsenic removal technologies.

2 Health Impact of Long Time Consumption of Arsenic Contaminated Water

Long terms consumption of arsenic contaminated water causes nausea, vomiting, and diarrhea. Chronic arsenic contaminated water consumption is causing the tell-tale appearance of dark spots on the skin and lesions on the palm (Fig. 3), sole, and trunk which can progress to skin cancer. Drinking arsenic contaminated water for a long time has been linked to increased risk of internal organ cancers [7, 8]. Prolonged consumption of arsenic contaminated water is linked to metabolic diseases such as heart disease, hypertension and type 2 diabetes mellitus [9].

Oxidation state of arsenic in water affects its toxicity. In general arsenic (III) is more toxic to arsenic (V) [10]. Drinking arsenic contaminated water is also linked to development growth of children along with low birth weight and still births [11–13].

Based on lifetime cancer risks the U.S. Environmental Protection Agency (US EPA) has established a national primary drinking water regulation, or maximum contaminant level (MCL), for arsenic of 10 μg/L [14]. The regulation came into

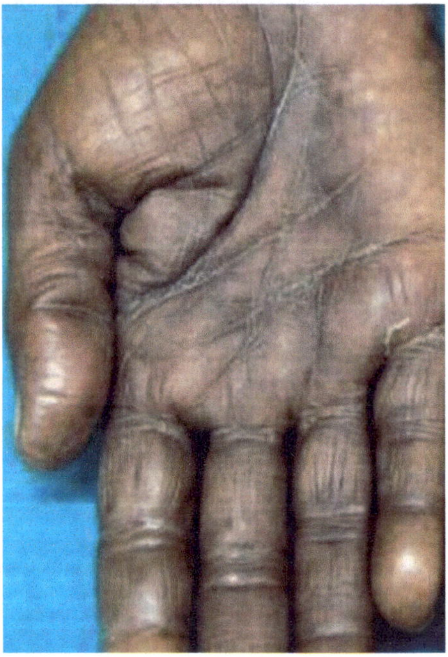

Fig. 3 Hardening of the palm due to chronic consumption of arsenic contaminated water

effect in 2006. The Government of Bangladesh MCL is set at 50 μg/L (Department of Public Health and Engineering (DPHE)) and the Indian Government MCL for arsenic is set at 50 μg/L [15].

3 Chemistry of Arsenic in Groundwater

Arsenic (As) is a group five element in the Periodic Table with the ability to exist in different oxidation states. Arsenic in groundwaters exists as arsenite (+3 charge), or arsenate (+5 charge) combined with multiple oxygen having a net negative charge existing in different forms in aqueous Solution. The pH and the redox potential of the water determine the forms of arsenite and arsenate that will predominate in aqueous solution and is shown in the Fig. 4 based on representation of existing speciation models and modified to illustrate the variation of existing aqueous species with changing pH at a fix redox potential [16]. The charge on the arsenic form greatly determines the ease of removal based on its interactions with surfaces and as earlier mention the form also affects the level of toxicity [16].

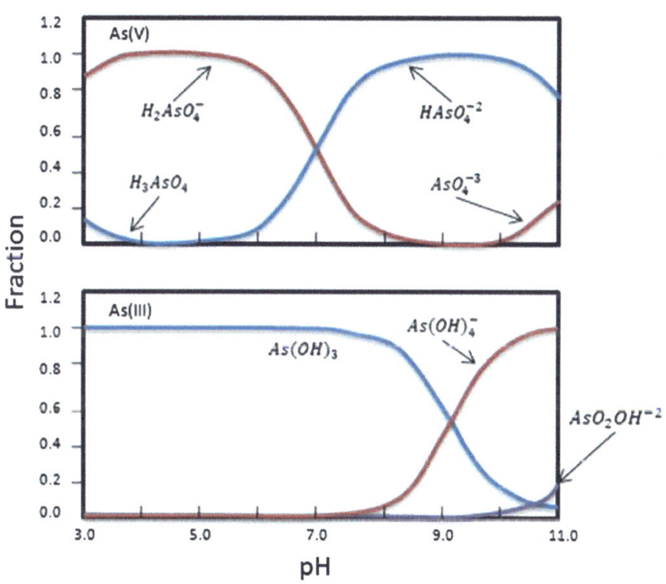

Fig. 4 pH-dependent speciation of As (V) and As (III) in aqueous solution (modified after [16])

4 Geogenic Occurrence of Arsenic in Groundwater

Global geogenic occurrence of arsenic in groundwater can be broadly associated with three groups of aquifers [1]:

1. Strongly reducing aquifers.
2. Oxidizing aquifers with high alkalinity and pH.
3. Aquifers with high arsenopyrite and other sulfide mineral content.

High groundwater arsenic is associated with aquifers having strongly reducing conditions is of the most concern in the context of public health due to its extensive frequent occurrence compared to other aquifers defined above. Aquifers with strongly reducing conditions exist in, e.g., Bangladesh, India, Vietnam, Cambodia, and Pakistan. The affected aquifers comprise micaceous sands, silts and clays deposited by the Ganges, Brahmaputra and Meghna river systems and their precursors during Holocene period. These aquifers are generally shallow (less than 100–150 m deep) [1]. Arsenic in the strongly reducing aquifers of Bangladesh and West Bengal is supplied most likely by the arsenic containing sulfide minerals in the Himalaya that are weathered, transported by the Ganges, Brahmaputra, and Meghna rivers, and deposited throughout the Bengal Basin. Geochemical analysis of the sediments from the Bengal Basin shows that a major portion of arsenic is strongly associated with secondary iron oxides, which were formed by the oxidation of primary and secondary sulfide minerals containing arsenic [17–19].

The most widely accepted release mechanism of arsenic into the anoxic groundwater is reductive dissolution of iron oxides in the soil of the aquifer to which the arsenic is attached to The reduction of the iron oxides decreases the binding sites for arsenic which causes the mobilization of arsenic into groundwater from the soil [17, 20–30]. These reduction processes are driven by microbial degradation of organic carbon sources present in the aquifer that serves as substrate and the source of electron for the microbes. The sources of the organic carbon in the aquifer include degradation of plant material deposited within the sediments during the process of landform development, dissolved organic matter from near-surface peat deposits, and dissolved organic matter recharged from the surface water bodies enhanced by irrigation pumping [21, 23, 25, 26, 30]. The aquifer sediments in most affected areas of Bangladesh and West Bengal are capped by a layer of clay or silt which restricts entry of air to the aquifers further ensuring an anoxic environment.

5 Testing as the First Step Towards a Sustainable Arsenic Mitigation Strategy

In a locality, as a first step towards a sustainable arsenic mitigation strategy, all the bore holes wells need to be tested for presence of arsenic or not (Fig. 5). The geology of a locality can vary greatly and as such would affect the release of arsenic into

Fig. 5 Testing and marking of test well as safe with green paint

the groundwater. Take the example of Araihazar [31–33] in greater Dhaka district of Bangladesh where even in this small municipality, the geology is different from locality to locality, and hence the wells in some areas of the municipality have high arsenic concentration and some wells have arsenic concentrations that are within the MCL. This necessitates the testing of all wells in a municipality in an arsenic affected area. The process is simple and costs less than a dollar a test using field test kit and marking the wells as safe or unsafe using a coding system. In Bangladesh the coding system is painting the snout of the well RED if arsenic concentration is greater than the MCL and GREEN if arsenic level in the well water is within the limit.

The involvement of the civic society is an essential part of the mitigation strategy that calls for sharing of water from the safe wells. The population needs to be motivated towards the concept of the precious recourse, safe drinking water. Dedicated community wells can be maintained to provide safe drinking water as a mitigation tools supported by the government or by the civic society. Having the availability of arsenic safe wells in a locality is the easiest form of mitigation measure that can be sustained but requires the involvement of the whole community.

6 Arsenic Removal Technologies as a Mitigation Strategy

In areas where there are no arsenic safe acquirers the only viable alternative is the promotion of arsenic removal technologies. The arsenic removal technologies that are being employed in the Subcontinent can be broadly classified as:

- Adsorption based arsenic removal technologies
- Coagulant based arsenic removal technologies
- Membrane systems, and
- In situ in ground based arsenic immobilization technologies.

6.1 Adsorption Arsenic Removal Systems

Adsorption is a mass transfer process where compounds in solution come to the solid liquid inter-phase by mass transfer principles and in the solid surface it is attached by physical or chemical forces. Arsenate and arsenite adsorbs to oxy-hydroxide surface of that are positively charged, such as a protonated surface hydroxyl group or through ligand exchange reaction in which the surface hydroxyl group is displaced by the adsorbing ion [34]. The adsorption reaction mechanism of arsenic species onto solid metal oxy-hydroxide surfaces by [35] Edwards, 1994 as:

$$= S - OH + H^{+} + H_2AsO_{4-} \rightarrow = S - H_2AsO_4 + H_2O \text{ (arsenate sorption)}$$

$$= S - OH + H^{+} + H_3AsO_{3-} \rightarrow = S - H_3AsO_3 + H_2O \text{ (arsenite sorption)}$$

The common filters that are being deployed use adsorption media of ferric oxy-hydroxide, activated alumina, ferric coated activated alumina, granular ferric hydroxide, and manganese dioxide coated sand [36, 37].

The efficacy of each media depends on water quality parameters like pH, the presence of other ions such as phosphates and silicates, and speciation of arsenic. System design and operational parameters such as empty bed contact time, loading rates, bed porosity effects system efficacy. The speciation of arsenic plays a significant role on the efficacy of an arsenic removal system. Arsenic (3) is uncharged in water and hence is difficult to remove from solution compared to charged arsenic (5) species from water [35].

They are flow through filters where influent arsenic laden water passes through the active media bed where the arsenic in solution is removed by the principles of adsorption outlined above. The filter bed effluent water is arsenic free. The systems deployed in the sub-continent can be highly engineered for a community or very rudimentary such a bucket system with the media bed in the bucket. The two extreme systems are shown in the Fig. 6.

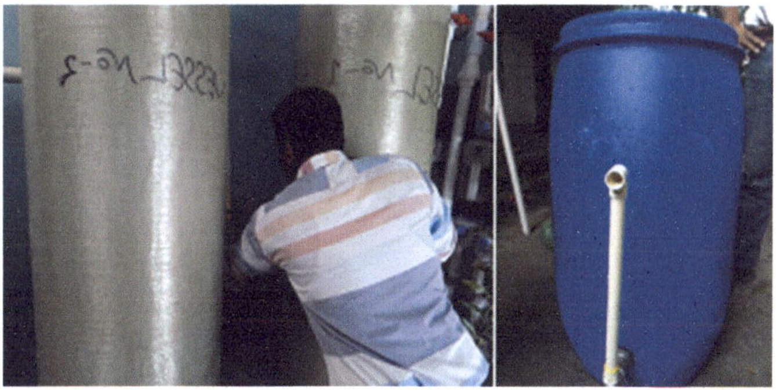

Fig. 6 Community scale and household scale adsorption bed with arsenic removal media

6.2 Community Systems Using Coagulation Followed by Filtration

In coagulation dissolved arsenic species are converted to insoluble products by the addition of a coagulant and the mechanism is thought to a combined mechanism of precipitation, co-precipitation, and adsorption around the formation of a solid oxy-hydroxide surface [35]. Trivalent metal salts are generally used as coagulating agent for arsenic removal systems. In general alum and ferric salts are used as the coagulant. System performance is affected by pH, coagulant dose and the speciation of arsenic. The coagulation treatment process for arsenic removal involves a chemical oxidation step to convert arsenic three to arsenic five followed by coagulant addition and pH adjustment to near neutral pH to remove arsenic from solution by the forming oxy-hydroxide solids followed by settling/microfiltration to remove the arsenic laden metal oxy-hydroxide solids.

A low-tech rural application of the coagulation filtration concept to remove arsenic is the use of zero valent iron as the source of the active coagulating agent. Rusted iron nails serve as the source hydrous ferric hydroxide that serves as the active ingredient to remove the arsenic from solution and that is then removed by filtration. This process has won the World Bank Prize (MIT filter) and the SONO filter (winner of the Granger Prize) [38].

6.3 Application of Membrane Systems

Both reverse osmosis (RO) and nano-filtration (NF) has a high rate of rejection for arsenic three and arsenic five species. The mechanism for rejection is prevention of diffusion as the mechanism of flow for RO and NF is solution diffusion. The efficacy of the system is not dependent on pH of the water or is not affected by the presence of other anions. For membranes with large pore size where the separation mechanism is straining based on molecule size (such as micro and ultra-filtration) coagulation with metal oxy-hydroxide as a prerequisite step is necessary [39]. The systems are gaining popularity of recently in the subcontinent with the advent of rural electricity coverage for the membrane systems are pressure driven and requires pumps to build the required pressure for the systems to operate.

6.4 In Situ in Ground Based Arsenic Immobilization Technologies

Subsurface arsenic removal (SAR) is an available treatment option that is relatively new. SAR technology was tested in Bangladesh by [40–43] and in West Bengal by [44]. SAR systems can be easily operated with a minimal level of waste generation [40–46]. The basic principle of SAR is to inject oxygenated water into an anoxic aquifer to create a subsurface iron oxide filtration zone for arsenic. SAR technology can be a relatively cheap treatment option to supply arsenic safe drinking water to the rural people of Bangladesh and West Bengal because SAR can be easily installed in preexisting shallow tube well with only minor retrofitting [43].

SAR technology involves three steps in removing arsenic from groundwater (e.g., [41]). The steps are as follows: (1) aeration of the extracted groundwater from the aquifer is performed in an aeration tank to increase dissolved oxygen concentration of the water. (2) after aeration the water enriched with oxygen is injected back into the same aquifer to ensure chemical reaction between oxygen ferrous iron in the aquifer so that hydrous ferric oxide (HFO) can be formed. (3) A larger volume of water with low dissolved arsenic can be extracted due to adsorption of arsenic on to HFO [41, 47, 48]. These three steps together is called a cycle of operation. Reductive dissolution of iron oxides releases a significant levels of iron and arsenic to the groundwater which results in co-occurrence of arsenic and iron in groundwater [20, 26, 49–53]. Thus the presence of dissolved Fe in groundwater is very important for immobilizing of arsenic in the subsurface by the processes of Fe(II) oxidation, followed by Fe oxide precipitation (forming the HFO surface), on which adsorption and co-precipitation of As, Fe, and many other ions occur. Episodic injection of aerated water into the aquifer maintains an oxidation zone within the aquifer and oxidize adsorbed Fe(II) to Fe(III)-oxide can exist. The outlined processes create new adsorption sites in the aquifer for dissolved Fe(II) and other elements such as arsenic to sorb. To reach the maximum removal efficiency, several injection-extraction cycles may be required [43]. The performance of SAR systems may be defined using Eq. 1 as described by [48].

$$Q_E = \frac{V}{V_i} \tag{1}$$

where V is the volume of extracted water meeting the drinking water standard and V_i is the volume of injected water.

SAR has the following advantages over other household and community arsenic removal systems, such as SONO and Alcan: [48, 54]

1. Aquifer materials in the subsurface can be used as filter media thus, no additional filter media is required.
2. Only minor maintenance is sufficient to operate a SAR system.
3. Locally available hardware can be used for the modification of an existing hand pump to build a new SAR system or for repairing purpose of any SAR system.
4. Fe is removed from the water along with arsenic, which significantly increase the potential for social acceptance.

6.4.1 Description of a Typical SAR System

A typical SAR unit is composed of a well, a tank for aeration, and pipelines for injection and extraction of water during SAR operation (Fig. 7). Separate pipelines with valves need to be connected to the SAR well for injection and extraction. The aeration tank is connected to the injection and extraction pipes to conduct SAR operation. A preexisting rooftop (approximately 5 m above ground level) is used normally to place the tanks to allow for gravity injection of aerated water. Showerheads are

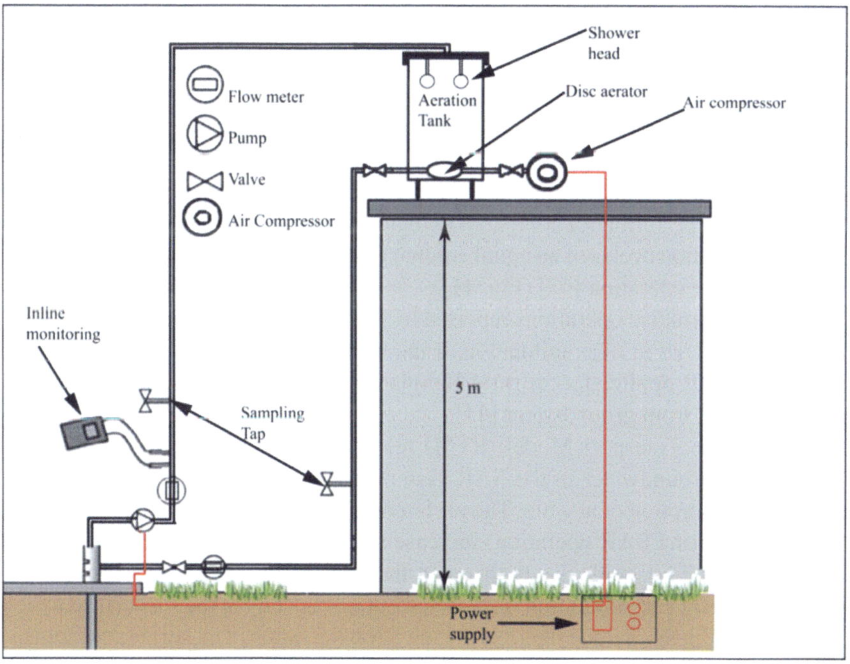

Fig. 7 Schematic diagram of a typical SAR system

installed at the top and disc aerators are placed at the bottom of the aeration tank to achieve high levels of dissolved oxygen in the water, air is supplied using an air compressor. Electrical suction pump is required for extraction of groundwater.

6.4.2 Efficiency of SAR Systems

SAR technology was tested in West Bengal by [44] and in Bangladesh it was tested by [40, 43] and more recently by [47]. Sen Gupta et al. [44] tested SAR at six different locations in Kasimpore, a village in North 24 Parganas District of West Bengal. The reported arsenic concentrations in the treated water by [44] were less that the WHO guideline value of 10 μg/L. In the study of [40], the concentration of arsenic in the extracted water never dropped below the Bangladesh guideline, while in the study of [43] the level of As in the extracted water dropped below the Bangladesh guideline value of 50 μg/L. The effect of different operational parameters on arsenic removal during SAR operation was investigated by [47] at two SAR units in Bangladesh. Arsenic concentration in the extracted water (for a certain volume) during all experiments conducted by [47] dropped below the Bangladesh guideline. SAR operation with a larger injection volume (5 m^3) performs better for arsenic removal in the subsurface than the use of a smaller injection volume (1 m^3) (Fig. 8). SAR performance (Q_E) for SAR well with 5 m^3 injection volume is approximately 1.5 and for SAR well with 1 m^3 injection volume the Q_E is approximately 1.1 (Fig. 8).

Higher arsenic removal for extraction with a lower pumping rate (13 L/min) compared to a higher pumping rate (50 L/min) was also reported by [47] (Fig. 9). Q_{Es} with lower pumping rate were always higher compared to higher pumping rate (Fig. 9).

Higher arsenic removal was reported by [47] for SAR operation with intermittent pumping compared to cycles with continuous pumping (Fig. 10).

SAR operation with repeated injection-extraction cycles of an equal volume produced better performance compared to regular cycles (Fig. 11). Q_{Es} with repeated injection-extraction cycles of an equal volume for all rounds were always higher than Q_{Es} with regular operation [47] (Fig. 11).

All three alternative operations reported by [47] resulted in better arsenic removal than regular operation. A combination of the three alternative operations is recommended for SAR application in rural Bangladesh to ensure substantial amount of arsenic removal from groundwater [47].

The Reactive Transport Model (RTM) results developed by [55] revealed that the pH of the groundwater in the SAR system during SAR operation significantly influenced the sorption of arsenic. The preferential removal of As(V) (after oxidation from As(III) during SAR operation) increase the overall arsenic sorption. Hydrous Ferric Oxide (HFO) gradually built up in the subsurface due to successive SAR cycles. These HFO accumulation increased sorption capacity in the subsurface which eventually facilitates arsenic removal during SAR operation. The surface complexation modeling suggests that simultaneous sorption of H_4SiO_4 is an important factor limiting arsenic removal during SAR operation [55].

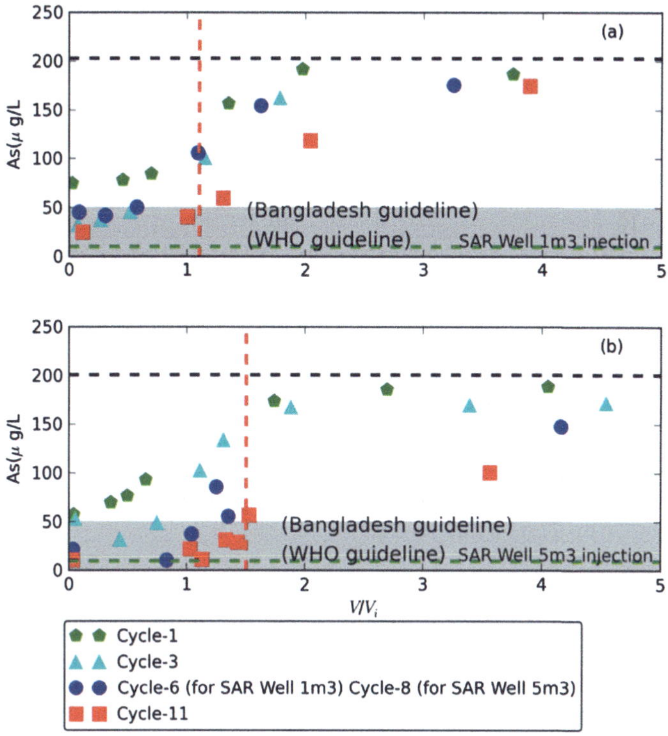

Fig. 8 Breakthrough curves showing the concentrations of arsenic (**a** and **b**) during SAR experimental cycles with different injection volumes (1 and 5 m^3). The background concentrations of As at the SAR wells are represented by the black dashed lines. The red dashed lines indicate the volume where As concentration in the extracted water exceeded the Bangladesh guideline during cycle 11. The Bangladesh guideline for arsenic is represented by the grey area (**a** and **b**). The WHO guideline for arsenic is represented by the green dashed line (**a** and **b**) (modified after [47])

The reactive transport model developed by [55] was used to assess the nationwide potential for SAR application in Bangladesh by [48]. A preliminary map was generated with suitable locations for SAR application in Bangladesh [48]. Results of the nationwide assessment for SAR application in Bangladesh revealed that the potential for SAR application in Bangladesh is substantial. [48] reported that a considerable performance can potentially be achieved with SAR technology for locations with As concentrations ranging between 50 and 150 µg/L. The lifetime of SAR systems for most of the locations in Bangladesh can be significant [48]. When a SAR unit produces 1000 L of arsenic safe water that can meet the water needs for 114 families for cooking and drinking water. The approximated cost for a SAR unit (with 1m^3 injection capacity) per family is US$13.00/year [48]. In contrast, a SONO filter (a household arsenic removal system) cost about US$45.00 to 50.00 with an average lifespan of 1 year [56]. The cost for a SAR unit per year per family (with a 1 m^3 injection capacity) is approximately half of the costs for a unit of SONO filter [48]

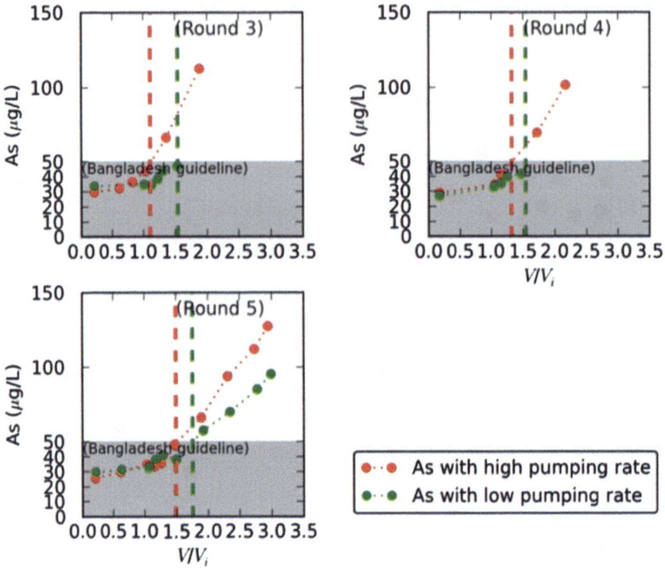

Fig. 9 Comparison of arsenic breakthrough between higher (red) and lower (green) pumping rates. The red dashed lines (cycles with high pumping rate) and green dashed lines (cycles with lower pumping rate) indicate the volume where As concentration in the extracted water exceeded the Bangladesh guidelines The Bangladesh guideline for arsenic is represented by the grey area (modified after [47])

and unlike the SONO filter does not have a finite life. Overall, based on the results reported by [44] suggests that SAR technology can be a potential option for potable water for rural communities exposed to high levels of geogenic groundwater arsenic in West Bengal. The preliminary map of suitable locations for SAR application reported by [48] can form the basis for future exploration of the possibilities of SAR in Bangladesh.

7 Concluding Remarks Towards Technology Application in the Subcontinent

Technology application ad its sustainability is greatly dependent on the chemistry of the groundwater and will dictate the operational life of the systems. Monitoring has to be a key feature of successful application of arsenic removal technologies. Another key feature is the ability of the community to successfully manage the high arsenic containing waste generated from the processes. The waste generated by the arsenic removal technologies are technology specific and as such each technology's environmental impact needs to be assessed with respect to the community's ability to safely handle the waste for long term sustainable application. There is no magic

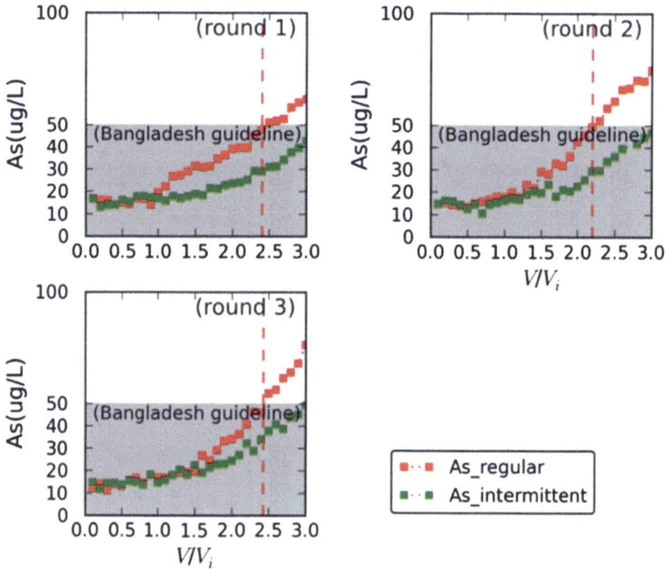

Fig. 10 Comparison of arsenic concentrations between regular cycles (red) and intermittent extraction cycles (green). The red dashed lines indicate the volume where As concentration in the extracted water exceeded the Bangladesh guideline during regular cycles. For intermittent cycles, the exceedance was beyond V/Vi = 3 Exceedance of the Bangladesh guideline and was not observed. The Bangladesh guideline for arsenic is represented by the grey area (modified after [47])

bullet, each technology has its benefits and drawbacks and as such the community has to be well informed about the technology for successful sustainable application. This paper we feel provides the background that will help a community understand the principles behind the arsenic removal technologies being deployed and promoted in the countries of the Indian Subcontinent.

8 Recommendations

Arsenic removal technology work however efficacy is greatly dependent on the type of technology and the chemistry of groundwater. They are not stand-alone systems. As such there has to be an on-ground monitoring system in place to monitor technology performance to ensure that the stakeholders are receiving arsenic safe water. Many of the treatment systems discussed generate arsenic rich spent media or high arsenic concentration waste liquid streams and the in fracture needs to be in place to handle the waste solids or the waste stream. Companies or organizations that intend to market arsenic removal technologies should be regulated heavily to ensure that a monitoring system is in place for treated water arsenic concentration monitoring and

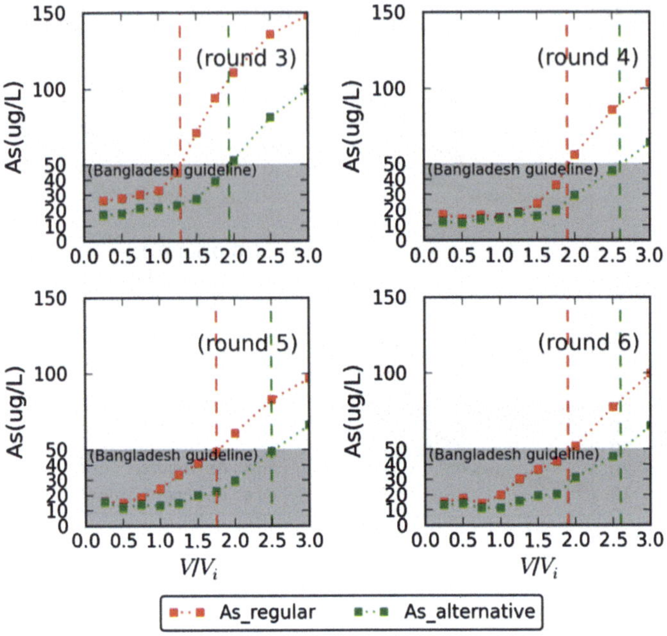

Fig. 11 Comparison of arsenic concentrations between regular cycles and alternative cycles (repeated injection–extraction cycles of an equal volume). The red dashed lines indicate the volume where As concentration in the extracted water exceeded the Bangladesh guideline during regular cycles. The green dashed lines indicate the volume where As concentration in the extracted water exceeded the Bangladesh guideline during alternative cycles. The Bangladesh guideline for arsenic is represented by the grey area (modified after [47])

the infrastructure in place to handle spent arsenic rich media or arsenic rich waste liquid stream.

Acknowledgements The authors would like to acknowledge the logistical support of North South University Bangladesh where both the authors are affiliated.

References

1. Smedley PL, Kinniburgh DG (2002) A review of the source, behaviour and distribution of arsenic in natural waters. Appl Geochem 17(5):517–568. https://doi.org/10.1016/S0883-2927(02)00018-5
2. Appleyard SJ, Angeloni J, Watkins R (2006) Arsenic-rich groundwater in an urban area experiencing drought and increasing population density, Perth, Australia. Appl Geochem 21(1):83–97. https://doi.org/10.1016/j.apgeochem.2005.09.008
3. Petrusevski B, van der Meer W, Baker J, Kruis F, Sharma SK, Schippers JC (2007) Innovative approach for treatment of arsenic contaminated groundwater in Central Europe. Water Supply 7(3):131–138. https://doi.org/10.2166/ws.2007.075

4. Smedley PL, Knudsen J, Maiga D (2007) Arsenic in groundwater from mineralised Proterozoic basement rocks of Burkina Faso. Appl Geochem 22(5):1074–1092. https://doi.org/10.1016/j.apgeochem.2007.01.001
5. Gunduz O, Simsek C, Hasozbek A (2009) Arsenic pollution in the groundwater of Simav Plain, Turkey: its impact on water quality and human health. Water Air Soil Pollut 205(1):43. https://doi.org/10.1007/s11270-009-0055-3
6. Jovanovic D, Jakovljević B, Rašić-Milutinović Z, Paunović K, Peković G, Knezević T (2011) Arsenic occurrence in drinking water supply systems in ten municipalities in Vojvodina Region, Serbia. Environ Res 111(2):315–318. https://doi.org/10.1016/j.envres.2010.11.014
7. ATSDR—Toxicological Profile: Arsenic (2000). https://www.atsdr.cdc.gov/toxprofiles/tp.asp?id=22&tid=3. Accessed 08 Sept 2019
8. Goshman LM (1985) Clinical toxicology of commercial products, 5th edn. In: Gosselin RE, Smith RP, Hodge HC (eds) Williams and Wilkins, 428 East Preston Street, Baltimore, MD 21202. 1984. 1987 + ix pp. 26 × 17.5 cm. ISBN 0-683-03632-7. $95.00. J Pharm Sci 74(10):1139. https://doi.org/10.1002/jps.2600741037
9. OEHHA (2004) California EPA Office of Environmental Health Hazard Assessment. Air toxic hot spots: adoption of a unit risk value for naphthalene [Online]. http://www.oehha.ca.gov/air/hot_spots/naphth.html
10. Stýblo M, Drobná Z, Jaspers I, Lin S, Thomas DJ (2002) The role of biomethylation in toxicity and carcinogenicity of arsenic: a research update. Environ Health Perspect 110(SUPPL. 5):767–771
11. Borzsonyi M, Bereczky A, Rudnai P, Csanady M, Horvath A (1992) Epidemiological studies on human subjects exposed to arsenic in drinking water in Southeast Hungary. Arch Toxicol 66(1):77–78
12. Yang C-Y, Chang C-C, Tsai S-S, Chuang H-Y, Ho C-K, Wu T-N (2003) Arsenic in drinking water and adverse pregnancy outcome in an arseniasis-endemic area in northeastern Taiwan. Environ Res 91(1):29–34. https://doi.org/10.1016/s0013-9351(02)00015-4
13. Ahmad SA et al (2001) Arsenic in drinking water and pregnancy outcomes. Environ Health Perspect 109(6):629–631
14. National Primary Drinking Water Regulations; Arsenic and Clarifications to Compliance and New Source Contaminants Monitoring (2021) Federal Register. https://www.federalregister.gov/documents/2001/01/22/01-1668/national-primary-drinking-water-regulations-arsenic-and-clarifications-to-compliance and new-source. Accessed 08 Sept 2019
15. Bureau of Indian Standards (2012) Indian standard for drinking water specification (Second Revision)
16. Khandaker NR, Brady PV, KRUMHANSL JL (2009) Bay of Bengal arsenic removal from drinking water: a handbook for communities
17. Islam FS et al (2004) Role of metal-reducing bacteria in arsenic release from Bengal delta sediments. Nature 430(6995):68–71. https://doi.org/10.1038/nature02638
18. Lowers HA et al (2007) Arsenic incorporation into authigenic pyrite, Bengal Basin sediment, Bangladesh. Geochim Cosmochim Acta 71(11):2699–2717. https://doi.org/10.1016/j.gca.2007.03.022
19. Kocar BD et al (2008) Integrated biogeochemical and hydrologic processes driving arsenic release from shallow sediments to groundwaters of the Mekong delta. Appl Geochem 23(11):3059–3071. https://doi.org/10.1016/j.apgeochem.2008.06.026
20. Nickson R, McArthur J, Burgess W, Matin Ahmed K, Ravenscroft P, Rahman M (1998) Arsenic poisoning of Bangladesh groundwater. Nature 395(6700):338. https://doi.org/10.1038/26387
21. BGS and DPHE (2001) Final report. Arsenic contamination of groundwater in Bangladesh, vol. 2. In: Kinniburgh DG, Smedley PL (eds) Brit Geol Surv (BGS) Technical report WC/00/19. British Geological Survey, Keyworth
22. Akai J et al (2004) Mineralogical and geomicrobiological investigations on groundwater arsenic enrichment in Bangladesh. Appl Geochem 19(2):215–230. https://doi.org/10.1016/j.apgeochem.2003.09.008

23. McArthur JM et al (2004) Natural organic matter in sedimentary basins and its relation to arsenic in anoxic ground water: the example of West Bengal and its worldwide implications. Appl Geochem 19(8):1255–1293. https://doi.org/10.1016/j.apgeochem.2004.02.001
24. Ravenscroft P, Burgess WG, Ahmed KM, Burren M, Perrin J (2005) Arsenic in groundwater of the Bengal Basin, Bangladesh: distribution, field relations, and hydrogeological setting. Hydrogeol J 13(5–6):727–751. https://doi.org/10.1007/s10040-003-0314-0
25. Meharg AA et al (2006) Codeposition of organic carbon and arsenic in Bengal Delta aquifers. Environ Sci Technol 40(16):4928–4935. https://doi.org/10.1021/es060722b
26. Harvey CF et al (2002) Arsenic mobility and groundwater extraction in Bangladesh. Science 298(5598):1602–1606. https://doi.org/10.1126/science.1076978
27. Harvey CF et al (2006) Groundwater dynamics and arsenic contamination in Bangladesh. Chem Geol 228(1–3):112–136. SPEC ISS. https://doi.org/10.1016/j.chemgeo.2005.11.025
28. Polizzotto ML, Kocar BD, Benner SG, Sampson M, Fendorf S (2008) Near-surface wetland sediments as a source of arsenic release to ground water in Asia. Nature 454(7203):505–508. https://doi.org/10.1038/nature07093
29. Polizzotto ML et al (2006) Solid-phases and desorption processes of arsenic within Bangladesh sediments. Chem Geol 228(1–3):97–111. SPEC ISS. https://doi.org/10.1016/j.chemgeo.2005.11.026
30. Neumann RB, Ashfaque KN, Badruzzaman ABM, Ashraf Ali M, Shoemaker JK, Harvey CF (2010) Anthropogenic influences on groundwater arsenic concentrations in Bangladesh. Nat Geosci 3(1):46–52. https://doi.org/10.1038/ngeo685
31. Van Geen A et al (2003) Spatial variability of arsenic in 6000 tube wells in a 25 km2 area of Bangladesh. Water Resour Res 39(5):HWC31–HWC316
32. Zheng Y et al (2004) Redox control of arsenic mobilization in Bangladesh groundwater. Appl Geochem 19(2):201–214. https://doi.org/10.1016/j.apgeochem.2003.09.007
33. Leber J, Rahman MM, Ahmed KM, Mailloux B, Van Geen A (2011) Contrasting influence of geology on E. coli and arsenic in aquifers of Bangladesh. Ground Water 49(1):111123
34. Amy GL (2000) Arsenic treatability options and evaluation of residuals management issues. American Water Works Association
35. Edwards M (1994) Chemistry of arsenic removal during coagulation and Fe–Mn oxidation. J AWWA 86(9):64–78. https://doi.org/10.1002/j.1551-8833.1994.tb06247.x
36. Banerjee K, Helwick RP, Gupta S (1999) A treatment process for removal of mixed inorganic and organic arsenic species from groundwater. Environ Prog 18(4):280–284. https://doi.org/10.1002/ep.670180415
37. Torrens K (1999) Evaluating arsenic removal technologies. Pollut Eng [Online]. https://pollutionengineering.com/archives/1999/po1070.199/pol9907works.html
38. Murtuz AKM, Chowdury MA, Khandaker NR (2013) Performance of arsenic removal filters in Bangladesh groundwaters. MIST J Sci Technol 2(2):58–88
39. Waypa JJ, Elimelech M, Hering JG (1997) Arsenic removal by RO and NF membranes. J AWWA 89(10):102–114. https://doi.org/10.1002/j.1551-8833.1997.tb08309.x
40. Sarkar AR, Rahman OT (2001) In-situ removal of arsenic-experiences of DPHE-Danida pilot project. Technologies for arsenic removal from drinking water. Bangladesh University of Engineering and Technology and The United Nations University, Bangladesh
41. Van Halem D, Bakker SA, Amy GL, Van Dijk JC (2009) Arsenic in drinking water: a worldwide water quality concern for water supply companies. Drink Water Eng Sci 2 [Online]. http://resolver.tudelft.nl/uuid:ac069d55-cb87-474b-9af6-421a07225339. Accessed 26 Dec 2018
42. Van Halem D, Olivero S, de Vet WWJM, Verberk JQJC, Amy GL, van Dijk JC (2010) Subsurface iron and arsenic removal for shallow tube well drinking water supply in rural Bangladesh. Water Res 44(19):5761–5769. https://doi.org/10.1016/j.watres.2010.05.049
43. Van Halem D et al (2010) Subsurface iron and arsenic removal: low-cost technology for community-based water supply in Bangladesh. Water Sci Technol 62(11):2702–2709. https://doi.org/10.2166/wst.2010.463
44. Sen Gupta B et al (2009) A simple chemical free arsenic removal method for community water supply—a case study from West Bengal, India. Environ Pollut 157(12):3351–3353. https://doi.org/10.1016/j.envpol.2009.09.014

45. Rott U, Meyerhoff R, Bauer T (1996) In situ treatment of groundwater with increased concentrations of iron, manganese and arsenic. Wasser-Abwasser 197:358–363
46. Appelo CaJ, de Vet WWJM (2003) Modeling in situ iron removal from groundwater with trace elements such as As. Arsenic in ground water Welch AH. In: Stollenwerk KG (eds) Springer US, pp 381–401
47. Rahman MM et al (2014) Exploratory experiments to determine the effect of alternative operations on the efficiency of subsurface arsenic removal in rural Bangladesh. Hydrogeol J 23(1):19–34. https://doi.org/10.1007/s10040-014-1179-0
48. Rahman MM (2017) Field experiments and reactive transport modeling of subsurface arsenic removal in Bangladesh
49. McArthur JM, Ravenscroft P, Safiulla S, Thirlwall MF (2001) Arsenic in groundwater: testing pollution mechanisms for sedimentary aquifers in Bangladesh. Water Resour Res 37(1):109–117. https://doi.org/10.1029/2000WR900270
50. Dowling CB, Poreda RJ, Basu AR, Peters SL, Aggarwal PK (2002) Geochemical study of arsenic release mechanisms in the Bengal Basin groundwater. Water Resour Res 38(9):121–1218
51. Swartz CH et al (2004) Mobility of arsenic in a Bangladesh aquifer: Inferences from geochemical profiles, leaching data, and mineralogical characterization. Geochim Cosmochim Acta 68(22):4539–4557. https://doi.org/10.1016/j.gca.2004.04.020
52. Postma D et al (2007) Arsenic in groundwater of the Red River floodplain, Vietnam: controlling geochemical processes and reactive transport modeling. Geochim Cosmochim Acta 71(21):5054–5071. https://doi.org/10.1016/j.gca.2007.08.020
53. Postma D et al (2010) Mobilization of arsenic and iron from Red River floodplain sediments, Vietnam. Geochim Cosmochim Acta 74(12):3367–3381. https://doi.org/10.1016/j.gca.2010.03.024
54. Van Halem D (2011) Subsurface iron and arsenic removal for drinking water treatment in Bangladesh
55. Rahman MM et al (2015) Reactive transport modeling of subsurface arsenic removal systems in rural Bangladesh. Sci Total Environ 537:277–293. https://doi.org/10.1016/j.scitotenv.2015.07.140
56. Shafiquzzaman M, Azam MdS, Mishima I, Nakajima J (2009) Technical and social evaluation of arsenic mitigation in rural Bangladesh. J Health Popul Nutr 27(5):674–683

Sequential Anaerobic–Aerobic Co-Treatment of Three Herbicides Mixture in Water: A Comprehensive Study on Biotransformation

G. B. Mahesh and Basavaraju Manu

Abstract Anaerobic reducing conditions tend the herbicides to undergo dehalogenation; dechlorination and demethylation reactions form simpler end products. Thus formed intermediate compounds can be mineralized in the aerobic reactor followed after anaerobic treatment. Therefore, this study was conducted to evaluate the sequential anaerobic–aerobic treatment of three herbicides mixture namely (2-ethylamino)-4-(isopropylamino)-6-(methylthio)-s-trazine) (ametryn), 3,6-dichloro-2-methoxybenzoic acid (dicamba) and 2,4-dichlorophenoxyacetic acid (2,4-d) with different formulations. The reactors were operated at hydraulic retention time (HRT) of 48 h, pH between 6.5–7.5 and at ambient reactor liquid temperature (27–32.2 °C). The long term study was conducted using two anaerobic reactors namely R1 (anaerobic control with no herbicide) and R2 (anaerobic reactor fed with 2, 4-d + ametryn + dicamba mixtures). The effect of increased herbicides concentration on the anaerobic–aerobic reactor during 400 days of the treatment period was evaluated. Two aerobic reactors were operated simultaneously to give post-treatment to the anaerobic effluent. The reactor performance was evaluated by monitoring herbicide removal efficiency of chemical oxygen demand (COD) and biogas production. The reactor's stability parameters pH, alkalinity, volatile fatty acids (VFA) and oxidation–reduction potential (ORP) were monitored on a daily basis. Both the anaerobic reactors were stabilized using 2 g/L of starch with total organic loading rate (OLR) of 0.21–0.215 kg-COD/m^3/d during 48 days, and aerobic reactors were stabilized in 14 days using anaerobic effluent as feed having OLR of 0.02 to 0.038 kg-COD/m^3/d. After achieving the quasi-state condition the influent was fed with known herbicide concentrations to the R2 reactor. The maximum COD removal efficiency obtained for different influent herbicide concentrations under anaerobic treatment from R2 reactor was 77–88%. Sequential anaerobic–aerobic removal efficiency for A2 reactor was found to be >85%. Addition of 5–10 mg/L of anthraquinone-2,6-disulphonate

G. B. Mahesh (✉)
Department of Civil Engineering, Madanapalle Institute of Technology and Science, Post Box No: 14, Kadiri Road, Chittoor District, Angallu, Madanapalle, Andhra Pradesh 517325, India

B. Manu
Department of Civil Engineering, Dakshina Kannada District, National Institute of Technology, Surathkal, Srinivasnagar (PO), Mangaluru, Karnataka 575025, India

© Springer Nature Switzerland AG 2022
S. Yadav et al. (eds.), *Wastewater Assessment, Treatment, Reuse and Development in India*, Earth and Environmental Sciences Library,
https://doi.org/10.1007/978-3-030-95786-5_6

(AQS) as a redox mediator enhanced the herbicides biotransformation efficiency in the anaerobic reactor (R2) by 5–10%. The samples were analyzed using GC-HRMS to identify the intermediate compounds.

Keywords Ametryn · Dicamba · 2,4-d · Biotransformation · SBR · AQS · Sequential anaerobic–aerobic treatment

Abbreviations

μg	Micro gram
ASBR	Anaerobic sequential batch reactor(s)
COD	Chemical oxygen demand
D	Day(s)
DO	Dissolved oxygen
G	Gram
GC-HRMS	Gas–chromatograph with high resolution mass Spectroscopy
HPLC	High performance liquid chromatograph
HRT	Hydraulic retention time
H	Hour(s)
KOH	Potassium hydroxide
Min	Minute(s)
mL	Milli litres
mV	Milli volts
MLSS	Mixed liquor suspended solids
MLVSS	Mixed liquor volatile suspended solids
$NaHCO_3$	Sodium bicarbonate
nm	Nanometer(s)
OLR	Organic loading rate
ORP	Oxidation reduction potential
SBR	Sequential batch reactor(s)
SRT	Solids retention time
STP	Sewage treatment plant
TP	Transformation product(s)
UASB	Upflow anaerobic sludge blanket
VFA	Volatile fatty acid(s)

1 Introduction

India is an agriculture-based country, having its 60–70% of the total population depends on agriculture contributing around 16–17% of total gross development product (GDP). Around 15–25% of crop production is lost due to pests, weeds, and diseases. India produces only 3 tons/hectare of yield, which is very less when compared to China, Brazil and the USA [1]. The total population of India is currently around 17.84% of the world population, has 2.4% of land area and 4% of water resources. The dependency on food increasing day by day as the population of the country increasing, which intern leading to adopt some advanced techniques to increase the crop yield and to protect the crops.

The advanced technology in the agricultural field focused mainly on increasing the crop yield by way of controlling weeds using chemicals, seed treatment agronomy, biotechnology, etc. India is the fourth-largest global producer of agrochemicals after the US, Japan, and China. Around 50% of the agrochemicals are consumed in the country, and the remaining is being exported, and this consumption rate may increase in the future [2]. India consumes around 0.5 kg/hectare of herbicide [3], which is very less in comparison with the quantity consumed by Japan (15 kg/hectare), United States (2 kg/hectare) and European Union (4 kg/hectare) during 2001–2003 [4].

The agrochemicals include pesticides, herbicides, fungicides, and bio-pesticides to eradicate pests, weeds, and fungus in the agricultural and non-agricultural fields. Herbicides are of prime importance as their usage is increasing due to a lack of available labor to remove weeds in various crop fields. The used herbicides and their derivatives have become a major concern in the field of environmental engineering as they join freshwater bodies through agricultural runoff and then join the downstream water bodies and increase its toxicity. The application of herbicides before rainfall has laid to the increase in herbicide load on the downstream water bodies [5]. The sources of herbicides include pesticide manufacturing units, soil leaching, surface runoffs, accidental spills, improper disposal, etc. Some of the primary herbicide compounds used are 2,4–dichlorophenoxy acetic acid (2,4-d), 2-ethylamino-4-isopropylamino-6-methyl-thio-s triazine (ametryn) and 3,6-dichloro-2-methoxybenzoic acid (dicamba) in different combinations [6]. Agricultural runoff contained up to 500 mg/L of pesticide [7], runoff from sugarcane fields, for instance, contained 24.5 mg/L of 2,4-d, 3.4 mg/L of ametryn and 93.7 mg/L dicamba [6]. The physico-chemical properties of these three herbicides are tabulated in the Table 1. The herbicide application in the crop fields is expected to rise in future due to the lack of availability of labors, increased food crisis, usage awareness, and cropland expansion.

Ametryn is strong toxicant for the dicots and toxicity level of the compound has been detected in terrestrial species that have fed grasses and broadleaf plants [8]. Its presence has been detected in surface and groundwater due to its low soil sorption capacity. Ametryn concentration 3.4 mg/L was found in the runoff water from agriculture land [6], and it was also found in wastewater treatment plants [9], and nearby the agricultural fields. Ametryn is a s-triazine group of herbicide, it has low water solubility of 209 mg/L at 25 °C, has a melting point (80 °C), and its pKa is 4 [10]. It is an endocrine disrupting compound [11], aquatic ecosystem disruptor [12], and cause different type of health issues to humans and other organisms [8]. The usage of herbicides has been banned in various countries including European Union (EU) from 2002 for environmental consequences [13].

Table 1 Physical and chemical properties of herbicides

Properties	2,4 –d ($C_8H_6Cl_2O_3$)	Ametryn ($C_9H_{17}N_5S$)	Dicamba (C_6H_2 $Cl_2(OCH_3)CO_2H$)
Structure	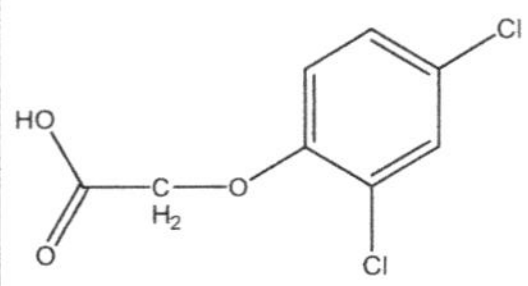	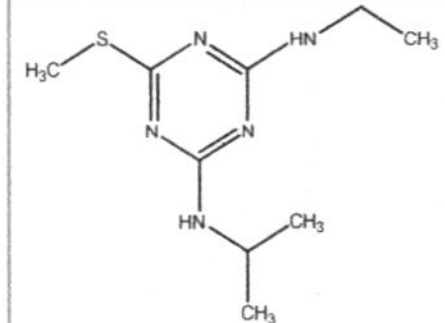	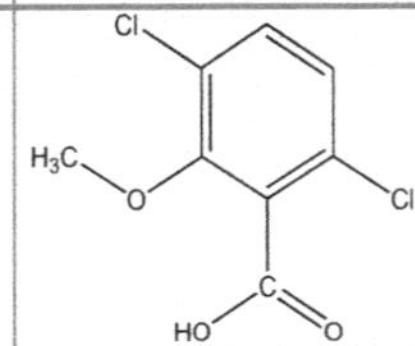
Synonym	2,4-Dichlorophenyloxy acetic acid	(2-ethylamino)-4-(isopropylamino)-6-(methylthio)-s-triazine	3,6-dichloro-2-methoxybenzoic acid
Molecular weight	221 g/mol	227 g/mol	221 g/mol
Solubility in water (mg/L)	890 at 20 °C	209 at 25 °C	4500 at 25 °C
Melting point	140.5 °C	85 °C	115 °C
Boiling point	160 °C	337 °C	200 °C

Source Manu and Sangami [6]

Dicamba is consumed to prevent the post-emergence of broadleaf weeds agriculture lands [14]. It has high water solubility (4500 mg/L), half-life (28.3 day), highly mobile in soil [15] and exists in water as anions, so that it is weakly adsorbed [16]. Dicamba can cause various health effects on aquatic life, animals, and also on human [17]. The application of a mixture of herbicides is being considered as an intensive agriculture practice to remove a different type of weeds effectively and efficiently [6]. A mixture of 2,4-d and ametryn are being used to remove weeds that are resistant to triazine herbicides in maize and sugarcane crops [18]. Use of different herbicide mixtures in the crop field to remove weeds that are resistant to individual herbicides may also effect on the non-target plants. The application of complex herbicide mixtures may accumulate over the soil due to poor solubility and volatilization.

Herbicides are highly mobile and form stable compounds during chemical hydrolysis and are well defined endocrine disruptors and exposure to such compounds causes various effects on to eyes, thyroid, liver, kidney and nervous system of the humans [19]. The used herbicides and transformation products have become a serious concern in the field of environmental sustainability as they join freshwater resources through crop land effluents and raise the toxicity of water. Persistence of these herbicides in the soil leads to contamination of both surface and groundwater. Various concentration of 2,4-d in surface and groundwater was detected even after the chemical was not used for a long time [20]. The effects of these three herbicides would be highly detrimental to human and also to other living organisms. Hence, herbicides have to be removed before discharging into water bodies. Therefore various authorities have prescribed standard limit for these herbicides in surface water, 2,4-d = 29 (surface water)–10 μg/L (groundwater) [21], ametryn = 14 (surface water)–1.4 μg/L (groundwater) [8] and dicamba = 200 μg/L (surface water)–14 μg/L (groundwater) [22].

The different physico-chemical.treatment methods have been adopted to remove herbicides including chemical oxidation processes, granular activated carbon adsorption, radiolytic degradation, advanced oxidation processes like Fenton's, electro-Fenton's, photo-Fenton's, photoelectro-Fenton's, electro-oxidation, photocatalysis, UV irradiation, electrolysis with UV irradiation photoelectrolysis and ozone treatment methods. Though the treatment mentioned above have shown considerable treatment efficiency, there is always a production of the complex intermediate product and which has to be treated further. Since most of the intermediate compounds are unstable and may not survive, they can be removed in the post-treatment itself.

Thus the physicochemical processes become tedious and uneconomical to operate. Some of the aforementioned physico-chemical methods have also been associated with biological methods to increase treatment efficiency and to reduce the cost. However, aromatic herbicides are removed partially due to their strong link between benzene ring and halogens, in general, physico-chemical treatment methods produce toxic intermediates which may pose toxicity than the parent compound. Biological treatment processes are cheaper than the physico-chemical methods in terms of investment and operation costs. The cost for biological treatment methods is range from 5 to 20 times less than chemical ones such as ozone and hydrogen peroxide in the case of the advanced oxidation process. The treatment cost can be reduced by 3 to 10 times in the case of biological methods [23].

To overcome some of the limitations of physico-chemical methods and to reduce toxic by-product formation, biological treatment processes are extensively being adopted in recent years. Apart from the shock loading effects and slow biomass stabilization, the biological methods are environmentally friendly, easy to operate and utilize locally available resources for the treatment. Several biological treatment methods have been developed by many researchers to treat herbicide present in water and wastewater.

Biological treatment methods to remove 2,4-d, ametryn and dicamba include the following methods: anaerobic biodegradation [24], biodegradation [25]; packed bed biofilm reactors [14, 18], membrane bioreactors [16, 26], UASB reactors [27], submerged biological anaerobic/aerobic filter [28], sequential batch reactors [29, 30]. Herbicide removal using pure cultures isolated from algae, fungi, and bacteria have also been used to remove herbicides efficiently [25]. But their suitability is limited to a particular type of herbicides, and practically it is difficult to maintain in pure form on a large scale at field conditions.

The conventional sequential batch reactors (SBR) are considered as an effective treatment option in the biological wastewater treatment methods because they are simple, flexible, and economically viable [31]. The mixed microbial consortia present in the reactor biomass can degrade different type of herbicides at various influent concentration levels even at different environmental conditions. SBR has been widely used to treat phenoxy herbicides including 2,4-d [32, 33], 2,4,6-trichlorophenol by modifying the existing SBR [34].

The herbicides containing halogens in the aromatic ring make them structurally stable and aerobically persistent can be treated efficiently under reductive conditions in anaerobic reactors, which can support the biotransformation of halogenated compounds [35]. Thus anaerobic sequential batch reactor (ASBR) can be used effectively in support of the reductive reactions. ASBR was used in the treatment of different pollutants including herbicides like dicamba [36], 2,4-d [27, 32], and refractory organic [37]. There were no significant researches have been reported to remove ametryn and dicamba by SBR method, but available studies are limited for the treatment of 2,4-d [33], dicamba [42] and ametryn [43].

Some studies reported that the biodegradation of herbicides primarily due to breaking up of bond between the benzene ring and the substituent group by methanogens in the presence of electron donor microbes [38]. The anaerobic reducing reactions can be enhanced by the addition of redox mediators like anthraquinone-2,6-disulphonate (AQS). They increase the rate of reaction by shuttling electrons from primary electron donors or from bulk electron donors to the electron-accepting organic compounds.

The TPs remained in the anaerobic reactor can be aerobically degraded at a longer acclimation period. Therefore, it becomes effective if the refractory halogenated aromatic compounds which cause recalcitrance are treated firstly by ASBR followed by aerobic SBR. The investigation reported on the sequential anaerobic–aerobic treatment of various environmental pollutants including azo dyes removal [39], textile wastewater treatment [40]. Removal of herbicides such as ametryn and dicamba separately using anaerobic–aerobic method has been reported previously in [41–45] but no such studies are reported for the treatment of mixture of herbicides.

In the present research, a lab-scale sequential anaerobic–aerobic reactor was established and evaluated for possible biotransformation and mineralization of herbicides. The simulated water contained a mixture of 2,4-d, ametryn, and dicamba. The reactors were stabilized using starch, and then after achieving the reactor stabilization, the simulated water with different herbicide concentration was fed. Impact of different concentrations of redox mediator on the treatment processes was evaluated.

2 Materials and Methodology

Figure 1 shows the laboratory reactor set up used in the present study. The anaerobic seed sludge was collected from the outlet of UASB reactor of sewage treatment plant (STP) and seed sludge for the aerobic reactor was collected from the primary settling tank of STP located in NITK campus, Surathkal, India.

The seed biomass was processed by passing through 250 μm sieve to get uniform solids and characterized for MLVSS and MLSS. Table 2 shows the various

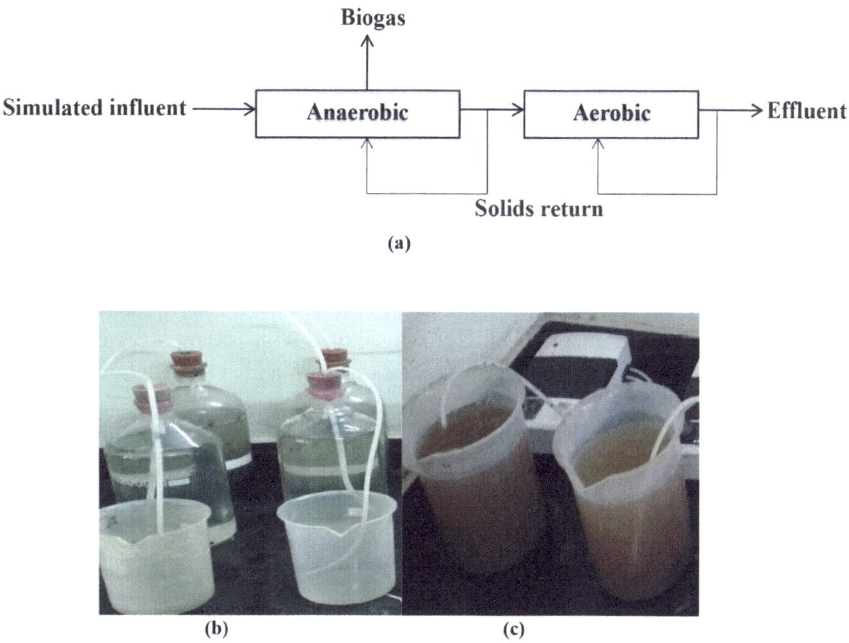

Fig. 1 **a** Flow diagram of the treatment process; **b** anaerobic, and **c** aerobic laboratory scale reactor set up

Table 2 The instruments and chemicals used in this study

Instruments	Particulars	Manufacturer
	Gas Chromatography–High-Resolution Mass spectrometry (GC-HRMS)	GC–Agilent, MS–Jeol
	Liquid Chromatography-Mass Spectrometry (LC–MS)	Shimadzu
	High Performance Liquid Chromatography (HPLC)	Agilent
	UV–VIS double beam Spectrophotometer	Systronics
	pH and ORP meter	Hanna
	High-Speed Centrifuge	Remi
Chemicals	2,4-d, ametryn, dicamba, methyl-tert-butyl ether	Sigma-Aldrich
	Starch and sodium hydrogen carbonate	Himedia
	Potassium hydroxide (99% purity), potassium dichromate, silver sulfate, mercuric sulfate, potassium iodide, sodium thiosulphate, ferrous ammonium sulfate, HPLC grade methanol, and ultra-pure water	Merck

instruments and chemicals used in this study.

2.1 Reactor Set-Up and Operation

The treatment process was carried out using a sequential batch reactor in anaerobic followed by an aerobic reactor, as showed in the flow diagram (Fig. 1). The reactors were operated manually with operating cycle include processes like feeding (10 min), reaction (23 h), settling (30 min), and decanting (20 min) [33]. Initially, the anaerobic reactors were operated with feed water containing 2 g/L of starch as carbon source and 4 g/L of $NaHCO_3$ as a buffer; aerobic reactors were fed with effluent from the anaerobic reactors as feed. One litre of seed sludge was inoculated with 9 g/L of MLVSS (sludge characteristics: MLSS = 76 g/L, MLVSS = 36 g/L) to each anaerobic reactor, (i.e., 250 mL of anaerobic sludge was diluted with 750 mL of water) and 1 L of simulated feed water containing 2 g/L starch was added to maintain the total volume of 2 L.

The aerobic laboratory-scale reactors were prepared by using 2 L capacity plastic beakers. One litre of seed sludge was inoculated to each aerobic reactor with 2500 mg/L of MLVSS (Sludge characteristics: MLSS = 6.3 g/L, MLVSS = 4.5 g/L), (i.e., 560 mL of aerobic sludge was diluted with 440 mL of water). And 1 L of tap water was added to maintain the total volume of 2 L on the first day and, the aerobic reactors were aerated using air diffusers. Starch was prepared by dissolving 10 g of starch powder in 250 ml of hot water. The composition of trace metal solution was prepared as per the protocols [39, 43]. The trace metal solution include in g/L; $COCl_2.6H_2O$:1.613, $FeSO_4$:8.39, $MgSO_4.7H_2O$:5, H_3BO_3:0.1, $ZnCl_2$:0.0473,

$CuSO_4.5H_2O$:0.0782, $NiSO_4.6_2O$:1.698, $(NH_4)_6MO_7O_{24}.4H_2O$:0.54, $CaCl_2$:7.776, $MnCl_2.4H_2O$:7.863.

The feed to anaerobic reactors was prepared using stock solution by diluting to the required concentration, with the addition of 4 g/L of $NaHCO_3$, 2 mL/L of trace metal solution and stock herbicide solution to give required herbicide concentration in 1 L volumetric flask containing water. The anaerobic reactors were operated at ambient temperature; 1 L of supernatant was decanted and fed with 1 L of fresh feed. The reactors were decanted manually by transferring 1 L of supernatant liquid to a beaker and capped instantly after feeding the fresh solution to avoid oxygen entry in to the reactor. The aerobic reactors were operated simultaneously by feeding 0.5 L of anaerobic effluent after 24 h by decanting 0.5 L supernatant. Influent and effluent characteristics of feed water and effluent water were analyzed for pH, temperature, ORP, alkalinity, COD, and herbicide removal efficiency.

The reactors were operated for several days till the quasi-steady condition was achieved. After achieving the quasi-steady-state condition, different concentrations of ametryn, 2,4-d and dicamba was fed to the respective reactors and treatment was carried out for about 400 days.

The removal efficiency of herbicide and COD was calculated using the following equation,

$$(\eta) = (C_{in} - C_f)/C_{in} * 100 \tag{1}$$

where,

(η): Removal efficiency (%).

C_{in}: concentration of herbicide or COD in the influent feed water (mg/L).

C_f: concentration of herbicide or COD in the effluent of reactors (mg/L).

2.2 Determination of Herbicides Intermediate Compounds

The sample extraction to detect transformation product using GC-HRMS was carried as per method 1699 [46]. Samples were analysed in GC-HRMS to detect the biotransformation products using the standard GC–HRMS method: column type: capillary; column class: standard non-polar; active phase: RTX-1; column length: 60 m; carrier gas: He; column diameter: 0.22 mm; phase thickness: 0.25 μm; data type: linear RI; program type: Ramp; start temp: 60 °C; end temp: 230 °C; heat rate: 10 K/min; end time: 35 min.

2.2.1 Water Quality Parameters

The alkalinity and chemical oxygen demand (COD, by closed reflux titrimetric method) were measured as per the protocols of standard methods [47]. pH and ORP values were measured in the reactors using the portable digital meter (edge®, Hanna Instruments, India). Dissolved oxygen (DO) in the aerobic reactors was measured using DO meter (HI 9741). Volatile fatty acid (VFA) concentration was measured using the standard operating methods [48].

3 Results and Discussion

3.1 Treatment of 2,4-D, Ametryn and Dicamba Mixtures in ASBR

The sequential anaerobic–aerobic treatment of mixture of ametryn, 24-d and dicamba was carried out in R2 and A2 reactor and the reactor operation condition is tabulated in Table 3. The reactor performance was monitored through COD removal efficiency and biogas production due to difficulty of detecting herbicide mixture concentration using HPLC. Figure 2a–c shows the profiles of COD removal, biogas production and ORP as a function of time during anaerobic treatment three mixed herbicides in R2 reactor. After the introduction of herbicides mixture on day 49, the COD removal reduced to 44% and biogas production <460 mL/d, which is less than the control (R1).

The COD removal efficiency remained lower than the R1, which indicated the presence of significant biodegradable organic compounds. ORP of R2 reactor was

Table 3 Operational conditions maintained during the sequential anaerobic–aerobic treatment of ametryn, 24-d and dicamba mixtures

Sl. No	Reactor operation (Days)	Experimental condition studied
1	1–48	Reactor start-up and acclimation to 2 g/L of starch (OLR = 0.21–0.215 kg-COD/m^3/d)
2	2 onwards	Anaerobic effluent fed to corresponding aerobic reactor
3	49–190	Influent herbicide concentration = (ametryn:2,4-d:dicamba) = (2:5:10) mg/L
4	116–142	Addition of AQS = 5 mg/L
5	143–190	Addition of AQS = 10 mg/L
6	191–400	Influent herbicide concentration = (ametryn:2,4-d:dicamba) = (4:10:20) mg/L
7	305–359	Addition of AQS = 5 mg/L
8	360–400	Addition of AQS = 10 mg/L

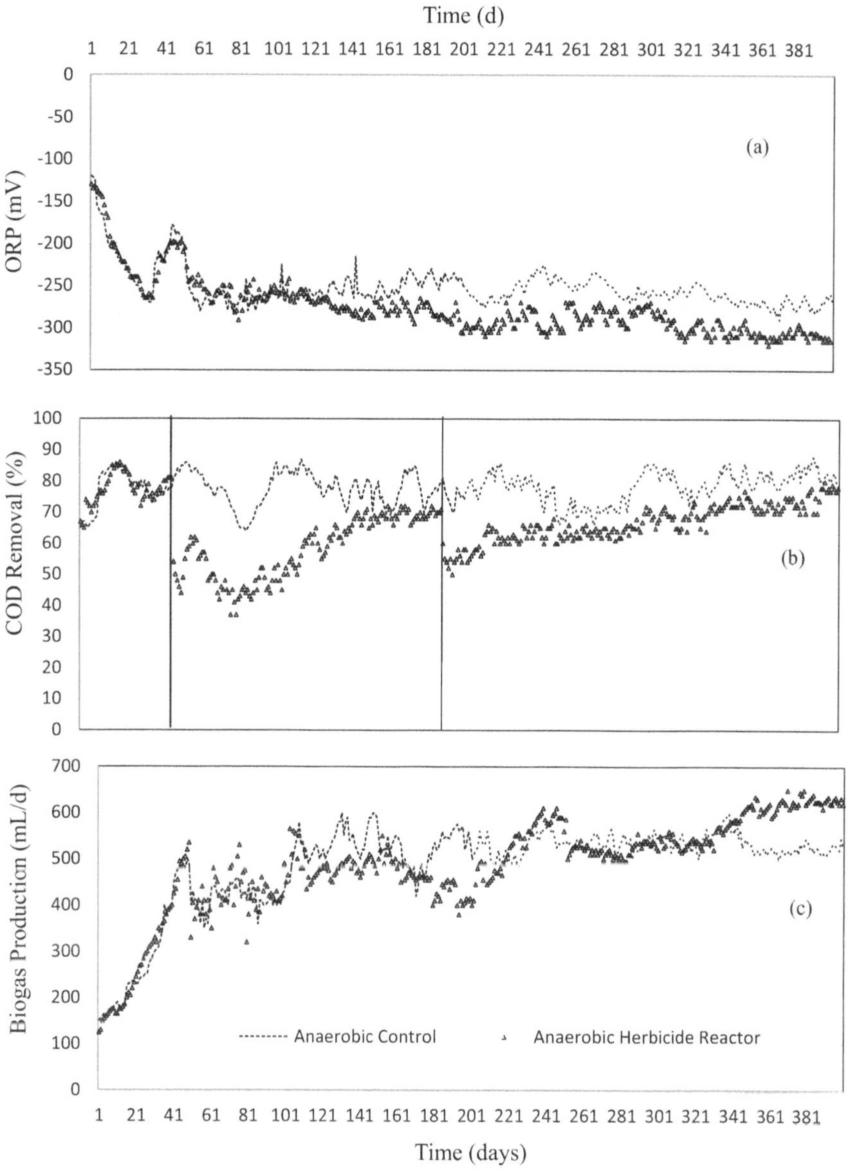

Fig. 2 **a–c** Variation of performance parameters during the anaerobic treatment of 2,4-d, ametryn and dicamba mixtures (R2) compared with anaerobic control (R1)

almost equal to the R1 reactor. To improve the reactor performance through enhanced redox reactions, 5 mg/L of AQS was added from 116 day. The reduction in ORP 10 to 20 mV than the control was observed in the R2 reactor, indicating an enhanced reducing reactions. AQS was raised to 10 mg/L from day 143 to improve the reactor efficiency then the ORP by 5 mV, but the maximum COD removal observed was about 70% and the biogas production was 550 mL/d.

Enhanced ORP indicates the increase in redox reactions in the reactor due to the mediating effect of AQS by shuttling the electrons [26]. The R1 reactor was able to yield high COD removal efficiency (>80%) and biogas production (>600 mL/d), and suggests the accumulation of toxic intermediate compounds within the R2 reactor. Hence, the mixture of three herbicides is considered to be more toxic and difficult to remove under anaerobic conditions. High effluent COD, VFA and reduced MLVSS concentration indicates the anaerobic sludge toxicity by the accumulated compounds.

It is believed that the mixture of three selected herbicides would cause more toxicity than the herbicides treated individually and also in mixture of two types. As expected the introduction of herbicides, reactor performance reduced and observed that which followed similar pattern as obtained previously. It has been reported that the maximum removal of 2,4-d [32, 33], dicamba [16] and ametryn [9] was achieved when they are treated separately. The treatment period required during the removal of individual herbicides was comparatively lesser than the number of days required during the treatment of mixture of three herbicides; this is observed even in the present study during ametryn and dicamba treatment.

The presence of high concentrations of chlorine groups would become toxic and reduces the biodegradability of chlorophenols [49]. After observing a stable COD removal efficiency, the influent herbicides concentration was raised to two fold from 191st day. The COD removal reduced to 45%, further increased gradually (65%) on a continued operation. The VFA and alkalinity observed was higher as expected during a toxic condition and the possible reasons are attributed to various parameters as discussed. High COD removal and biogas production observed than the control towards the end of the treatment period. This behaviour may be an indication of bacterial adoptability and hence it may be considered that the mixture of three selected herbicides have a tendency to get biodegraded anaerobically.

The biogas production in the R2 reactor was much lower than R1 as observed (Fig. 2). Continued operation with the similar herbicide and reactor conditions and in the presence of AQS (5–10 mg/L), the biogas production increased above the control. The continued operation supports the biomass to get acquainted with the feed condition and thus can develop metabolic pathway required for the herbicide degradation. The increased biogas production of R2 than the R1 reactor indicates the herbicides have been converted to intermediates, VFA and further in to end products as CO_2/methane gas as observed during the treatment of individual herbicides. The toxicity of herbicides can be minimised in the presence of a co-substrate like starch and thus converted to TPs and then to biogas.

3.2 Sequential Anaerobic–Aerobic Treatment of Three Herbicide Mixtures

Figure 3 shows the profile of COD removal as a function of time during aerobic treatment of R2 effluent in the aerobic reactor (i.e., A2 reactor performance). COD removal efficiency is observed to be >80% till day 130 and it was comparatively lower than the control. The liquid content of the reactor turned to dark grey colour even at continued aeration (DO: 3–4 mg/L), indicates accumulation TPs. The poor sludge settling, low COD removal efficiencies observed till 80th day due to sludge toxicity. The grey colour was disappeared and high COD removal efficiency was achieved after 100 days. The continued operation supported the aerobic bacteria to slowly acquaint with system and allowed the activation of inactive biomass to metabolize the organic matter leading increased COD removal efficiency. Biomass adaptability is an important criterion in any biological treatment process, a proper biomass activity during the treatment of toxic compounds like herbicides supports the efficient treatment. In our previous studies biomass acclimation has been achieved for treating the herbicides like dicamba and ametryn with high efficiency [42, 43]. It has been observed the acclimation of biomass to the single type herbicide compound is comparatively faster than the reactor treating the mixture of herbicides. Such observations suggest that the biomass has to be acclimated for a long period to achieve maximum removal efficiencies of selected herbicides mixture.

After raise in influent herbicide concentration, COD removal was around 82%, which suggests that the bacteria adapted to the existing influent condition to degrade particular compounds. Hence, a further rise in herbicides concentration promotes enhanced removal efficiency than the previous stage. Such observations have been discussed previously by [50], where 2,4-d was removed >99% in the presence of

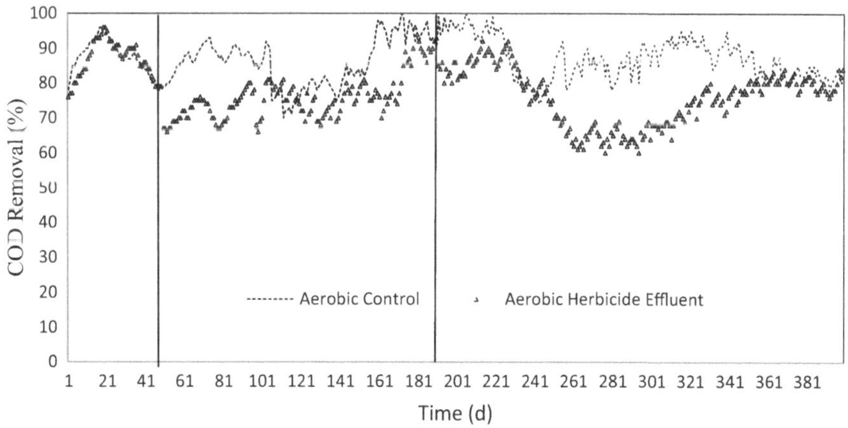

Fig. 3 Performance parameters of aerobic reactor treating 2,4-d, ametryn and dicamba mixtures (A2) compared with the aerobic control (A1)

high influent chlorophenol concentrations. The significant amount of TPs of herbicides mixture leaving the system was indicated by high effluent COD and VFA (>900 mg/L). High VFA concentration in the effluent indicates the presence of excess non-degraded organic matter, which has to be removed.

The COD removal efficiency of both A1 and A2 remained the same for few days up to 243 days and A2 COD removal reduced to 60%, a further gradual increase in the COD removal was observed over 75 days. The low COD removal efficiency in the A5 reactor than the A1 reactor by 15–22% would indicate the inability of aerobic biomass to degrade the TPs, which had contributed to high VFA concentration up to 1600 mg/L. This may be a toxic condition as discussed previously, and the alkalinity observed during this period was 2400–2800 mg-$CaCO_3$/L. The continued operation supported the degradation of VFA thereby reducing the effluent COD level. A stable COD removal efficiency of 80–85% was achieved from A1 and A5 reactors after 360 days of operation.

The similar COD removal efficiencies obtained for both control and herbicide treatment reactor and high VFA concentration indicate that anaerobic transformation products of three mixed herbicides have not completely degraded. High alkalinity observed on the initial days of herbicide introduction and also rises in the influent concentration for the second time. Eventually, it was observed that the reactor responded positively with increased COD removal, with a reduction in alkalinity below 2400 mg-$CaCO_3$/L. High alkalinity in the effluent may indicate that the compounds inhibitory impacts on the biomass [41]. R2 reactor is actually able to remove high COD removal than the R1; this may be an indication of biotransformation of herbicide compounds at different dosages and their further mineralization. Thus the sequential anaerobic–aerobic system may be considered as an efficient method to detoxify and also to remove mixture of herbicides that are potentially significant.

4 Recommendations for Future Study

The studies may be conducted to.

- Long operation periods for treating high influent dosages as required for manufacturing industrial effluents.
- Quantify the concentrations of intermediate compounds formed treatment.
- Find out the type of bacteria present in the reactor sludge by polymer chain reaction (PCR) amplification and 16S rRNA sequencing.
- Reduce the treatment time required by modifying the existing anaerobic–aerobic method.

References

1. Deshpande T (2017) State of Agriculture in India. PRS Legislative Research. http://prsindia.org/sites/default/files/parliament_or_policy_pdfs/State%20of%20Agriculture%20in%20India.pdf
2. Tata Strategic Report (2016) Next generation Indian agriculture—role of crop protection solutions, a report on Indian agrochemical industry FICCI. http://ficci.in/spdocument/20744/Agrochemicals-Knowledge-report-2016.pdf
3. De A, Bose R, Kumar A, Mozumdar S (2014) Targeted delivery of pesticides using biodegradable polymeric nanoparticles. Springer, New Delhi, India
4. OECD (2008) Environmental performance of agriculture at a glance. Organisation for economic co-operation and development. http://www.oecd.org/greengrowth/sustainable-agriculture/40953155.pdf
5. Conte LO, Schenone AV, Alfano OM (2016) Photo-Fenton degradation of the herbicide 2, 4-D in aqueous medium at pH conditions close to neutrality. J Env Manag 170:60–69
6. Sangami S, Manu B (2017) Fenton's treatment of actual agriculture runoff water containing herbicides. Water Sci Technol 75:451–461
7. Chiron S, Fernandez-Alba A, Rodriguez A, Garcia-Calvo E (2000) Pesticide chemical oxidation: state-of-the-art. Water Res 34:366–377
8. USEPA (2010) Reregistration eligibility decision (red) for ametryn. https://archive.epa.gov/pesticides/reregistration/web/pdf/Ametryn_red.pdf
9. Navaratna D, Shu L, Jegatheesan V (2016) Evaluation of herbicide (persistent pollutant) removal mechanisms through hybrid membrane bioreactors. Biores Technol 200:795–803
10. Frías S, Sánchez MJ, Rodríguez MA (2004) Determination of triazine compounds in ground water samples by micellar electrokinetic capillary chromatography. Anal Chim Acta 503:271–278
11. Sanderson JT, Seinen W, Giesy JP, Van den Berg M (2000) 2-Chloro-s-triazine herbicides induce aromatase (CYP19) activity in H295R human adrenocortical carcinoma cells: a novel mechanism for estrogenicity? Toxicol Sci 54:121–127
12. Velisek J, Stara A, Zuskova E, Kouba A (2017) Effects of three triazine metabolites and their mixture at environmentally relevant concentrations on early life stages of marbled crayfish (Procambarus fallax f. virginalis). Chemos 175:440–445
13. Liu H, Han P, Liu H, Zhou G, Fu B, Zheng Z (2018) Full-scale production of VFAs from sewage sludge by anaerobic alkaline fermentation to improve biological nutrients removal in domestic wastewater. Biores Technol 260:105–114
14. González-Cuna S, Galíndez-Mayer J, Ruiz-Ordaz N, Murugesan S, Piña-Escobedo A, García-Mena J, Santoyo-Tepole F (2016) Aerobic biofilm reactor for treating a commercial formulation of the herbicides 2, 4-D and Dicamba: biodegradation kinetics and biofilm bacterial diversity. Int Biodeter Biodeg 107:123–131
15. Comfort SD, Inskeep WP, Macur RE (1992) Degradation and transport of dicamba in a clay soil. J Environ Qual 21:653–658
16. Ghoshdastidar AJ, Tong AZ (2013) Treatment of 2,4-D, mecoprop, and dicamba using membrane bioreactor technology. Env Sci Pollut Res 20:5188–5197
17. Shin EH, Choi JH, Abd El-Aty AM, Khay S, Kim SJ, Im MH, Shim JH (2011) Simultaneous determination of three acidic herbicide residues in food crops using HPLC and confirmation via LC-MS/MS. Biomed Chromat 25:124–135
18. Sandoval-Carrasco CA, Ahuatzi-Chacón D, Galíndez-Mayer J, Ruiz-Ordaz N, Juárez-Ramírez C, Martínez-Jerónimo F (2013) Biodegradation of a mixture of the herbicides ametryn, and 2,4-dichlorophenoxyacetic acid (2,4-D) in a compartmentalized biofilm reactor. Biores Technol 145:33–36
19. United States Environment Protection Agency (2005) Prevention, pesticides and toxic substances (7508C). Registration Eligibility Decision for 2,4-D. http://archive.epa.gov/pesticides/reregistration/web/pdf/24d_red.pdf

20. Laganà A, Bacaloni A, De Leva I, Faberi A, Fago G, Marino A (2002) Occurrence and determination of herbicides and their major transformation products in environmental waters. Anal Chimi Acta 462:187–198
21. WHO (2003) 2,4-D in drinking-water. 2,4-D Drink
22. Hamilton DJ, Ambrus A, Dieterle RM, Felsot AS, Harris CA, Holland PT, Wong SS (2003) Regulatory limits for pesticide residues in water (IUPAC Technical Report). Pure Appl Chem 75:1123–1155
23. Marco A, Esplugas S, Saum G (1997) How and why combine chemical and biological processes for wastewater treatment. Water Sci Technol 35:321–327
24. Milligan PW, Häggblom MM (1999) Biodegradation and biotransformation of Dicamba under different reducing conditions. Environ Sci Technol 33:1224–1229
25. Szewczyk R, Kuśmierska A, Bernat P (2018) Ametryn removal by Metarhizium brunneum: biodegradation pathway proposal and metabolic background revealed. Chemos 190:174–183
26. Navaratna D, Elliman J, Cooper A, Shu L, Baskaran K, Jegatheesan V (2012) Impact of herbicide Ametryn on microbial communities in mixed liquor of a membrane bioreactor (MBR). Biores Technol 113:181–190
27. Sponza DT, Uluköy A (2006) Treatment of 2,4 dichlorophenol (DCP) in a sequential anaerobic (upflow anaerobic sludge blanket) aerobic (completely stirred tank) reactor system at increasing organic loading rates. Desal 195:235–250
28. Nasseri S, Baghapour MA, Derakhshan Z, Faramarzian M (2014) Degradation of atrazine by microbial consortium in an anaerobic submerged biological filter. J Water Health 12:492–503
29. He X, Wareham DG (2011) 2,4-D removal via denitrification using volatile fatty acids. Water Sci Technol 63:178–183
30. Mangat SS, Elefsiniotis P (1999) Biodegradation of the herbicide 2, 4-dichlorophenoxyacetic acid (2, 4-D) in sequencing batch reactors. Water Res 33:861–867
31. Irvine RL, JrLH K, Asano T (1989) Sequencing batch reactors for biological wastewater treatment. Critical Rev Environ Sci Technol 18:255–294
32. Celis E, Elefsiniotis P, Singhal N (2008) Biodegradation of agricultural herbicides in sequencing batch reactors under aerobic or anaerobic conditions. Water Res 42:3218–3224
33. Chin H, Elefsiniotis P, Singhal N (2005) Biodegradation of 2, 4-dicholophenoxyacetic acid using an acidogenic anaerobic sequencing batch reactor. J Env Eng Sci 4:57–63
34. Taraban RH, Berry DF, Berry DA, Walker HL (1993) Degradation of dicamba by an anaerobic consortium enriched from wetland soil. Appl Env Microbiol 59:2332–2334
35. Field JA, Stams AJM, Kato M, Schraa G (1995) Enhanced biodegradation of aromatic pollutants in cultures of anaerobic and aerobic bacterial consortia. Antonie Van Leeuwenhoek 67:47–77
36. Khorsandi H, Ghochlavi N, Aghapour AA (2018) Biological degradation of 2, 4, 6-trichlorophenol by a sequencing batch reactor. Env Process 5:907–917
37. Weinberg B, Teodosiu C (2012) Monitoring and assessment of herbicides removal by industrial wastewater treatment. Env Eng Manag J 11:215–224
38. Ghattas AK, Fischer F, Wick A, Ternes TA (2017) Anaerobic biodegradation of (emerging) organic contaminants in the aquatic environment. Water Res 116:268–295
39. Manu B, Chaudhari S (2002) Anaerobic decolorisation of simulated textile wastewater containing azo dyes. Bioresour Technol 82:225–231
40. Abiri F, Fallah N, Bonakdarpour B (2017) Sequential anaerobic-aerobic biological treatment of colored wastewaters: case study of a textile dyeing factory wastewater. Water Sci Technol 75:1261–1269
41. Mahesh GB, Manu B (2019a) Biodegradation of ametryn and dicamba in a sequential anaerobic-aerobic batch reactor: a case study. Water Pract Technol 14:423–434
42. Mahesh GB, Manu B (2019b) Biological treatment of 3,6-dichloro-2-methoxybenzoic acid using anaerobic-aerobic sequential batch reactor. Env Process 6:493–509
43. Mahesh GB, Manu B (2019c) Removal of ametryn and organic matter from wastewater using sequential anaerobic-aerobic batch reactor: a performance evaluation study. J Env Manag 249
44. Mahesh GB, Manu B (2020) Enhancement of ametryn biodegradation efficiency using anthraquinone-2,6-disulphonate in anaerobic-aerobic treatment. Env Eng Man J 19(7):1125–1236

45. Mahesh GB, Manu B (2020) Anaerobic co-digestion of 2, 4-dichlorophenoxyacetic acid with starch followed by aerobic post-treatment and identification of dominant bacteria. Springer Singapore
46. USEPA (2007) Method 1699: Pesticides in Water, Soil, Sediment, Biosolids, and Tissue by HRGC/HRMS
47. APHA (2005) 21st edition standard methods for the examination of water and wastewater. American Public Health Association/American Water Works Association/Water Environment Federation, Washington, DC
48. Terry BE (2014) Standard operating procedure: approximate volatile acids by titration. https://www.cefns.nau.edu/~teb/ambl/ambl_SOPs.html
49. Annachhatre AP, Gheewala SH (1996) Biodegradation of chlorinated phenolic compounds. Biotechnol Adv 14:35–56
50. Ma JY, Quan XC, Yang ZF, Li AJ (2012) Biodegradation of a mixture of 2, 4-dichlorophenoxyacetic acid and multiple chlorophenols by aerobic granules cultivated through plasmid pJP4 mediated bioaugmentation. Chem Eng J 181:144–151

Wastewater Management and Treatment Technologies with Recycling and Reuse Issues in India Leading to Zero Liquid Discharge (ZLD)

Ashim Kumar Bhattacharya and S. N. Roy

Abstract Wastewater treatment and its recycling issues are considered to be as one of the greatest challenges in today's perspective leading to sustainable water management practices. Wastewater generated from industrial, municipal and communities is contaminated significantly our available water resources and, in some cases, make it toxic as well. It has been estimated that by 2025 India will face severe water scarcity when the water availability in India will drop below 1,000 m^3 per capita unless adequate and sustainable water management practices are initiated on immediate effect. So, these wastewaters must be treated properly before they are discharged. Currently, a very small percentage of domestic and industrial wastewater is treated in India. All large cities in India together generate over 30 billion litres of sewage of which only 6.2 billion litres is treated. In Class-I cities only 26% of the wastewater is treated while in Class-II cities a mere 4% of the wastewater is treated. In case of Industry, around 60% of wastewater is remaining untreated. Recently the Central and State regulatory bodies has implemented stringent discharge norms for wastewater to protect the quality of Nation's water. Based on the fact, suitable technology solution must be taken into consideration to treat the polluted and sometime toxic wastewater generated from industries and municipalities. At the same time wastewater recycling and reuse issues must be given priority to protect country's freshwater resources. Further, online monitoring of effluent quality standards has also been implemented by Central Board for close monitoring of wastewater related issues which at the same time helps to protect our fresh water resources nationwide. The Indian wastewater Industry presently facing acute problems with respect to treatment with recycling and reuse issues across the country as the existing facilities do not function up to the mark

A. K. Bhattacharya (✉)
Academy of Water Technology and Environ Management, Kolkata 70008, India

Environment Pollution Monitoring Cell, National Institute of Technological Teachers Training and Research, Salt Lake, Kolkata, India

S. N. Roy
Ceramic Membrane Division, CSIR-Central Glass & Ceramic Research Institute, Kolkata 700032, India

Adamas University, Kolkata 700126, India

© Springer Nature Switzerland AG 2022
S. Yadav et al. (eds.), *Wastewater Assessment, Treatment, Reuse and Development in India*, Earth and Environmental Sciences Library,
https://doi.org/10.1007/978-3-030-95786-5_7

as desired in revised regulatory norms. The main reasons for the above problems are (a) selection of incorrect and some time backdated technology, (b) faulty design of the system, (c) poor operation and maintenance and (d) lack of operator training for skills and knowledge development. As per estimation, the metros and the largest cities in India are collectively responsible for producing around 38 billion litters of waste water every day, which will eventually need to be recycled. Treated waste water could be the answer to the future of a secure water source in the country. They can be used for industries like Thermal plants, Coal sector, Mining and Mineral processing which often not required so high purity water and thereby help to meet the other industrial and agricultural needs of the nation. Long term challenges of depleting ground and surface water resources can be addressed if waste water recycling is effectively implemented across the country. In the present paper, wastewater management practices in industrial and municipal sector with treatment methodology and various technological aspects have discussed. Different Biological Treatment technology based on conventional, fixed and fluidized bed using media as a carrier of Biological population have been covered. Advance waste water treatment technology like membrane bioreactor; application of membrane-based separation technology during tertiary treatment for recycle and reuse issues have also been taken into consideration. Comparison of Performance efficiency, space availability and cost economic factor of different technologies during primary, secondary and tertiary/advance treatment have been highlighted. Finally, concept of zero liquid discharge leading to sustainable water management practices has also been incorporated with one or two case study examples.

Keywords Wastewater treatment · Water scarcity · Regulatory discharge limits · Fresh water resources · Online monitoring of effluent quality standards · Wastewater recycling · Membrane based separation technology · Tertiary/Advance treatment · Zero liquid discharge · Sustainable water management

1 Introduction

Due to unprecedented growth of population with rapid urbanization as well as industrialization, the entire world specifically developing countries like India is facing challenges on clean and safe water supply. Wastewater generated from industrial, municipal and communities are increasing day by day. It is contaminated significantly our available surface and ground water resources and, in few cases, make it toxic as well. India accounts for 2.4% of Global land area, contributing 16% of world population but represents only 4% of world fresh water resources.

Total water resource availability in the country estimated to be about 1123 BCM (690 BCM from surface and 433 BCM from ground). Irrigation sector seems to be in the top which consumed around 85% of total water availability (Table 1), which again reported to be increase to 1072 BCM by 2050. Ground water is the major source for irrigation in the country. Annual groundwater recharge is about 433 BCM

Table 1 Water demand (billion cubic meters by sector wise) in India till 2050

Sector	2010	2025	2050
Irrigation	688	910	1,072
Industry	12	23	63
Drinking water	56	73	102
Others	57	87	210
Total	813	1,093	1,447

of which irrigation covered 212.5 BCM and 18.1 BCM covered by domestic and industrial sector [1]. Based on present population growth-rate (1.9% per year), the Indian population is expected to cross the 1.5 billion mark by 2050. Due to increasing population with infrastructural and industrial growth, the per capita average annual freshwater availability of the country has been reducing significantly since 1951 from 5177 m^3 to 1588 m^3 in 2010. India is already declared as a water stressed country being per capita availability falls below 1700 cum per person per day. And it is expected to further reduce to 1341 m^3 in 2025 and 1140 m^3 in 2050. Therefore, there is an urgent need for efficient water resource management through enhanced water use efficiency and waste water recycling [2]. Table 1 depicts the water demand in India till 2050.

During recent years, urban and infrastructure development in India is moving mostly in a unplanned manner. Further, infrastructure needed to cater the development process is being stretched to its breakeven point. This leads to water management to be an inherent part of planning and development. The municipalities and industries are observed to be continuously making substantial investment in water and wastewater treatment sector in the country. And this results in opening wide opportunities for technology and equipment solution providers in the said sector [3].

Currently, a very small percentage of domestic and industrial wastewater is treated in India. All large cities in India together generate over 30 billion litres of sewage of which only 6.2 billion litres is treated. In Class-I cities only 26% of the wastewater is treated while in Class-II cities a mere 4% of the wastewater is treated. In case of Industry around 60% of wastewater is being treated [4]. Further, as per the UNESCO and WWAP (2006) estimates [5], the productivity of industrial water usage for India (IWP, in billion constant 1995 US$ per m^3) is seems to be the lowest reported to be just 3.42 and about 1/30th of that for Japan and Republic of Korea. Further by 2050, it has been reported that about 48.2 BCM per day of wastewaters having potential to meet 4.5% of the total irrigation water demand of the country [6]. So, the wastewaters generated by industrial and municipal sectors must be treated and managed properly to make it suitable for discharge and possible reuse [7].

The reducing per capita availability of water and deteriorating water quality has forced the country to look for sustainable and effective water technologies to meet clean and quality water demand [3]. Wastewater treatment and its recycling issues are considered to be as one of the greatest challenges today in a country like India that will lead to sustainable water management practices.

Recently the Central and State regulatory agencies has implemented stringent discharge norms for wastewater to protect the quality of Nation's water resources. Based on this fact, suitable technology solution must be bringing into the picture to treat the polluted and sometime toxic wastewater generated from industries and municipalities. At the same time wastewater recycling and reuse issues must also be given high priority to meet country's freshwater demand. Further, recently online monitoring of effluent quality standards has also been implemented by Central board for close monitoring of wastewater discharge parameters and compliance issues which ultimately helps to protect our fresh water resources nationwide to a great extent.

1.1 Common Effluent Treatment Plants

The Small-Scale Industries in the country has been reported to be increased to large numbers in recent years which again raise the concern on increased volume of wastewater generation in the country. Further, most of the SSI units due to space constrain, lack of skills and knowledge, and limited financial capacity individually unable to install and operate their own wastewater treatment plant, resulting in inability to meet the regulatory issues. To address the pollution coming out from these categories of industries, Common Effluent Treatment Plants (CETPs) seems to be one of the viable solutions for small and medium scale enterprises for effective wastewater management.

In India, Ministry of Environment and Forest (MoEF) during 1991 initiated an innovative financial support scheme for CETPs to ensure growth of the small and medium entrepreneurs (SMEs) in an environmentally compatible manner. While initially the scheme was launched for first ten years, but based on evaluation of need base demand, it was extended further. Accordingly, the MoEF instructed various State Pollution Control Boards (SPCBs) to examine the possibilities of establishing CETPs in various industrial estates in the respective states. As reported, in India more than 150 CETPs have been set up so far under this scheme. The CETP facilitates in reduction of number of discharge points in an industrial estate for better enforcement by regulatory bodies. The investment of substantial government finances in the CETP scheme was justified on the basis of potential benefits in terms of pollution reduction and environmental improvements by protecting our fresh water resources [8].

1.2 Problem Statement

The Indian wastewater Industry presently facing acute problems with respect to treatment with recycling and reuse issues across the country as the existing facilities do not function up to the mark as desired in revised regulatory norms.

The main reasons for the above problems are (a) selection of incorrect and some time back dated technology, (b) faulty design of system, (c) poor operation and maintenance and (d) lack of operator training for skills and knowledge development.

As per estimation, the metros and the largest cities in India are collectively responsible for producing around 38 billion litters of waste water every day, which will eventually need to be recycled. Treated waste water could be the answer to the future of a secure water source in the country. They can be used for industries like Thermal plants, Coal sector, Mining and Mineral processing which often not required so high purity water and thereby help to meet the other industrial and agricultural needs of the nation. Long term challenges of depleting ground and surface water resources can be addressed if waste water recycling is effectively implemented across the country.

Due to improper design, poor maintenance, frequent breakdown of electricity and lack of availability technical man power, the wastewater treatment facilities do not function properly, in fact, often remain closed for most of the time [9]. One of the major problems with many waste water treatment facilities is that none of the implemented technologies has been assessed by cost benefit analysis with payback period. Due to this, most of the local authorities are often became less interested in smooth running of waste water treatment plants. Based on a performance evaluation studies of STPs in some selected Indian cities carried out by CPCB, it has been reported that though the waste water treatment capacity in the country has increased by about 2.5 times during last three decades, still hardly 10% of the sewage generated has been reported to be treated properly and effectively. The rest finds its own way to contaminate the rivers and ground water resources in the country and thereby disturbing the natural dynamic equilibrium of our ecosystems [7, 10].

1.3 Technology Trends

The knowledge on wastewater treatment for reuse has evolved quite a long time before and it has been advanced with human history [11]. Reuse of untreated municipal wastewater has been practiced for many centuries to divert the human waste outside of urban settlements. Likewise, use of domestic wastewater for land application is an old and common practice. This under different stages of development has led to better understanding of process, treatment technology with simultaneous development of water quality standards. In countries like Europe and United States of America wastewater treatment and it's recycling and reuse has been reported to be practised for quite some time. However, in recent years, with enhance demand for drinking water resources; the rate of water reuse is becoming a priority issue in the Asia Pacific regions as well, specifically in a country like India [3].

Majority of the Sewage Water Treatment plants (STPs) in India were developed under various river action plans (from 1978–79 onwards) and are located in Class-I and class II cities/towns along the banks of major rivers [12]. Oxidation Pond and activated sludge process reported to be the mostly commonly employed technology followed by Up-flow Anaerobic Sludge Blanket technology. Later quite a number of Waste Stabilization Ponds Technology has also been reported to be implemented successfully. A World Bank Report [13] came out strongly in favour of stabilization ponds as the most suitable wastewater treatment system for municipal sewage treatment plants in developing countries like India, where land is often available at reasonable opportunity cost and skilled labour is in short supply. But with recent trends of urbanization with infrastructural growth in the country together with land availability plays one of the pivotal roles for shifting from selection and implementation of aforesaid land-based technology. Technology trends from then onwards focused move towards more space saving technology, lower installation and commissioning time with performance oriented improved effluent quality suitable for recycle and reuse.

Wastewater treatment technology mainly involve a preliminary treatment in the initial stage to remove coarse /bulk materials with substances like oil and grease and then follows three stage of treatment process namely Primary, Secondary and Tertiary Treatment process. Primary and Tertiary treatment process are mostly based on physico-chemical treatment. Different technologies involve in the process includes coagulation-flocculation, de-emulsification for treatment of oily wastewater, sedimentation/clarification, dissolve air flotation, flash mixer, clariflocculator, dual media filter, activated carbon filter, pressure sand filtration and tank stabilization etc. Secondary treatment process mainly known as biological treatment process is responsible for degradation of organic matter relying on bacterial population. The dominating technologies in the biological treatment process are mainly based on mechanism of suspended and attached growth process. More details of the aforesaid technologies have been discussed in the later sections.

With increasing water scarcity and demand along with stringent regulatory norms forced Industries to search more on wastewater recycling technologies for sustainable growth. Recycling of wastewater involves a combination of technologies to treat the wastewater and make it reusable with attainment of safe disposal standards. As energy-efficient cost effective processes are need of the hour membrane based separation technologies, hybrid/solar desalination and advance oxidation technologies will gain importance during the years to come.

1.4 Regulatory Trends

With the increase of urbanization and industrial activity, availability of fresh water resources in the country is in declining trend. The regulatory and compliance issue as set by the government authority become more stringent to maintain the wastewater discharge quality requirements. There are total 46 categories of industry have been identified specific to each industry segments in India which includes industry like distilleries, pulp and paper, power sector, refineries etc. Recently, based on the trend of water resource scenario in the country, the policy on Zero Liquid Discharge (ZLD) has been drafted by Ministry of Environment and Forests (MoEF) which strives the industry for gradual move to ZLD status specifically water intensive industries like

Textile, Pulp and paper, Steel and Power etc. The implementation part of the policy will be governed by individual State Pollution Control Boards. Presently, only a few states in the country have been advised to specific industry end users like Textile, Paint, Pulp and Paper, Pharma, Automobile manufacturer, Breweries to achieve the ZLD status as a mandatory requirement. In fact, this is expected to be implemented thorough out the country during the coming years [3].

In the present paper, wastewater management practices in industrial and municipal sector with treatment methodology and various technological aspects will be discussed. Different Biological Treatment technology based on conventional, fixed and fluidized bed using media as a carrier of biological population have been covered. Advance waste water treatment technology like membrane bioreactor; application of membrane-based separation technology during tertiary treatment for recycle and reuse issues have also been taken into consideration. Comparison of Performance efficiency, space availability and cost economic factor of different technologies during primary, secondary and tertiary/advance treatment have highlighted. Finally, concept of zero liquid discharge leading to sustainable water management practices has also been incorporated with one or two case study with examples.

2 Wastewater Management and Treatment Practices in India in Industrial and Municipal Sector

2.1 Treatment Methodology

Wastewater depending on its source and characteristics may subject to different treatment options. The wastewater management and treatment practices followed by industrial and municipal sector as discussed in the previous Sect. 1.3 are based on collection and preliminary treatment followed by primary and secondary treatment process. Finally wastewater is allowed to pass through tertiary and advance treatment process to remove residual solids (mostly suspended), organic matters and sometimes nutrients as well. Advance treatment process and operation often makes it suitable for recycle and reuse meeting the regulatory norms set by central and state authorities (Fig. 1). Clear understanding, in-depth knowledge of all these treatment methodologies, technological aspects, selection process and factors regulating the treatment mechanism is very important for better management of wastewater treatment plants (ETPs/STPs/CETPs).

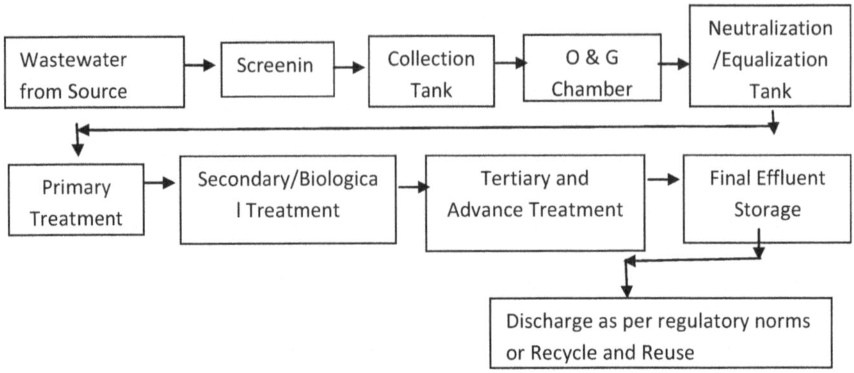

Fig. 1 Treatment methodologies for wastewater treatment

2.2 *Technological Aspects*

Based on treatment methodology, the availability of different types of technologies and their applicability to Indian wastewater sectors are discussed in detail in the following sections:

2.2.1 Primary Treatment Technologies (Physico-chemical Treatment of Wastewater)

Primary Treatment Methodology is considered as the 1st stage of wastewater treatment process. It is mostly based on physico-chemical unit operations. The treatment process followed largely depends on the operations being carried out in the production/utility process in the plant. Physical operations such as screening, sedimentation, skimming and flotation are used. The steps involved screening, removal of oil and grease by skimming mechanism or by application of de-emulsification process prior to skimming in case of emulsified oily waste treatment. Next process involves equalization/neutralization processes followed by sedimentation/clarification with or without coagulation-flocculation. It has been observed that in many of the industry sectors like Pharmaceutical, Food processing, Paints, Dairies the removal of oil and grease is carried out manually which seems to be ineffective and there is every possibility of carryover of oil and grease in next stage of treatment process. As a result, post treatment performance got affected. Installation of proper skimming mechanism with belt type skimmer as shown in Fig. 2 may strongly be recommend for solving the issue in an effective manner.

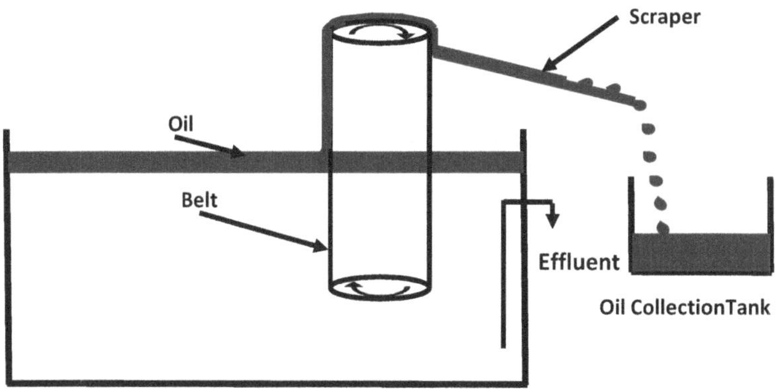

Fig. 2 Belt type skimmer for removing oil and grease from wastewater

2.2.2 Coagulation-Flocculation Technology

For removal of suspended solids, coagulation–flocculation technology (Fig. 3) is the most dominating one in globally and also in Indian Industry sector. Here selection of chemicals specifically polymers plays a most crucial role for obtaining optimum removal efficiency of the suspended solids. It's a three-stage treatment process i.e. rapid mixing (1st stage) followed by slow speed mixing (2nd stage) and lastly sedimentation/clarification process (3rd stage).

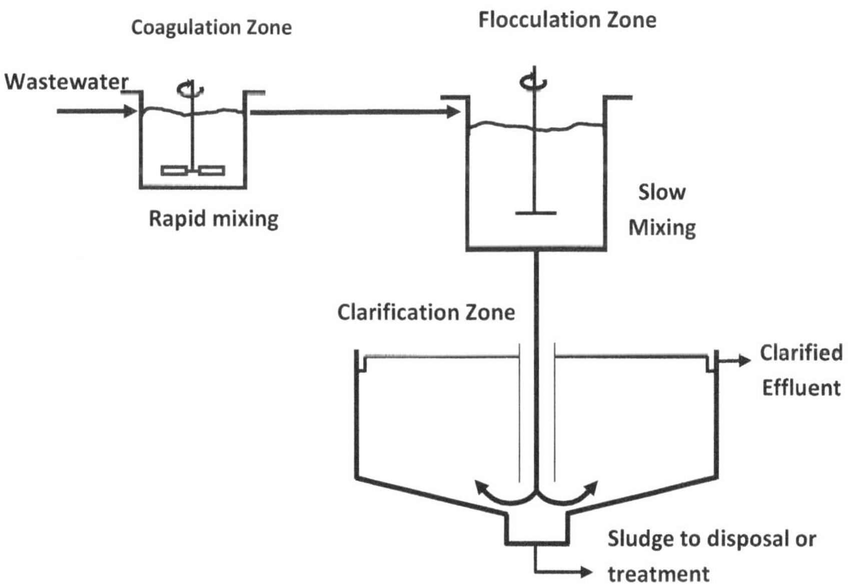

Fig. 3 Chemical coagulation-flocculation process in wastewater treatment

The equipment used in the 1st stage of the treatment process is known as Flash Mixer which involve high speed mixing (150–200 rpm) of influent with the chemicals commonly termed as coagulant. The examples are Alum, Poly Aluminium Chloride (PAC), Lime, Ferric Alum etc. In this stage agglomeration of suspended solids and colloidal impurities occurred by charge neutralization process. Retention time provided here typically vary from 30 s to 2 min.

2nd stage is the floc building stage where after agglomeration. Here, flocculants are added in the flocculation tank with a stirring arrangement with a speed of 30–50 rpm and retention time of 20–45 min. Here the particles come closer together to build large size heavier particles commonly known as floc having very high settling rate.

3rd stage is the sedimentation or say clarification process where the suspended/colloidal impurities settle down after floc building stage and effluent got clarified in the sedimentation tank or clarifier. In this zone, 1–3 h retention time is provided. The clarified effluent then allows passing to the next stage of treatment process namely Biological Treatment process.

Some design considers coagulation, flocculation and sedimentation in a single unit often termed as Mixing-Flocculation-Sedimentation (MFS) Unit.

During primary treatment process, selection of coagulant plays a crucial role to obtain system efficacy and it depends upon the nature of suspended solid to be removed, wastewater matrix, design facility, and chemicals cost. Quantity required depend wastewater flow and suspended solids loading rate. However, final selection of coagulant (or coagulants) should be made with jar testing experiments at lab level first followed by plant scale evaluation. Consideration must be given to required treated effluent quality, effect upon downstream treatment process performance, cost factor and sludge generation/handling cost and its disposal through effective treatment [14].

In coagulation–flocculation process during clarification stage clarifier generally chosen circular type of clarifier where retention time provided is 2–3 h. But sometimes the clarifier has been found to replaced by tube settler or say plate separator type of mechanism which requires much less space, lower installation commissioning time and faster settling time as well with retention time varies from 1 to 2 h maximum.

2.2.3 Advantages of Tube Settler

Tube settlers offer a cost-effective approach for upgrading existing wastewater treatment plant clarifiers and sedimentation basins to obtain enhanced performance with energy efficient manner. It often helps to reduce the tank age/footprint required in new installations or improve the performance of existing settling basins by reducing the solids loading on downstream filters. Tube settlers generally made of light weight PVC and can be easily supported with minimum structures that often incorporate the effluent trough supports [15].

One great advantage of tube settlers is that they are often available in a variety of module sizes and tube lengths to fit any tank geometry, with customised design and engineering offered by the manufacturer.

It uses multiple tubular channels placed at an angle of 60° and adjacent to each other, which combine to form an increased effective settling area. This provides for a particle settling depth which is significantly less than the settling depth of a conventional clarifier, reducing settling times. The channel collects solids into a thickened mass which promotes the solids to slide down along the tube channel. It can be installed as needed to fit in a new or existing clarifiers/basin of any size [15].

Only in case of high TSS loading and large capacity wastewater treatment plant, tube settler type of mechanism often lead to chocking problems which required frequent cleaning-making it operation/maintenance intensive. In that case circular clarifier may be looked for a better choice.

2.2.4 Dissolve Air Flotation Technique

Flotation is a unit operation that separate simultaneously oil and grease with suspended solids from the wastewater stream. Separation is carried out by introducing fine gas (usually air) bubbles into the liquid phase. In this section it has been discussed in detail considering it as one of the most effective treatment systems for wastewater containing suspended solids with oil and grease. The most common technique is that of dissolved air flotation (DAF), in which the waste stream is first pressurized with air in a closed tank. Then it is allowed to pass through a pressure-reduction valve, after that wastewater enters the flotation tank (Fig. 4) where, due to the sudden reduction in pressure, minute air bubbles in the order of 50–100 microns in diameter are formed. As the bubbles rise to the surface, the suspended solids and oil or grease materials adhere to them and are carried upwards. Most of the cases, chemicals are used to enhance flotation performance with coagulants, flocculating agents, emulsion breaking agents etc. [16].

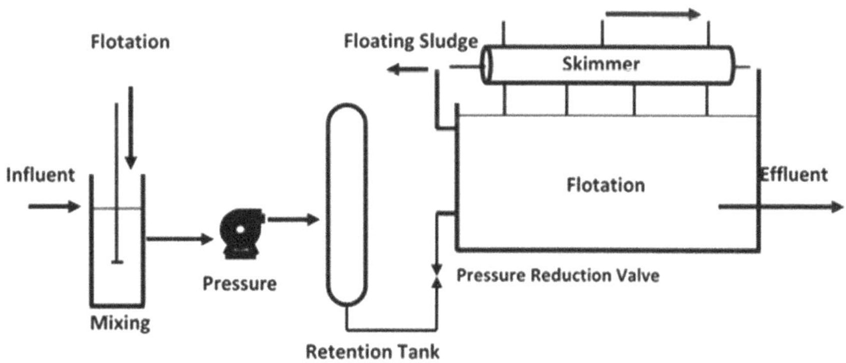

Fig. 4 Process design of a dissolved air flotation system

The main advantage claimed for a DAF system is the faster rate at which very small or light suspended solids can be removed in comparison with settling [16].

Performance of DAF system has been reported to be dependent on several factors like inlet solids concentration and air to solid ratio, maintenance of pH of the system.

Also proper flow rates and the continuous presence of trained/skilled operators are found to be key performance criteria for any DAF system. The principal advantages of this dissolve air flotation technique over sedimentation are that very minute or light particles that settles very slowly in sedimentation process can be removed almost completely and also in a less time. Once the particles floated in the surface, can be removed by skimming operation [17].

In one study, oil removal was reported to be 90% [18]. Another study, in first case a unit of tuna processing wastewaters, the DAF able to removed 80% of oil and grease and 74.8% of suspended solids in one case and a second case showed removal efficiencies of 64.3% for oil and grease and 48.2% of suspended solids. The main difference between these last two effluents was the lower solids usually content of the second [19]. Although considered very effective, DAF systems are not generally recommended for treating oily wastewater from small scale industrial units due to its high cost [20] and also requirement of skilled/trained operators [14].

2.2.5 Biological Treatment Technologies

Biological treatment is considered as an essential and integral part of any wastewater generated from industrial or municipal source. The wastewater generally contains soluble organic impurities or a mix of the both types of wastewater sources. The obvious treatment performance with cost benefits, both in terms of capital investment and operation/maintenance costs of biological treatment, over other treatment processes seems to be quite significant in any integrated wastewater treatment plant.

Biological treatment process often referred to as secondary treatment process where the microorganisms used are responsible for the degradation of the organic matter with the stabilization of organic wastes. The biological treatment processes are broadly classified as aerobic (in which aerobic and facultative micro-organisms predominate) or anaerobic (which use anaerobic micro-organism. Further, if the micro-organisms are suspended in the wastewater during biological operation, the operations are "called suspended growth processes", while the micro-organisms that are attached to a surface act as a carrier over which they grow are called "attached growth processes".

In most of the cases aerobic treatment process is followed. In case of effluent with high organic load e.g. in distillery, dairies, food processing, slaughter house effluent—the biological treatment process carried out first by anaerobic treatment followed by aerobic treatment to reduce the load to aerobic process and also to achieve the desired output in terms of BOD removal efficiency. The end product is methane in case of anaerobic process while in aerobic treatment the end product is carbon dioxide and water.

The low biomass yield does mean the nutrients requirement of an anaerobic process is lower than that of the aerobic process. In terms of BOD:N:P, the aerobic process would have required 100:5:1 while the anaerobic process only require 100:3.5:0.5. Notwithstanding this lower requirement, nutrients supplementation may still be considered in many industrial wastewaters treatment where wastewater are seems to be nutrients deficient even for anaerobic processes [21].

The conventional sewage treatment technologies such as Activated Sludge Process (ASP), Waste Stabilization pond (WSP) [22] up flow Anaerobic Sludge Blanket (UASB) Reactor etc., are commonly adopted in sewerage system to treat wastewater up to secondary level as per the effluent standards.

Biological treatment using aerobic activated sludge process has been followed widely well over a century both in municipal and industrial wastewater treatment. However, in recent years due to stringent regulatory norms forced the industry and municipal sector to follow various advance biological treatment technologies based on suspended and attached growth process.

Different dominating technologies in the Biological Treatment field in the country as per present practices applicable to industrial and municipal wastewater treatment are mentioned below.

2.2.6 Conventional Activated Sludge Process (ASP)

This is the most common and widely used aerobic biological treatment process to treat the wastewater generated from Industrial and Municipal services. The process is based on suspended growth of bacterial population where effluent after removal of suspended solids in primary treatment enter into the system comprising of a Aeration tanks and a secondary clarifier (Fig. 5). Diffused aeration technique is now followed to maintain Dissolve Oxygen level typically in the range of 2 mg/l. Biological growth is monitored by measuring Mixed Liquor Suspended Solids (MLSS) in the range of 1800- –4,500 mg/l typically or Mixed Liquor Volatile Suspended Solids (MLVSS)

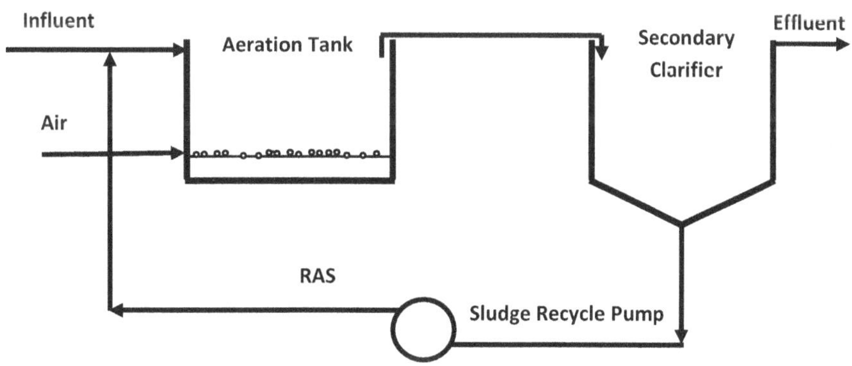

Fig. 5 Schematic of conventional activated sludge process

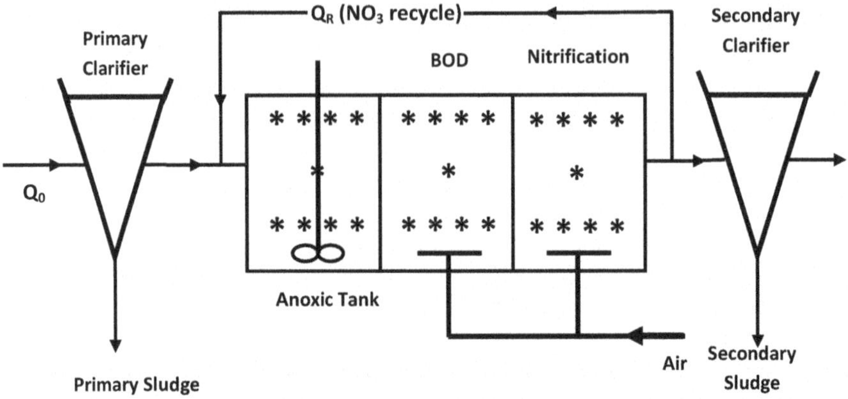

Fig. 6 Schematic of fluidized bed bio-reactor (FAB)/moving bed bio-reactor (MBBR) process

which should be 70% of the MLSS value for effective degradation of COD and BOD of the wastewater by removing soluble organics present.

2.2.7 Fluidized Aerobic Bio-reactor (FAB) More Popularly Known as Moving Bed Bio-reactor (MBBR)

This technology is based on the attached growth process. Here bacteriological growth is higher compare to conventional activated sludge process due to additional surface area availability provided by the media. Diffused aeration is essential and the reactor carrying the media acts as a fluidized bed. Thus, the process often referred to as Fluidized Aerobic Bioreactor (FAB) or also termed as Moving Bed Bioreactor (MBBR) as well. Schematic flow diagram of the process is presented in Fig. 6. This technology seems to be preferable option than the conventional activated sludge process due to higher surface area availability resulting in enhanced organic load reduction in the wastewater [22].

2.2.8 Submerged Aerobic Fixed Film Bio-Reactor (SAFE)

The key features of this process are in the following which represents in Fig. 7.

- Tube settlers inside aeration tank offer space economy
- Essentially a fixed film media as the name suggest with enhanced oxygen supply through submerged aeration
- Reactors up to 6 m deep enabling low land requirements
- Large biomass and long solid retention time in the reactor leading to low 'food to micro-organism ratio and higher organic removal based two stage biological oxidation

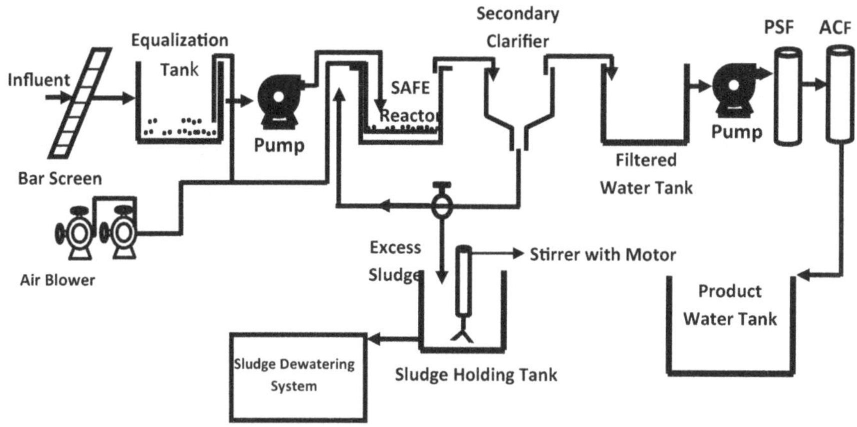

Fig. 7 Schematic of submerged aerobic fixed film bio-reactor (SAFE) process

- Many Pharmaceutical, food processing plants based on such technology are functioning in industrial wastewater application in India.

2.2.9 Sequential Batch Reactor also Known as Cyclic Activated Sludge System (SBR/CASS)

This is one of the newer technologies recently using in Indian industrial and municipal wastewater treatment sector. The treatment methodology of this technology may be looked as a variation of ASP technology which is essentially a batch treatment by combining, primary settling, aeration, secondary settling and decanting the treated effluent in a series of sequenced and or simultaneous reactions in the same basin on a time deferred cycle. Thus, multiple basins are considered for the design basis whereby when one tank is in one part of the cycle such as aeration, another tank will be settling and discharging the treated effluent in a cyclically repeated operation. For the aeration purpose high efficiency fine bubble non-clog membrane type diffusers are preferred. The rate kinetics for the bio reaction in this process is non steady type, not like conventional activated sludge process where reaction kinetics is based on continuous flow steady state. Close monitoring and optimization is essential for achieving the higher rate. Schematic diagram of Sequencing Batch Reactor process is presented in Fig. 8.

The process has the ability to remove simultaneously BOD with N and P from the wastewater. The technology got popularized and seems to produce very effective results in India in municipal sewage treatment. The consideration of this technology for industrial wastewater treatment specifically for Distillery, Dairy, Slaughter House, Textile, Tannery wastewater is yet to be established.

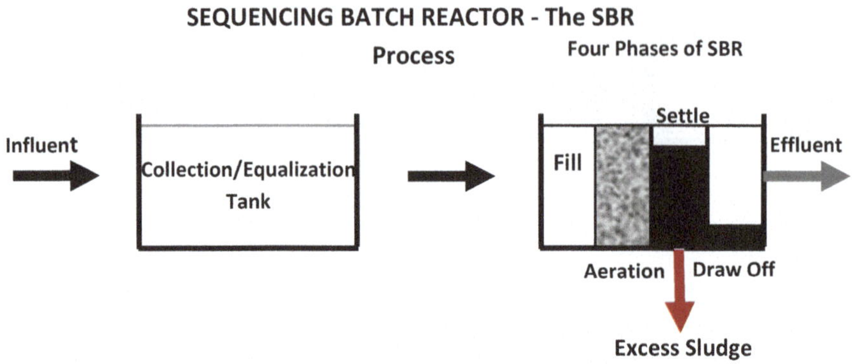

Fig. 8 Schematic of sequential batch reactor (SBR) also known as cyclic activated sludge system (CASS)

2.2.10 Integrated Fixed Film Activated Sludge System (IFAS)

During the initial stage of wastewater treatment plant installations where two stage biological treatment is followed-comprising trickling filter containing mostly stone or sometime plastic media as biological growth carrier (often referred to as packed bio-tower) [23]. The process is then followed by activated sludge process based on suspended growth, followed by secondary clarifier. But with technological advancements, space constraints, stringent regulatory norms this technology is presently rarely followed in Indian Industrial and Municipal wastewater treatment.

The improved and advance version of above configuration has been developed now for implementing in newer industrial as well municipal wastewater treatment systems. The process consists of fluidized aerobic bioreactor often termed as moving bed bioreactor (MBBR) instead of bio-tower followed by activated sludge process. This process has been reported to be produce very successful results in some of the industries like refineries and petrochemical, where the existing wastewater treatment system based on single stage conventional activated sludge process followed by secondary clarification tank is undergoing a capacity expansion plan or forced to meet stringent discharge norms. This hybrid process comprising of fluidized/moving media and activated sludge process is known as Integrated Fixed Film Activated Sludge (IFAS) process [23].

Some major advantages of the aforesaid process are providing below:

- The IFAS system provides enhanced surface area for biomass to grow and degrade the organic contaminants that are reluctant to biodegradation or also may sometime toxic to the environment as well.
- The overall efficiency of IFAS process is higher and more effective in nitrification of wastewater than conventional activated sludge process alone.
- Due to less wastage of sludge, the IFAS process provides much improved sludge management and dewatering facility as compared to the activated sludge process [23].

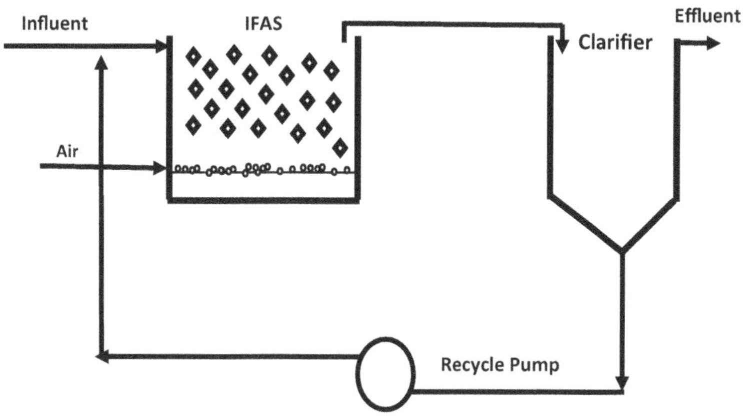

Fig. 9 Schematic for integrated fixed film activated sludge (IFAS) system

The sequences are shown in Fig. 9.

- It can be easily incorporated in the existing activated sludge system to meet additional load due to increased production and process capacity requirement and/or stricter regulations without the need of constructing additional concrete tanks.
- Foot-print of IFAS has a smaller with lower capital and operating cost.

2.2.11 Membrane Bioreactor (MBR)

Membrane Bioreactor technology is now considered to be the most advance technology producing high quality water almost ready for recycle and reuse. This technology combines the aeration and secondary clarifier in one and the same tank by sucking out the aerated mixed liquor through membranes instead of settling in a separate downstream tank. Thus, it yields a treated sewage with practically almost no BOD and suspended solids. Virtually clear and transparent treated liquid obtained from the treatment. The membrane is a matter of proprietorship and the throughput per membrane module offered by various companies are different and also each one advocates different shapes of the membranes like flat sheet, cross flow, dead end flow etc., which makes it difficult for common validated standard design criteria.

MBR technology has been in extensive usage for treatment of domestic sewage, but for industrial waste treatment applications, its use has been somewhat limited or selective. Recently as the wastewater recycling scenario growing in a faster rate the industries are become more prone to this technology. Figure 10 showed the difference in the two processes lies in the method of separation of bio-solids [23].

In the MBR process, the bio-solids are separated by means of a polymeric membrane based on microfiltration or ultra filtration unit, as against the gravity clarification process in the secondary clarifier in conventional activated sludge process.

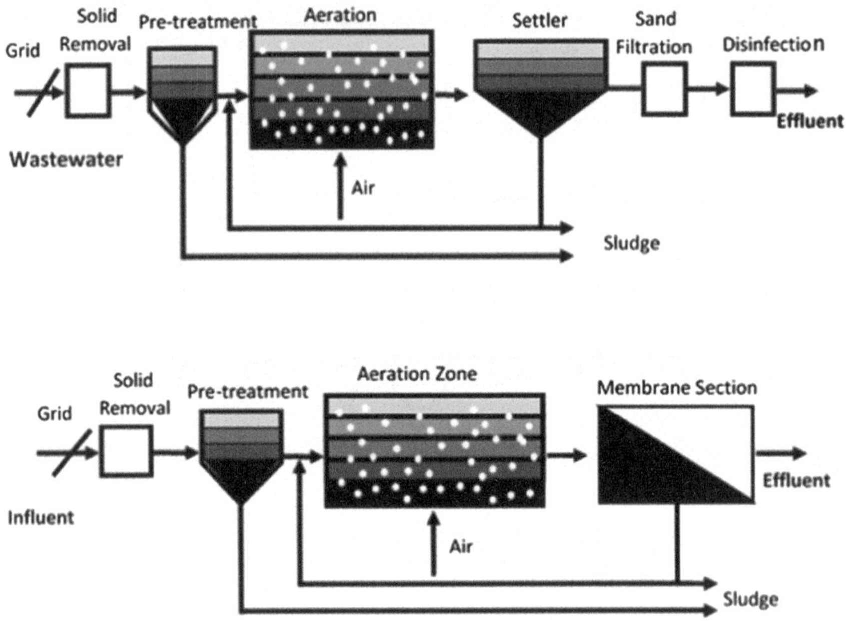

Fig. 10 Schematic of conventional activated sludge process (top) and external (side stream) membrane bio-reactor (bottom)

Therefore, the advantages of MBR system over conventional activated sludge system are obvious as listed below:

- Membrane filtration provides a positive barrier to suspended bio-mass that they cannot escape the system unlike gravity settling in activated sludge process, where the bio-mass continuously leaving the system along with clarified effluent and sometimes a total loss of solids is also occurred due to process upsets and sludge-bulking in the secondary clarifier. As a result, much higher bio-solids concentration measured as MLSS/MLVSS can be maintained (around 3 to 4 times) in a MBR process (around 10,000–12,000 mg/l) in comparison to the activated sludge process (around 2500–4,500 mg/l).
- Due to the above aspect of MBR, aeration tank volume in the MBR process can be one-third to one-fourth the volume of the aeration tank in an activated sludge system. Further, instead of gravity settling based clarifier, a much more compact tank is needed to house the membrane cassettes in case of submerged MBR and skid mounted membrane modules in case of non-submerged, external MBR system.
- In fact, in MBR system requires 40–60% less space as required for activated sludge system. The scope of civil work and overall foot-print as a result reduced to a great extent.

- Due to membrane filtration (micro/ultra filtration), the treated effluent in case of MBR process is much higher quality compared to conventional activated sludge process. Further it avoid the installation of tertiary filtration systems like sand filter and carbon filter which seems to be very common in activated sludge process after secondary clarification. The treated effluent produce is almost ready for reuse like cooling tower make-up, floor washing or for gardening/horticultural use etc. [23].

Typical treated water quality from MBR system is:

- BOD_5 <5 mg/L
- Turbidity <0.2 NTU
- TSS levels: <0.5 mg/l.

Such high-quality effluent is considered to be an important factor for recycle and re-use based on present scenario of fresh water availability.

2.2.12 Comparison of Different Biological Treatment Technologies Based on Performance with Cost Benefit Analysis

Table 2 Illustrate the comparison of Fluidized bed bioreactor (MBBR) process with Membrane Bioreactor process (MBR) for biological treatment of wastewater.

The performance of a biological treatment technology plays a very significant role with respect to effluent quality achievable for possible recycle and reuse. Tables 3 and 4 provide a technological summary with respect to some effluent quality parameters as mentioned therein.

Table 2 Comparison of MBR and MBBR technology

Sl No	Parameters	MBBR	MBR
1	Type of process	Aerobic fixed film continuous-attached growth	Aerobic continuous
2	Operating parameters	Automatic-PLC based	Automatic PLS based
3	Odour	Odour possibility-as Sludge not fully stabilized	Sludge fully stabilized-no odor
4	Treatment efficiency achievable	90–95% removal of BOD	95% removal of BOD
5	Post treatment requirement for industrial or indirect potable reuse	Filtration-preferable ultrafiltration followed by disinfection, If TDS is not an issue	Only minimal disinfection; If TDS is not an issue
6	Ease of operation during shutdown/maintenance	Maintenance of media is difficult	Membrane cassette can be isolated even treating the full flow
7	Level of operator attention	Medium	Low

Table 3 Different biological treatment technologies comparison summary

Parameters	SBR	MBBR	MBR
BOD in mg/l	10–20	10–20	3–5
TSS in mg/l	<30	<30	<5
Nh3-N in mg/l	2–5	2–5	<1
TP in mg/l	<0.5	<0.5	<0.5
TN in mg/l	<10	<10	<10
Direct use application	Only for landscaping	Only for landscaping	Possible

2.2.13 Tertiary Treatment Technologies

Tertiary Treatment generally includes physical and chemical treatment process followed after secondary treatment to meet the desired treatment objective. Here, final polish is provided to the effluent that improves wastewater quality before it is reused, recycled or discharged to the environment to meet regulatory norms as defined by Central or State Pollution Control Boards.

The major treatment technologies follow in Indian industrial and municipal sector are based on flocculation/sedimentation, pressure sand filtration, activated carbon adsorption, membranes filtration, ion exchange, de-chlorination and reverse osmosis.

Pressure Sand filtration and Activated Carbon Filtration are mostly used during tertiary treatment after secondary clarification mainly to remove fine residual insoluble impurities including trace organics, colour, odor etc.

Depending on financial criteria and effluent quality requirements sometimes Dual Media Filter, Multigrade filter is also provided after secondary clarification. These types of filters contain sand, gravels and Anthracite based media in a single unit and often plays the role of both sand and carbon filter in a single unit operation reducing operation, maintenance cost, backwash frequency etc. to a great extent.

Adsorption

Globally carbon has been recognised as an adsorbent for centuries. Recent changes in wastewater discharge standards regarding toxic pollutants have placed additional emphasis on this technology. Adsorption is particularly effective during tertiary treatment process for treating low concentration waste streams. After secondary treatment process it acts as a polisher in meeting stringent regulatory norms. The adsorption process utilized to remove colour, odour and other soluble organic pollutants from wastewater [24].

The activated carbon adsorption also helps to removes toxic chemicals such as insecticides, pesticides, phenols and its related compounds, cyanides and organic dyes which are difficult to treat by conventional treatment methods. Dissolved organics are adsorbed on carbon surface as waste water containing contaminants is allowed to pass through the adsorbent. Granular activated carbon is reported to be a very good adsorbent compare to powder activated carbon due to its high surface area to volume ratio [23].

Table 4 Comparison of different biological technologies with respect to cost (capital and operational), treated effluent quality and space requirement

Parameters	Conventional ASP	SBR	MBBR	IFAS	MBR
Treated effluent quality	Meets specified discharge standards with additional filtration step	Meets/exceeds specified discharge standards without additional filtration step	Meets/exceeds specified discharge standards with additional filtration etc	Meets/exceeds specified discharge standards with additional filtration etc	Exceeds specified discharge standards without additional filtration step. Very good for recycle provided TDS level permits
Ability to adjust to variable hydraulic and pollutant loading	Average	Very good	Very good	Very good	Very good
Pre-treatment requirement	Suspended impurities e.g. oil and grease and TSS removal	Suspended impurities e.g. oil and grease and TSS removal	Suspended impurities e.g. oil and grease and TSS removal	Suspended impurities e.g. oil and grease and TSS removal	Suspended impurities e.g. oil and grease and TSS removal with fine screening
Secondary clarifier requirement	Needed	Aeration Basin acts as clarifier	Needed	Needed	Clarifier is replaced by membrane filtration
Complexity to operate and control	Simple, but not operator friendly	Operator friendly	Operator Friendly	Operator friendly	Skilled manpower required
Capital investment	Low	Low	Medium	High	Very High
Operating cost	Low	Low	Medium	High	Very High
Space requirement	High	Low	Average	Average	Low

For many water treatment applications it has been proved to be the least expensive treatment option to remove wide variety trace organic contaminates during tertiary treatment process. Its suitability on any specific application will found to be depending on type of contaminants and costs as they directly relate to the quantity of carbon consumed.

Bio-physical Process of Adsorption

The biophysical process is considered as an augmentation and modification of Activated Sludge Process. In this process, powdered activated carbon is added in activated sludge process. Adsorption and Bio degradation take place as a combined process in a single tank, resulting in synergistic effects. The process has been developed in early seventies named as PACT$^{(R)}$ process and widely applied in USA. Later it spread in Europe and other parts of the world. In this process powdered activated carbon is encapsulated in the biological flocs and two different mechanisms takes place. One is adsorption in to powder activated carbon and next is biological removal-both are being complementary to each other. It helps to tackle specifically reluctant, poorly and sometimes non biodegradable organic compounds. Advantages are better sludge sedimentation and dewatering ability, less foaming, increased nitrification, enhanced removal efficiency with improved process stability [25]. Finally, in many cases disinfection process is often applied to make it suitable for further advance treatment with an objective to recycle and reuse.

2.2.14 Disinfection Process

Disinfection refers to the destruction of mainly pathogenic or say disease causing organisms from waste water. In the field of wastewater treatment, the categories of microorganisms which are human centric and responsible for the disease are bacteria, protozoa, cysts, helminths and viruses. The three-disinfection processes which are popular and widely followed in industrial and municipal applications namely chlorination, ozonation and UV disinfection are discussed in the following section.

Chlorination

Chlorination is a process of disinfection most widely used in treatment of both industrial and domestic wastewaters. Other kinds of use reported in wastewater treatment as advance oxidation process e.g. cyanide oxidation. Usually the effluents are treated with chlorine just before their final discharge to the receiving streams or water bodies. Chlorine gas or hypochlorite solutions found to be used the latter being easier to handle [16]. Chlorine forms hypochlorous acid which in turn forms hypochlorite in water solutions.

$$Cl_2 + H_2O \rightarrow HOCI + H^+ + Cl^-; HOCI \rightarrow H^+ + OCl^-$$

During chlorination process often wastewaters may contain considerable amounts of ammonia or volatile amines resulting in an increased demand of chlorine to achieve a desired degree of disinfection as they react with chlorine to produce chloramines. The amount of products depends on the factors like pH, ammonia concentration and the concentration of organic amines present in the wastewater.

The degree of disinfection is attributed to the residual chlorine present in treated water. During initial phase of chlorine dosing residual chlorine is negligible due to presence of reducing agents in the wastewater. With further addition of chlorine residual will appear as combined chlorine residual due to formation of chloramines. A decreasing phase of chlorine residual then observed due to complete reaction of ammonia and amines present with the added chlorine with disappearance of chloramines. Free residual chlorine will then appear on further addition of chlorine which is also referred to as "breakpoint chlorination". The main objective of obtaining some free chlorine residual is to ensure complete disinfection [16].

Chlorination units are generally simple, consisting of a vessel in which the wastewater and the chlorine are brought into contact to produce a good mixing condition. Residence time recommended not less than 30 s. Prior to final discharge, sufficient time (about 15 min) must be provided for the chlorine to react to meet the objective. This may be done in the ducts which carry the wastewater to the discharge point, provided the residence time exceeds 15 min. A typical configuration is shown in Fig. 11 [16].

The levels of free available chlorine should comply with the local regulations that usually vary between 0.2 and 1 mg/l. Some cases excess chlorine as residual in wastewater effluents was identified as the main toxicant suppressing the diversity, size, and quantity of fish in receiving streams or ecosystems [26]. To achieve effective disinfection, although 15 min is very common, retention times of up to 30 min are also used. The chlorine dosage needed to achieve the residuals will vary with the wastewater considered: 2–8 mg/l is common for an effluent from activated sludge plant, and can be as high as 40 mg/l in the case of septic wastewater [16, 17, 27].

Ozonation

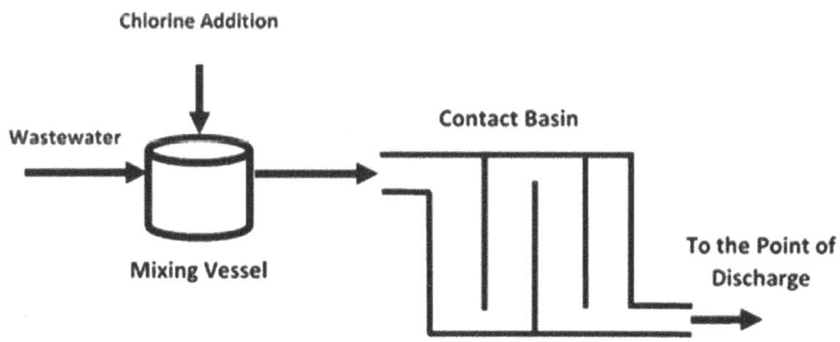

Fig. 11 A typical configuration of chlorination system

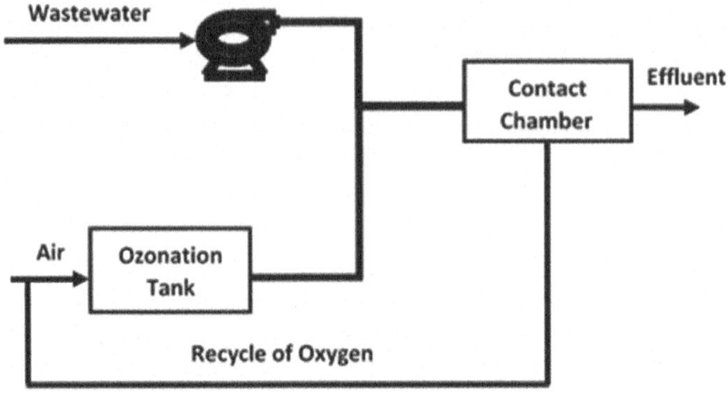

Fig. 12 Schematic diagram of an ozonation system

Ozonation is considered to be a very effective disinfection process applicable to water and wastewater treatment plant operation. It is a unit operation applied at the end of the treatment similar to chlorine or uv disinfection. Ozone seems to be as a very effective oxidizing agent useful for disinfection purpose to kill the bacteria and viruses from the wastewater. Ozone (O_3) is produced when a high voltage is discharged across a narrow gap in the presence of air or oxygen. Ozonation systems in situ generate O_3. From safety point of view ozonation is a much better choice than chlorination. Figure 12 represents a schematic diagram of an ozonation system.

Once ozone has been added and reacted, it enhances the dissolved oxygen level of the effluent to be discharged, which often beneficial to the receiving water stream. Also, as an advance treatment process it often helps to remove traces of any oxidizable impurities present in the wastewater in terms of COD and BOD as well. Contact tanks are usually closed to recirculate the oxygen-enriched air to the ozonation unit. Advantages over chlorination are that it does not produce dissolved solids and is affected neither by ammonia compounds present nor by the pH value of the effluent. It is also reported to be used to oxidize ammonia and nitrites present specifically in fish culture facilities [16, 28].

Ultraviolet (UV) Disinfection

This technique is primarily employed as a disinfection process that inactivates waterborne pathogens without use of chemicals. Additionally, UV is also effective for residual TOC removal, destruction of chloramines and ozone. The process is not considered as a popular one during industrial and municipal wastewater treatment plant application for disinfection purpose like chlorine and ozone.

In Indian wastewater sector the use of disinfection process follows the sequence:

Chlorination > Ozonation > UV disinfection. Considering cost as another factor which follow the sequence Ozonation > UV > Chlorination.

2.2.15 Advance Treatment Technologies for Recycle and Reuse

After tertiary treatment the requirement of advance treatment technology arises especially when the objective is to recycle and reuse. Advance treatment often may be considered to further stabilize chemical and biological oxygen demand in wastewater including removal of nitrogen and phosphorus. Sometimes the wastewater is also treated with substances like ozone or hydrogen peroxide as an advance oxidation process to destroy many trace and harmful contaminants that may contribute to compliance issue in the treated water to be discharged.

While considering the objective of recycle and reuse, advance treatment mainly focused on application of membrane-based separation technologies.

Based on the effluent quality obtained after tertiary treatment, selection of membrane-based technologies has been taken into consideration. Further purification of organics and dissolved salts in tertiary treated wastewater is achieved by use of reverse osmosis. The reverse osmosis is based on the certain specific polymeric membranes, usually cellulose acetate, polyamide or nylon to pass clean and pure water at fairly high rates and salts are rejected by the membranes.

RO membranes are very often susceptible to fouling due to presence of organics, colloidal impurities and microorganisms. A selection of proper pre-treatment systems like Sand filtration, Carbon filtration, Micro filtration and Ultra filtration systems often found suitable to get rid of this fouling issues.

Ultrafiltration (UF)

Ultrafiltration membrane often considered as an important pre-treatment process before reverse osmosis. Also it has been now widely applied during wastewater recycling as an advance treatment process when total dissolve solids in final treated water is not an issue.

It is responsible for retaining only macro molecules, viruses, traces of oily substances, colloidal silica and suspended solids. Thus, components like salts, solvents and low molecular weight organic solutes pass through ultrafiltration membrane with the permeate water [24]. The pressure differences across UF membrane are considered to be negligible due to non-retaining the salts by the membrane. But flux rates through the membranes are seems to be quite high allowing the use of lower pressures.

Nanofiltration (NF)

In the membrane spectrum Nanofiltration is positioned after Ultrafiltration but before Reverse Osmosis with a typical operating range of 0.001–0.01 μm [29]. This is essentially a lower pressure version of membrane capable of removing bivalent hardness causing ions such as calcium or magnesium together with bacteria, viruses, and colour. The treatment cost of nanofiltration is generally lower than reverse osmosis treatment including operational cost. Nanofiltration is preferentially recommended when permeate with some residual TDS but without colour, COD and hardness is acceptable. Feed water system needs to be maintained similar to reverse osmosis. Turbidity and colloidal impurities need to be maintained at minimum level. In case of high turbidity specifically when colloidal silica is present, pre-treatment through Microfiltration and Ultrafiltration is needed to be included in the treatment scheme. Disinfection of feed will be necessary to remove micro-organism from permeate.

3 Zero Liquid Discharge Concept and Case Study Example

Zero liquid discharge (ZLD) is a process developed to provide beneficial effects to industrial and municipal organizations for tackling the stringent environmental norms applicable to liquid and solid waste management. The process enables to protect our environment to its highest level with no effluent discharge or say left over leading to sustainable environmental practices. ZLD systems employ the implementation of advanced wastewater treatment technologies to recover and recycle virtually all of the wastewater produced.

Some of the salient features of a ZLD systems:

- A Zero liquid discharge facility (ZLD), is an industrial or municipal plant means without discharge of any wastewaters.
- Target ZLD is normally achieved by maximum waste water recovery
- Separation achieve by evaporation or boiling of water or part of waste water not reusable after passing through membrane process, in evaporators, crystallizers and recovery. The process produces solid waste and high purity water ready for reuse.
- Finally, ZLD facility enhances corporate image of an organization with a improved marketing image by providing better water resource optimization on sustainable basis.

Presently the major drivers of ZLD in the country are

(i) Water scarcity with growing fresh water demand along with still negligible rate of waste water recycling and water conservation practices following by the industries
(ii) Economics: recycled water becomes more affordable as the usable water supply from conventional sources becomes more expensive in many cases especially coastal and water stress areas.
(iii) Growing social responsibility and as an education towards awareness of environmental issues
(iv) ZLD cost is generally high in most cases but it might be a more economic solution when waste needs to be arranged and transported in large volumes over long distances on daily basis [30].

3.1 *Process Schematic (in General) for a ZLD Project*

Phase-I

Influent → ETP → Treated water from ETP → Ultrafiltration → Reverse Osmosis → Permeate Water → Recycle/Reuse.

Phase-II

Reject from Reverse Osmosis systems → Evaporation/Crystallization → Solid Waste.

Some major components of ZLD Systems

- Ultrafiltration Systems
- Reverse Osmosis (RO) mostly are of two stages
- Mechanical/Multi-effect Evaporators (MEE)
- Agitating Thin Film Digester (ATFD), etc.

3.1.1 Multi-effect Evaporators

This is the most popular and widely accepted unit operation when the desired objective is to achieve a zero liquid discharge facility to meet the stringent regulatory norms in wastewater treatment for an industry (Fig. 13).

The following points seem to be important while considering a Multi-Effect Evaporator system:

- For optimizing the design of an evaporator, one of the most significant consideration is the steam economy (kg of liquid evaporated per kilogram of steam used). In fact steam economy ultimately governs the successful implementation of the process.

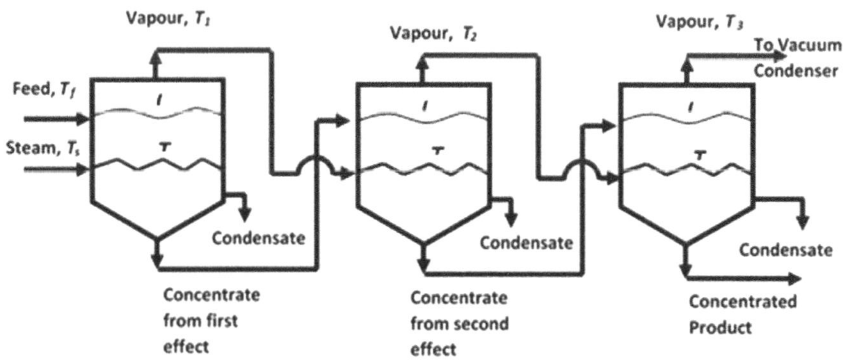

Fig. 13 Multi-effect evaporators for zero liquid discharge facility

- The most effective way to obtain high economies is to use a **Multiple Effect Evaporator**, whereby the vapor from one effect—is used to heat the feed in the next effect, where boiling occurs at lower pressure [31].
- Further, by thermo-compression the vapors will condensate at a temperature high enough to be reused for the next effect through compression, will always help to enhance efficiency of the evaporator.

3.1.2 Agitating Thin Film Digester (ATFD)

This is another unit operation/equipment which often considered during designing a zero liquid discharge system.

Operating Principle

- ATFD stands for evaporation of water/solvents to convert concentrated liquid to dry powder or flakes. Selection is important based on application -either recovered solvent or dry product or dry sludge.
- Agitated Thin Film Dryer is considered to be the ideal equipment for continuous processing of concentrated liquid for drying. The design basis consists of cylindrical, vertical body with heating jacket and a rotor inside of the shell which is equipped with rows of pendulum blades all over the length of the dryer.
- Here, heat will transfer from jacket to main shell by smooth agitation while water/solvent will evaporate and liquid will convert gradually to slurry, to cake and finally to either dry powder or flakes [32].

3.2 Solar Based Application in ZLD

This application is possible where land is not an issue and easily available. The reject from RO systems is directly sent to solar based evaporation pond where under sunlight evaporation take place with subsequent crystallization of the solid. The dry solid is taken out and stored for safe disposal as per solid waste management rules/guidelines. This is seems to be quite effective with respect to energy savings point of view and also significantly low treatment cost with minimum manpower requirement. Considering the climate scenario in India this type of project may be looked for southern and western states of the country.

3.3 Application Areas

- *Pharmaceutical Industries*
  - Drying of API and its intermediates to eliminate product the conventional process of drying.

  - Recovery of important solvents from the feed.
- *Textile Industries*
  - Drying of concentrated liquid for s recovery.
- *Agro Chemicals-Dyers and Pigment Industries*
  - Drying of product for recovery of inorganic product, to eliminate other conventional systems
- *Chemical and Petro-Chemical Industries*
  - Drying of Chemical and petro-chemical products to recover the powder or cakes.
- *Sugar Industries and Distilleries*
  - Drying of concentrated effluent to make the Zero Liquid Discharge System.
- *Effluent Treatment Plants*
  - One stop solution for drying of effluent to make Zero Liquid Discharge.

4 Case Study of Recycle and Reuse Leading to Zero Liquid Discharge

Case Study-1

A pharmaceutical industry manufacturing formulation-based product generates waste water at a capacity of 100 KLD. The influent sources are from (i) Utility, process and kitchen waste and (ii) another source is from sewage water from sanitation activities with a capacity of 10 kl per day. The effluent characteristics from the two sources are provided below:

Influent Characteristics:

(i) pH—6.2–7.2, TSS—145–240 mg/L COD: 750–1300 mg/l, BOD: 630–850 mg/l, Oil and Grease: 45–60 mg/l; TDS: 350–430 mg/L
(ii) pH: 6.5–7.8, TSS: 130–280 mg/l, Oil and Grease: 25–40 mg/L, COD: 450–550 mg/L, BOD 260–300 mg/l.

Desired Treated water quality for recycling and reuse:

The process design has been carried out based on following Treated water quality for possible recycle and reuse as secondary purpose (non potable use):

pH: 7.0-8.0, TSS: <5 mg/L, BOD: <10 mg/l; COD: 50 mg/L (max), O&G: BDL, TDS: <500 mg/L.

Process schematic

Influent → Bar screen → Collection Chamber → Oil and Grease Chamber → Equalization Tank → Flash Mixer → Flocculator → Primary Clarifier → Aerobic

Treatment (ASP) → Secondary Clarifier → Filter Feed Tank with Chlorine Dosing System → Pressure Sand filter → Activated Carbon Filter → Treated Water Tank → Ultrafiltration Membrane → Final Treated water Storage → CT Make UP and Gardening use.

STP Process Design: Influent → Bar screen → Collection Chamber → Oil and Grease Chamber → Equalization Tank → Aerobic Treatment → Secondary Clarifier → Treated water with Chlorine Disinfection → Send to ETP Treated water storage for Ultrafiltration Treatment Plant.

Sludge from the secondary clarifier returns back to aeration tank to maintained Active Biomass as Mixed Liquid Suspended Solids (MLSS).

Final treated water parameters

pH—7.3-7.8; TSS <3mg/L, Oil and grease—BDL; COD—30-40 mg/l ; BOD—6-8 mg/l, TDS—<400 mg/L; Total Bacteriological Count-Nil.

As Total Dissolve Solids are well within regulatory limits (below 2,000 mg/l) the Reverse osmosis systems are not considered in the treatment scheme and thereby minimizing the treatment as well as operation and maintenance cost.

The treated water is recycling reuse for cooling water make up, gardening; road /pavement washing, fire fighting storage. No water is discharging to any streams or river leading to zero-discharge facility and thereby meeting successfully the regulatory board requirements.

Case Study-2

Textile Mill Wastewater Treatment with recycle and reuse:

Typical influent characteristics: pH: 10.5–11.5, Temp: 55–60 °C; TSS: 230–285 mg/L, COD: 750–930 mg/L, BOD: 380–470 mg/L, Pt Co Color: 1200–2200, Flow: >300 KLD.

Process schematic: Influent → Influent → Bar screen → Collection Chamber → Cooling Tower system → Equalization Tank → Flash Mixer → Flocculator → Primary Clarifier → Aerobic Treatment (ASP) → Secondary Clarifier → Filter Feed Tank with Chlorine Dosing System → Pressure Sand filter → Activated carbon Filter—Treated water Tank → Feed to two stage RO systems.

1st stage:

Treated water → Microfiltration → Ultrafiltration Membrane →RO Membrane →Permeate (60% Recovery) → Final Treated water → storage for Recycle and Reuse.

Reject (40%) from the 1st stage →

2nd Stage RO Membrane system → Permeate for recycle and reuse (20% Recovery),

Overall recovery combining 1st and 2nd stage: 80% (60% from 1st stage plus 20% from 2nd stage).

Reject from the 2nd stage: Multi-effect Evaporator → Vapour → Condenser → Condensate → Treated water storage.

Solid from MEE: Crystallization, Collection and Storage and disposal as per solid handling rules.

Final Treated Water Quality: pH: 6.7-7.5, TDS: <10 mg/L, BOD: <15 mg/L, COD: <50 mg/l, TSS: <1 mg/L, O&G: Nil.

The final treated water used in process side application and feed to DM (demineralised) plant for boiler application.

5 Conclusion

Every one of us must accept the fact that water is a basic necessity for the survival of humans. There is interplay of various factors that govern access and utilisation of water resources and in light of the increasing water demand it becomes important to look for holistic, people centred and industry specific approaches for water and wastewater management. It needs the combined initiative and action of all levels, for the socioeconomic development of the country. Wastewater Management with recycle and reuse must be taken as a key performance indicator to protect our fresh water resources. Looking into Industry, it is the second largest user of water after agriculture. The amount of water used varies widely from one industry type to another. Many businesses, notably the textile, food, beverages, and pharmaceutical sectors are highly water intensive and consume water by using it as an ingredient in processing finished products. Many businesses discharge of wastewater or wash water into natural fresh water ecosystems. Each litre of water moving through a system represents a significant energy cost. Water losses in the form of inefficient treatment, delivery leakages, theft, and consumer waste all directly affect the amount of energy required to deliver water for its proper use. Wastage of water directly leads to a waste of energy. Industries and municipalities must focus on effective water uses by reducing demand, maximizing recycling opportunities; minimizing water losses due to breakdowns and leakages; and investing in new water storage infrastructure. For Sustainable development optimal management of water and wastewater leading to zero discharge is now another challenge before the industry. Finally, in terms of working methodology, it may be looked as a systematic approach of Identifying, Measuring, Monitoring and Reducing the Water Consumption by different activities by Industries and municipalities. It should be an exercise of stewardship of water resources through deployment of cost effective, performance oriented appropriate/advance technology focusing on recycle and reusing the wastewater for the greatest good of society and the environment on sustainable basis.

6 Recommendations

(i) Development of strategy for systematic utilization of water resources in various industrial processes reducing water demand and losses.

(ii) Implementation of energy efficient, performance oriented and cost-effective technologies that helps to achieve sustainable environmental practices with opportunities for recycle and reuse.
(iii) Conducting water balance study in a regular interval to understand the water consumption pattern and improvement needs.
(iv) Any better approach and improved technology solution may give a new pathway to the wastewater management in industrial and municipal sector in India.

References

1. CGWB (2011) Ground Water Year Book, India 2010–11, Central Ground Water Board, Ministry of Water Resources. Government of India. http://www.cgwb.gov.in/documents/Ground%20Water%20Year%20Book-2010-11.pdf
2. CWC (2010) Water and related statistics, Water Planning and Project Wing, Central Water Commission, India
3. Water and wastewater treatment, AECINFRA. http://aecinfra.com/water-and-wastewater-treatment/
4. CPCB (2009) Status of water supply, wastewater generation and treatment in Class I cities and Class II towns of India. Series: CUPS/70/2009-10. Central Pollution Control Board, India
5. Van-Rooijen DJ, Turral H, Biggs TW (2008) Urban and Industrial Water Use in the Krishna Basin. Irrig Drain. https://doi.org/10.1002/ird.439
6. Bhardwaj RM (2005) Status of wastewater generation and treatment in India, IWG-Env. Joint Work Session on Water Statistics, Vienna, 20–22 June.
7. Wastewater production, treatment, and use in India. http://www.ais.unwater.org/ais/pluginfile.php/356/mod_page/content/111/CountryReport_India.pdf
8. CPCB (2006) Performance status of common effluent treatment plant in India.
9. CPCB (2007) Advance methods for treatment of textile industry effluents, resource recycling series: RERES/&/2007, Central Pollution Control Board, India
10. Trivedy RK, Nakate SS (2001) Treatment of hospital waste and sewage in hyacinth ponds. In: Trivedy RK, Kaul S (eds) Low cost wastewater treatment technologies, ABD, Jaipur, India, pp 132–163
11. Andreas and Shane (2015) Waste water treatment & reuse: past, present and future. Water 7:4887–4895
12. CPCB (2005) Parivesh sewage pollution—News Letter, Central Pollution Control Board, Ministry of Environment and Forests, Govt. of India, Parivesh Bhawan, East Arjun Nagar, Delhi 110032. http://cpcbenvis.nic.in/newsletter/sewagepollution/contentsewagepoll-0205.html
13. Shuval HI, Adin A, Fattal B, Rawitz E, Yekutiel P (1986) Wastewater irrigation in developing countries: health effects and technical solutions. Technical paper no. 51. World Bank, Washington DC
14. Coagulation and flocculation mrwa.com. https://www.mrwa.com/WaterWorksMnl/Chapter%2012%20Coagulation.pdf
15. Tube settler systems for clarification—Water online https://www.wateronline.com/doc/tube-settler-systems-for-clarification-0001
16. Wastewater treatment in the fishery industry—Primary treatment. http://www.fao.org/3/V9922E/V9922E04.htm
17. Metcalf and Eddy Inc (1979), Waste-water engineering: treatment, disposal, reuse, McGraw-Hill Book Co, New York, USA.

18. Ilet KJ (1980) Dissolved air flotation and hydrocyclones for wastewater treatment and by-product recovery in the food process industries. Water Serv 84:26–27
19. Ertz DB (1980) Dissolved air flotation treatment of seafood processing wastes: an assessment. In: Proceedings of 8th national symposium on food processing wastes, Cincinnati, Ohio, USA
20. Anon (1986) Environmental assessment and management of the fish processing industry, Sectoral studies series Nr 28, UNIDO, Vienna, Austria
21. Wun Jern NJ (2006) National University of Singapore, industrial wastewater treatment, Published by Imperial College Press, London
22. Sewage Treatment Plant (STP)—HAR Wassertec. https://www.harwassertec.com/sewage-treatmentplantstp.
23. Biological wastewater treatment. www.watertoday.org. https://www.watertoday.org/Article%20Archieve/Aquatech%2012.pdf
24. Effluent treatment in textile industry-2. www.chemarc.com https://www.chemarc.com/content/effluent-treatment-in-textile-industry-2/5b0d6cc8d04ee50d542a11a6
25. Boere JA, Berg EVD, Mudde CG, Amersfoort NV (1992) Industrial wastewater treatment by bio physical process, The Netherlands, BETIEE conference, Toulouse
26. Paller MH, Lewis WM, Heidinger RC, Wawronowicz JL (1983) Effects of ammonia and chlorine on fish in streams receiving secondary discharges. J Water Pollut Control Federat 55:1087–1097
27. Eckenfelder WW (1980) Principles of water quality management. CBI Publishing Co., Boston, USA
28. Monroe DW, Key WP (1980) The feasibility of ozone for purification of hatchery waters. Ozone Sci Eng 2:203–224
29. Schoeberl P, Brik M, Braun R, Fuchs W (2004) Treatment and recycling of textile waste water. Desalination 171:173–183
30. Zero liquid discharge (ZLD) concept, evolution and technology options. https://gpcb.gov.in/images/pdf/ZLD_PRESENTATION_3.PDF
31. Industrial evaporator manufacturer & supplier in Hyderabad. https://kepengg.com/evaporator-plant.html
32. Agitated Thin Film Dryer (ATFD) Manufacturer in Hyderabad. https://kepengg.com/agitated-thin-film-dryer.html

Wastewater and Its Impacts in India

Bharati Brahmacharimayum and Pranab Kumar Ghosh

Abstract Management of wastewater generated by the cities has gained a lot of prominences nowadays due to rapid urbanization and increased awareness about environment protection. However, wastewater sector is one of the sectors of urban infrastructure which has been characterized by huge gap between the facilities created for treatment and the waste being generated. The result of this glaring gap is highlighted in the form of huge quantity of wastewater being left untreated. With the objective to identify strategies on how to bridge this gap, review of the Indian wastewater sector is being undertaken. The review focused on alternative treatment technologies, innovative operational system for sustainable treatment of wastewater, and political and legal framework for wastewater management. In order to complement the conventional three stages treatment process of wastewater, sustainable treatment options have been reviewed to gain insights on cases of successful implementation of these innovative technologies in treatment of wastewater in India. The legal framework and policy initiatives have also been reviewed to ascertain the enabling framework that has been established for holistic wastewater management. The policy initiatives have been undertaken by the government with the motive to bring flexibility to the system and equip the decision-makers with the options to devise site-specific solutions for treatment of wastewater for cities with different sizes and socio-economic characteristics. The major thrust has been towards preference of decentralized wastewater system which has been equipped with features to fine-tune the system to meet the challenges being faced by the sector.

Keywords Wastewater management · Decentralized wastewater system · Wastewater recycling and reuse · Policy framework for wastewater

---

B. Brahmacharimayum
State Climate Change Cell Manipur, Directorate of Environment, Govt. of Manipur, Imphal East 795005, India

P. K. Ghosh (✉)
Department of Civil Engineering, Indian Institute of Technology Guwahati, Guwahati, Assam 781 039, India
e-mail: pkghosh@iitg.ac.in

© Springer Nature Switzerland AG 2022

S. Yadav et al. (eds.), *Wastewater Assessment, Treatment, Reuse and Development in India*, Earth and Environmental Sciences Library,
https://doi.org/10.1007/978-3-030-95786-5_8

1 Introduction

The management of waste has gained a lot of prominence in recent times in urban areas. The increased generation of waste due to rapid urbanization and the concern about protecting the environment from the adverse effects of unscientific waste management practices could be attributed to being the reason behind renewed interest in waste management. Amongst the urban services, wastewater management is one of the sectors which have been characterized by wide infrastructure gap for service provision. The reason for the failure to completely treat the wastewater generated by the city could be attributed to various factors. One of the factors is the lack of budgetary resources to create/rejuvenate the ageing infrastructure. Practice to charge for the service through adoption of polluters' pay principle is a concept which is yet to be implemented in totality by the various cities. Adoption of this concept could resolve the funding gap to a certain extent. Another critical issue faced by the sector is the selection of either a centralized or decentralized wastewater system. Decentralized wastewater system has the flexibility of fine-tuning the treatment system to the site-specific conditions and dispense with the need to make huge investment in creating the sewer network. However, this system has not been adopted widely by the various cities as most of the cities still prefer to adopt the conventional centralized wastewater system. Because of this, stringent environmental protection standards, legal reforms, and policy initiatives have been undertaken by the government with the objective to establish a sustainable system for management of wastewater.

This chapter presents a brief overview of the Indian wastewater sector along with the various challenges that need to be overcome to ensure a holistic wastewater management system in India. The chapter comprises of eight sections. The current status of the wastewater management sector is discussed in the second section. The third section focuses on the common treatment options while the various sustainable treatment options that are nowadays being used in place of the conventional techniques is discussed in the fourth section. The legal framework and the policy initiatives that have been undertaken to ensure a holistic wastewater management system in India is discussed in the fifth and sixth section, respectively. The key challenges being faced by the sector and the strategies can be adopted to overcome these challenges is the focus of the final section. The chapter concludes with the key conclusions that have been drawn from the analysis of the Indian wastewater sector.

2 Wastewater Production and Treatment—Current Status

The key sources for the production of wastewater can be considered to be broadly classified into domestic and industrial sources. As per CPHEEO, around 70–80% of the water supplied to the households for domestic consumption is later released as wastewater. It has been estimated that class-I cities and class-II towns, representing 72% of the urban population of India generates per capita wastewater of around

Table 1 Sewage generation in Class-I cities and Class-II towns of India [5]

Parameter	Class I cities		Class II towns	
	2006	2009	2006	2009
Sewage generation (in MLD)	26,164	35,558	2,965	2,696
Sewage treatment capacity (in MLD)	6,047	11,554	200	234
Untreated (%)	77	68	93	92

98 L per capita per day (lpcd). Table 1 shows the sewage generation from class I and II cities and towns of India during the period 2006 and 2009. During 2015, the quantity of sewage produced in the country was estimated to be around 61,754 million litre per day (MLD) while the sewage treatment capacity was around 22,963 MLD [1]. As a result, around 62% of the sewage (i.e. around 38,791 MLD) remains untreated and this is discharged directly into nearby water bodies. It has also been observed that more than 50% of the sewage in India is generated from the states of Maharashtra, Tamil Nadu, Uttar Pradesh, Delhi and Gujarat. Besides the domestic sources, another source of wastewater is from a large number of small scale industries which are located haphazardly in crowded locations such as residential areas. In the context of industrial wastewater treatment, around 5,500 MLD generated by small-scale industries are left untreated [2]. This sector on account of lack of adequate financial resources is not able to establish their own treatment facilities. Therefore, in order to reduce the pollutant load of the receiving water bodies, there is a need to established common effluent treatment plants (CETPs) for such industries across the country.

The large gap in the sewage generation and sewage treatment is distinctly visible. One of the main reasons for the failure to ensure 100% treatment of the generated wastewater could be attributed to the assumption that unless underground sewers are built and unless the sewage is conveyed via these sewers the wastewater in open drains is treated as 'unofficial' and need not be taken into account for treatment before disposal [3]. The treatment of only meagre percent of the wastewater generated is not a surprising factor when this assumption is taken into account [4]. The objective of the government towards the fulfillment of 100% sanitation will not be possible as long as the sewage is classified into 'official' and 'unofficial' category for treatment.

3 Common Treatment Options

The natural cleansing system used to be sufficient enough to treat the generated wastewater in olden days. The wastewater discharged into the water bodies easily get diluted by a large amount of clean water and, with passage of time, microbial communities present in nature slowly break down the sewage as well other pollutants resulting in clean treated water with its passage through the natural water system. However, this natural cleansing system works when the quantity is meagre but the wastewater discharged nowadays is overloading the natural system. Alternative treatment options have been devised to treat the wastewater and these options can be classified into the following three stages:

i. The preliminary stage in which coarse solids and other large materials are removed by screening and grit removal.
ii. Primary stage in which the settleable organic and inorganic solids are removed after allowing it to settle down and skimming off the floating scum.
iii. Secondary stage in which biological processes come into play for a breakdown of the contaminants.

Most sewage treatment plants operated by the state governments in the country have been established under schemes financed by the Government of India, and these are functioning at sub-optimal efficiency level. In most of the class I cities, majority of the sewage treatment plants use activated sludge process (ASP) as the preferred treatment technology, followed by up flow anaerobic sludge blanket (UASB) technology [6]. However, in Class II towns, waste stabilization ponds (WSP) technology is widely used, followed by UASB. The main reason for the preference of ASP over other technologies for treatment in large cities is because of the space constraint as ASP requires less space in comparison with UASB and WSP.

In conventional ASP technology, primary and secondary sludge is treated in anaerobic reactors thereby producing very less amount of processed sludge and substantially reducing the aeration costs of the primary and secondary sludge. The biogas generated from the anaerobic reactors is an added advantage which indirectly minimizes the operational cost. UASB technology utilizes grit chamber as the preliminary treatment unit and requires only one day retention time. It is one of the most economical technologies as it requires only the passage of the sewage through the treatment system and biogas generated is considered as a value-added by product. However, all the anaerobic processes are highly environmentally sensitive and a slight shock load of the feed is enough to hamper the overall performance of the treatment system. In the case of waste stabilization ponds (WSP), waste is treated through three stages of ponds. The first pond is anaerobic in nature, second pond being facultative in nature followed by the third pond as the maturation stage with a total hydraulic retention time of 5 to 7 days.

Owing to the large-scale industrialization and development, the pollutant load has greatly varied from domestic sewage with high organic content to the one which contains various chemicals such as heavy metals, chemical compounds along with toxic and radioactive wastes. These wastes cannot be removed by ordinary settling and need more complex and advanced treatment processes. Some of the advanced techniques employed for treating this wastewater ranges from biological treatment processes to physico-chemical methods such filtration, carbon adsorption, distillation, and reverse osmosis [7]. The preferred approach for treating the wastewater could be one of the above-mentioned treatment option or a combination of it for remediating it to the extent that the effluents can be reused for various industrial, agricultural, or recreational purposes.

4 Wastewater Recycling and Reuse—Sustainable Treatment Options

Conventional centralized sewage treatment technologies, which are currently being used widely in India, are expensive, needs complex technology and, most importantly, unable to meet the demand for treatment of all the generated wastewater. As a result, the untreated or partially treated wastewaters reach the water bodies and pose several health and environmental problems. The focus should now shift to adoption of sustainable treatment technologies to treat the wastewater locally and reuse it. One of the potential solutions towards sustainable treatment technologies is the decentralized wastewater treatment system. In a decentralized wastewater system, besides lower operation and maintenance cost the capital investment requirement is also lower in comparison with a conventional centralized wastewater system. The need for a shorter network for wastewater collection, requirement for less quantity of wastewater for treatment, and the use of simpler and natural treatment options are the main highlights of the decentralized system. The wastewater can be managed and treated at the generation source itself through adoption of new emerging natural processes which can be designed according to the site-specific requirement and are much more cost-effective than the conventional treatment processes. The treated wastewater even has the potential for local reuse in non-potable purposes such as toilet flushing, car washing, horticulture/gardening, groundwater recharge, and such other uses.

Furthermore, the utilization of wastewater effluent in agriculture will result in conservation of water quality as well as serve the irrigation purposes. Similarly, if municipal wastewater is managed in a proper manner could reduce the problem of surface water pollution to a great extent. The nutrients contained in the sewage such as nitrogen and phosphorus can be utilized thereby reducing the consumption of chemical fertilizers [8].

Some of the successful case studies on various treatment options are highlighted in the following section [9].

1. *Decentralized wastewater treatment system*: An example of successful decentralized wastewater treatment is the case of Aravind Eye Hospital, Puducherry, India. Around 340 KLD of wastewater including greywater generated from sinks, bathrooms and kitchens and black water from toilets of the hospital building is treated at their treatment plant. Treatment of the wastewater has been done using a combination of both aerobic and anaerobic processes. The wastewater after treatment has been utilized for gardening as well as flushing within the hospital premises.
2. *Wetland system for wastewater treatment*: This treatment system normally consists of a primary settling tank followed by a planted filter bed. The inlet and outlet zones of the bed are filled up with larger gravels of 50–100 mm diameter. In addition to this, a water regulation chamber is also constructed before the final collection tank. The quality of the water is monitored at various points throughout the bed via the monitoring ports. Similar kind of system has

been successfully implemented on a pilot scale in IIT Bombay campus on an experimental basis.

3. *Soil biotechnology*: A combination of physical processes like sedimentation, infiltration and biochemical processes is utilized in this technology to remove suspended solids as well as the organic and inorganic pollutants from the wastewater. The key component of the system is the use of a culture containing native micro-flora. An example of treating the wastewater using this technology is the case of Chaudhary Charan Singh International airport, Lucknow. In this plant, the wastewater trickles down the layers of stone or rubble, soil media (weathered rock) containing micro-flora culture, which aid in the removal of the pollutants in the process.
4. *Green bridge technology*: This technology uses a combination of Ecofert, which is an active microbial consortium, biomats, sand, gravels, and plants. The treatment involves a series of green bridge filter which traps the floatable and suspended solids while microbial bioremediation by Ecofert reduces the organic and inorganic content present in the wastewater. This technology was developed by Shristi Eco Research Institute (SERI), Pune. Treating the polluted stretch of Ahar River in Udaipur is an example of successful implementation of this technology.
5. *Nualgi technology*: Nualgi powder is nanoparticles containing silica and micronutrients such as iron, manganese, and magnesium, which trigger the growth of diatoms. The growth of diatoms, in turn, degrades the organic content of sewage thereby increasing the dissolved oxygen content. This technology was used by local farmers to solve the problem of fish death due to high pollution of Madivala Lake of Bangalore by adding around 5–6 kg of Naulgi powder. Improved water quality has resulted in reviving the aquatic food chain and improved fish production.

5 Legal Framework for Wastewater Sector in India

The Indian government has enacted several legislations for the wastewater sector. The key legislations that have been enacted are as follows:

- *Water (Prevention and Control of Pollution) Act, 1974*: This act, which was amended in 1988, aims to provide for prevention and control of water pollution and specifically prohibits the discharge of any poisonous, noxious or polluting matter into any stream or well. This is the Act that also established the Central and State Pollution Control Boards, and as per this act consents from the State Pollution Control Board are required for any new discharge into any new stream or well. Under this Act, it is mandatory to obtain "Consent to Establish" from Pollution Control Board before starting work on the sewage treatment plant and also obtain "Consent to Operate" after construction [10].
- *Water (Prevention and Control of Pollution) Cess Act, 1977*: This Act, which was amended in 1999, provides for levy and collection of a cess on water consumed

by establishments with a view to augment the resources of the Central Pollution Control Board and State Pollution Control Boards for prevention and control of pollution of water under the Water Act, 1974. Cess is payable for activities as outlined in Schedule II to the State Government.

- *Environment (Protection) Act, 1986*: The discharge standards for treated sewage, the noise standards governing the sewage treatment plant (STP), the air emission standards governing the STP are prescribed in this Act. The PCB is empowered to tighten these standards wherever it is needed. As per clause (1) of Sect. 3, the Central Government is given the authority to take measures to protect and improve the quality of the environment, and prevent and control environmental pollution.
- *Municipal Byelaws*: These laws mandate the owner of any property to dispose of sewage in accordance with the provisions. For example, when municipal sewers are within a certain distance from the property, the owner is obligated to connect the establishment to those sewers as per the bye-laws.
- *Town & Country Planning Acts*: Every project must adhere to the applicable town &country planning Act.

The first three Acts have been promulgated within the power granted to the Union Government under the entry 56 on the Union List of the Seventh Schedule of Indian Constitution. Entry 56 stresses on inter-state rivers and river valleys because most Indian rivers, which are the major receptors of pollution, are not confined to the boundaries of a single state. There is an element of national and public interest, pursuant to which, these Acts have been framed with the intention to preserve these supra-statal water bodies through provisions specified in those Acts. The purpose of the Water Act (1974) is to promulgate the Central Pollution Control Board and State Pollution Control Boards into existence for the function of "Prevention and Control of Water Pollution." The Water Cess Act (1977) provides for a source of revenue for the boards to enable them discharge their functions laid out in the Water Act (1977) and is done by instituting a set of levies and penalties. Levies are instituted upon all types of consumers of water releasing the used water (polluted water) into the environment and penalties are for those consumers breaching the standards laid down in the Environment (Protection) Act (1986). These acts, which are complementary and supplementary in nature, thus define the regulatory framework for wastewater in India.

6 Policy Framework for Wastewater Sector in India

National Urban Sanitation Policy is one of the key policy initiatives that form the policy framework for the wastewater sector in India. It delegates the responsibilities for sanitation and wastewater management amongst the three tiers of the government [11]. As per this policy, Government of India is envisaged to assist with generating awareness, dividing institutional responsibilities, providing assistance for funding projects as part of City Sanitation Plans, national-level monitoring and evaluation, and mainstreaming sanitation into natural investment in urban infrastructure and housing. While the state governments are expected to assist with assigning institutional responsibilities, resources, and capacities; setting standards at the state level within the overall framework of the national standards; resolving issues of tenure and space in providing sanitation facilities for the poor; monitoring and evaluating cities' performance; and conduct capacity building and training exercises. Finally, the urban local bodies are to undertake preparation of city sanitation plans, carry out planning and financing schemes, creating assets and managing systems to meet service norms, fixing tariff and carry out revenue collection, and engaging stakeholders in ensuring 100% sanitation.

The National Urban Sanitation Policy (2008) has mandated a different approach to the management of wastewater services in India. Some of the notable changes include wider participation of public, NGOs and CBOs in wastewater management. The policy has advocated taking the assistance of NGOs and CBOs in mobilizing communities, raising awareness and working out affordable and community-managed solutions to sanitation. Furthermore, the services of NGOs and private parties may be sought by ULBs for (i) preparing and disseminating materials for IEC, (ii) conducting baseline surveys and stakeholder consultations, (iii) maintaining a comprehensive GIS-based database, (iv) implementing physical works, and (v) letting out and supervising O&M management contracts. It has been proposed to even take the help of citizen's groups, NGOs, and CBOs for effective monitoring and evaluation of wastewater services.

With respect to the choice of technologies, the preferred approach should be up-gradation and retrofitting of the difficult existing solutions in order to fully tap its potential. While devising solutions for existing areas, the approach should be towards the selection of technological options which can be upgraded and those technologies that recycle and reuse treated wastewater should be encouraged. In addition to this, state governments are being urged to set standards for ULBs on use of low energy-intensive onsite/decentralized wastewater treatment technologies and distributed utilities. The concept of recovery of operation & maintenance cost through introduction of user charges has become a key cornerstone of the new policy. Finally, the role of private sector through PPP has been proposed to be an alternative route for implementation of key projects and activities in wastewater sector.

7 Challenges of Wastewater Sector in India

Indian wastewater sector is one of the few sectors that has been neglected and has not been given enough attention by the various stakeholders resulting in several challenges which need to be overcome to improve its service level. One of the challenges is relating to whether to adopt a centralized or decentralized approach to wastewater treatment. The National Urban Sanitation Policy has focused on decentralized wastewater treatment approach for sewerage and sewage treatment while the conventional model of service delivery has been on treating the sewage through a network of large and long underground sewers conveying sewage to a distant treatment plant with a capacity to ideally treat all of the wastewater generated in the city. Service delivery through centralized projects entails high capital investment in sewers, pumping stations and plants and subsequently expenditure on material, energy, and labor for operation and maintenance. Such model will become a financial burden for small, fiscally-constrained cities, which cannot afford sewage drainage systems, let alone sewage treatment ones [3]. Furthermore, collection costs of the total budget for wastewater management are high in a centralized system, particularly in small communities with low population densities. In such a scenario for small cities, decentralized model of service delivery could be a feasible and viable alternative. Decentralized systems keep the collection component of the wastewater management system as minimal as possible and focus mainly on necessary treatment and disposal of wastewater [12].

Wastewater treatment plants have long lives, and the technology adopted at the time of their commissioning depends on the quality of raw wastewater and the nature of the receiving medium. Amongst the technologies used in wastewater treatment plants, conventional Activated Sludge Process (ASP) or a variation of it is the technology used in almost 60% of the plants in India. The efficacy of ASP, which is a biological process, depends on the quality of the wastewater in terms of its biological content. However, nowadays, the sources of wastewater are from domestic and commercial establishments. The wastewater thus contains both biological and chemical constituents, and its composition varies from one part of the city to another part. In such case, adopting a centralized approach will mean selecting a single technology for the entire city that will lead to reduced efficiency of treatment process in some areas with lower biological content, thereby resulting in incurring unnecessary expenses for some areas of the city. The preferred approach, however, would be to adopt area-specific approach for treatment. Decentralization approach embodies area specificity in its model so project can be implemented based on the ground realities and preferences of users of the area under consideration.

Low collection efficiency is another key challenge of the Indian wastewater sector. One of the reasons for this could be attributed to the practice of not considering the wastewater in open drains while treating the sewage in the underground sewers as the quantity of wastewater to be treated within the concerned city. Underground sewerage is no doubt a non-negotiable characteristic of centralized wastewater sector while the decentralized system has the inherent flexibility to respond to ground realities of wastewater management issues of the concerned area. In a decentralized system, the open drains conveying the wastewater or the user-specific septic treatment system can be easily incorporated into the existing system to arrive at an optimal solution. Furthermore, the option to explore for innovative solution for using the open drains as a treatment zone can be explored wherein bioremediation techniques can be used to partially treat the waste and convey this partially treated waste to the treatment plant. This will certainly reduce pollution and also turn the drain, from a stinky and dirty sewer to more aesthetically pleasing waterway, which will become a part of the city's landscape.

Lack of adequate funds by the ULBs is one of the challenges being faced by them and this has limited the ability of ULBs to provide the basic urban services. One of the suggestions is to levy user charges. The user charges should be designed to cover the operation and maintenance cost, depreciation cost and debt servicing requirement, in case the project has been funded with debt financing. An important step in this direction could be through strict implementation of the polluter pays' principle. Premises and establishments, which release wastewater into their surroundings, should be obligated to pay in accordance with the investment they have to make for treatment and disposal on their own, in case of absence of a sewerage service. The new sanitation policy has empowered the ULBs to set the tariff, and in case of adoption of decentralized system the ULB can set area-specific tariff structure for that area depending on the socio-economic characteristics of that area. For example, affluent urban areas, owing to the affordability of the users of that area, may include wastewater recycling and reuse in their sewage management system with a view to conserve the scarce resource such as clean water. And, in fact, high user charges do not deter use amongst those that can afford to pay [13].

8 Conclusions

The treatment of wastewater is an essential process for the conservation of our environment and needs to be encouraged in India. The sector has been characterized by a huge gap between the infrastructure facilities and the amount of waste being generated, thereby leaving behind a significant portion of wastewater left untreated. The failure to ensure holistic wastewater management could be attributed to the key challenges being faced by the sector. Some of the key challenges worth mentioning are lack of budgetary resources, practice of treating only the wastewater conveyed through underground sewers leading to low collection efficiency, preference for centralized system which entails huge capital investment, lack of interest to adopt innovative sustainable treatment technologies.

There is a need for a paradigm shift in treatment of wastewater in India. The focus should be to embrace new sustainable technologies for treatment in place of the conventional treatment process. The practices of opting conventional centralized wastewater system also need to be relooked in the light of the need to build in flexibility in the system and making its cost-effectiveness. The feasibility of opting decentralized wastewater system over the conventional system should be explored to overcome the challenges being faced by the sector.

9 Recommendations

New sustainable technologies for treatment of wastewater should be explored in place of the conventional treatment process. The practices of opting conventional centralized wastewater system also need to be relooked in the light of the need to build in flexibility in the system and making its cost-effectiveness. The feasibility of opting decentralized wastewater system over the conventional system should be explored with the objective to overcome the challenges being faced by the sector.

References

1. CPCB (2016) CPCB Bulletin—vol I. Central Pollution Control Board, New Delhi
2. ENVIS (2017) Newsletter. [cited 2018 August 09, 2018]; Available from: http://cpcbenvis.nic.in/cpcb_newsletter/sewagepollution.pdf
3. Planning Commission (2011) Report of the working group on urban water supply and sanitation for 12th Five Year Plan (2012–17). Planning Commission, Government of India, New Delhi
4. MoUD (2011) Report on urban infrastructure services. High-Powered Expert Committee, Ministry of Urban Development, Government of India, New Delhi
5. CPCB (2009) Status of water supply, wastewater generation and treatment in class I cities and class II towns of India. In: Control of urban pollution series: CUPS/70/2009. Central Pollution Control Board, New Delhi
6. CPCB (2005) Status of sewage treatment in Inda. Central Pollution Control Board, New Delhi
7. EPA (1998) How wastewater treatment works—the basics. United States, Environmental Protection Agency, Washington, D.C.
8. FAO (1992) Wastewater treatment and use in agriculture. Food and Agriculture Organisation, Rome
9. CSE (2014) Decentralised wastewater treatment and reuse: case studies of implementation on different scale—community, institutional and individual building. Centre of Science and Environment, New Delhi
10. CPHEEO (2013) Manual on sewerage and sewage treatment systems: Part A—Engineering. Central Public Health & Environmental Engineering Organisation, Ministry of Housing and Urban Affairs, Government of India, New Delhi
11. MoUD (2008) National urban sanitation policy. Ministry of Urban Development, Government of India, New Delhi
12. Massoud MA, Tarhini A, Nasr JA (2009) Decentralized approaches to wastewater treatment and management: applicability in developing countries. J Environ Manage 90(1):652–659
13. Varghese S (2012) Corporatizing water. India's draft national water policy. Institute for Agriculture & Trade Policy, Minnesota

Wastewater Remediation

Nutrient Loading Impact on Remediation of Hydrocarbon Polluted Groundwater Using Constructed Wetland

Abhishek, Pankaj Kumar Gupta, and Brijesh Kumar Yadav

Abstract Pollution of soil–water resources by the release of hydrocarbons is a major concern because of the wide coverage of groundwater table by these pollutants under varying subsurface conditions. Amongst several remediation technologies, bioremediation has been proved to be the most environmental benign technology. The efficiency of bioremediation technique can be amplified by controlling the prevailing environmental conditions and by modifying the constituents using plants, the process termed as biostimulation. In this chapter, treatment wetlands have been employed and investigated at laboratory and field scales in order to replicate the real field conditions. The performance of an engineered wetland planted with *Canna generalis* is evaluated to decontaminate hydrocarbon polluted groundwater. Practical and numerical experiments are performed for a duration of two months under controlled conditions to investigate the impact of nutrient loading on the bioremediation of Toluene, the selected hydrocarbon. The observed breakthrough curves show a slow initial removal rate which reached up to a level of 99% at the end of the experiment. Results presented in this chapter can be used for designing treatment wetland for decontaminating hydrocarbon polluted water in the fields.

Keywords Treatment wetland · Hydrocarbon pollution · Bioremediation · Nutrient loading

---

Abhishek
School of Environment and Society, Tokyo Institute of Technology, Yokohama 226 8503, Japan
e-mail: abhishek.a.ab@m.titech.ac.jp

P. K. Gupta (✉)
Faculty of Environment, University of Waterloo, 200 University Ave W, Waterloo, ON N2L 3G1, Canada
e-mail: pk3gupta@uwaterloo.ca

B. K. Yadav
Department of Hydrology, Indian Institute of Technology Roorkee, Uttarakhand 247667, India
e-mail: brijkfhy@iitr.ac.in

© Springer Nature Switzerland AG 2022

S. Yadav et al. (eds.), *Wastewater Assessment, Treatment, Reuse and Development in India*, Earth and Environmental Sciences Library,
https://doi.org/10.1007/978-3-030-95786-5_9

1 Introduction

Rapid urbanization and increasing population have caused unprecedented stress on the surface- and sub-surface water resources globally. Shallow regions of Indian aquifer have been reported the most vulnerable to hydrocarbon pollutants amid the intense industrial development, especially petrochemical refineries [14]. Bhattacharyya and Shekdar reported that more than twenty-eight thousand tons of oily sludge are generated by Indian refineries annually. Leakage from storage tanks and pipelines (underground) transporting the raw- and processed petroleum hydrocarbons cause these landscapes to be even more vulnerable to hydrocarbon contamination. A spilled hydrocarbon spreads over the groundwater system forming a pure phase pool [12]. The water-soluble compound like toluene starts dissolving to underlying groundwater resources and finally form a dissolved plume [12]. The use of this contaminated groundwater for drinking propose may cause a significant negative impact on microbiota [13] and human health, as toluene is carvcinogenic. For this reason, more attention of environmental scientist and geochemist is needed to remediate such pollutant from groundwater systems to fulfill the drinking water demand of increasing population and to minimize the impact of hydrocarbon pollution.

Various bioremediation approaches have been developed by many researchers in the last decades, in which plant-assisted bioremediation using constructed wetlands have been reported among the most cost-effective techniques [2]. Constructed wetland gained widespread acceptance since the 1980s due to its high performance and low installation and maintenance costs. Considerable information on the design and operation of these treatment wetlands has accumulated since that time [11]. However, very little attention has been paid to the performance evaluation of plant-assisted bioremediation and Horizontal Flow Treatment Wetland (HFTW) for the treatment of hydrocarbon contaminated groundwater resources. Most of these studies focus on the treatment of stormwater and wastewater containing several pollutants, including hydrocarbons [9]. The outcomes/results of these studies have restrictions to be used directly to design a bioremediation system for hydrocarbon polluted groundwater because (a) wastewater contains high background nutrient load, which significantly affect the removal rate, (b) enhanced co-metabolism due to high organic pollutants play important role performance of wastewater treatment. Yadav et al. [15] reported high biodegradation of dissolved BTEX from batch having 50% wastewater and 50% polluted groundwater at room temperature. While the biodegradation rate was very low in batch having 100% polluted groundwater resources.

Laboratory scale experiments examining the removal of BTEX compounds under aerated condition revealed the higher removal rates in planted mesocosms [2, 5, 6]. Domestic wastewater (mixed with the contaminated groundwater collected from the refinery site) was added to grow the root zone microbes in different mesocosms. The results fortified that the positive role of plant (root zone) on the biodegradation of toluene from contaminated groundwater resources. However, several questions remain unanswered, and further work is required to be done to fill the knowledge gap. Additional research could be conducted to examine the effect of (inorganic) nutrient loading on toluene removal from polluted groundwater in HFTW.

Amidst all the recent advancements, the strategies relevant for the realistic agrarian field conditions where the nutrients' loading followed by their leaching to the vadose zone are major causes of groundwater pollution and hence needs to be studied in detail. Therefore, in this chapter, extensive laboratory work has been performed to investigate the effect of nutrient loading in a two-dimensional experimental setup. This research focused on designing cost-effective treatment wetlands to remediate hydrocarbon (toluene in the current study) polluted subsurface using integrated practical experiments-numerical modeling. The simulation runs using HYDRUS CW2D model was solved with and without plant under three different level of nutrient loadings. This coupled effect has not been addressed adequately for toluene in the previous studies.

2 Constructed Wetlands-Literature Updates

Biodegradation of petroleum hydrocarbons is a complex process that is strongly influenced by physical, chemical, and biological factors such as temperature, oxygen, nutrients, salinity, pH, physical state, the concentration of the contaminant and the composition and adaptability of the microbial population [2]. Many studies have shown success with enhanced bioremediation and treatment wetlands which accelerates the natural process by providing nutrients, electron acceptors, and competent degrading microorganisms [3]. Constructed wetlands are the treatment systems that use natural processes involving wetland vegetation, soils, and their associated microbial assemblages to improve water quality. The functionality of the system is illustrated in Fig. 1a–c.

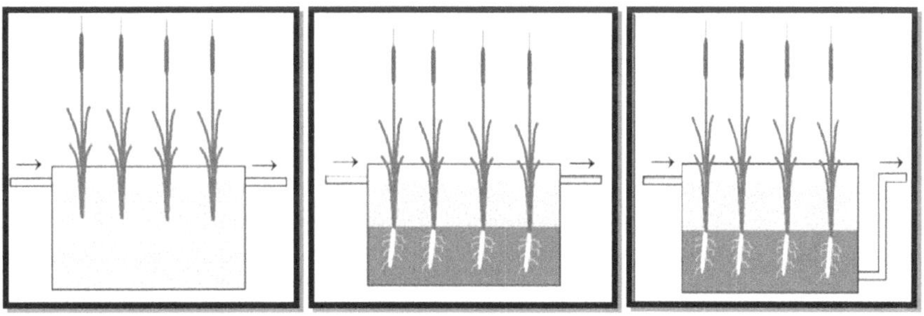

(a) Planted with Macrophytes sp. Without gravel bed. Free water surface flow.

(b) Planted with Macrophytes sp. With gravel bed. Free water surface flow.

(c) Planted with Macrophytes sp. With gravel bed. Free water surface. Effluent from bottom.

Fig. 1 Different types of constructed wetlands that are generally used for contaminant removal from subsurface resources

Constructed wetlands outperform the conventional treatment methods because of their low operational costs, unskilled manpower- and smaller land area requirements, and inherent habitat enhancement [4]. CWs, typical natural (mimics the natural systems the most) and eco-friendly systems, are considered as 'black boxes' where the interactions between the constitutive components are not well understood. The performance of the wetland systems depends on the various physical, chemical, and biological components and their interactions with biotic and abiotic habitats. This technique has been proved promising since the rhizosphere is abundant in microbes capable of hydrocarbons' natural degradation, which further depends partly on the prevailing ecosystem (temperature, oxygen, nutrients, salinity, pressure, pH, chemical composition, physical state, and contaminant concentration, etc.) and partly on the operating conditions. Research shows that the former factor plays a vital role in most cases. Hence, to enhance a time-bound effective removal of the pollutant (hydrocarbon), the design criteria become the vital user-defined parameter (Table 1).

The hydrocarbons (say, non-aqueous phase liquids: NAPLs) get introduced to the subsurface through poor storage practices, inadequate disposal facilities, or accidental spills. As a hydrocarbon migrates downward through the subsurface, a part of it remains in the soil matrix as residual, which, along with the mobile phase, causes a prolonged water quality degradation Yadav et al. [15]. Toluene was used for the current case study because of the high thermodynamic stability (and high water solubility, 1780 mg L^{-1}) followed by enhanced environmental persistence of Benzene ring.

The freshwater wetlands are typically nutrient-limited due to the heavy nutrient demand by the plants. Moreover, they could also be nutrient traps in the form of

Table 1 Recent experimental and numerical studies performed to investigate the removal of various contaminants from the subsurface

Target compounds	CW types/Domain size, planted with	Key findings	References
Toluene	28 cm inner diameter × 30 cm height, *Canna generalis*	Reduction of removal time of toluene was reduced by 25% in the presence of the plantation	[2]
Benzene	1 m L × 0.6 m W × 0.8 m H, *Phragmites karka*	Toxic and inhibitory effects are system-dependent	[1]
Benzeze	75 cm H × 10 cm in diameter, *Phragmites ustralis* (Cav.) Trin. ex Steud	The seasonal removal efficiencies for benzene decreased with time and were much higher for the indoor compared to the outdoor wetlands	[12]
Diesel oil	10 m × 1 m × 60 cm, Typha seedlings at a density of 7.5 plants/m^2	Plantation enhanced the removal efficiencies attributed to the phytoremediation	Omari et al.

biomass, thus necessitating the additional nutrient load (not excessive) to the system to enhance the biodegradation [7]. Therefore, the wetland system under consideration was subjected to the nutrient loading to foster the removal rates in engineered wetlands by altering the wetland matrix to possess (a) concurrence of aerobic and anaerobic to enhance (de)nitrification, (b) an internal carbon source. The CW2D module for the HYDRUS-2D software package [8] was used to model the degradation kinetics of the organic matter in subsurface flow Horizontal Flow Treatment Wetland (HFTW). Achieving high treatment performances in CWs within a short time is critical and hence adapting wetland design involves a trade-off between CW efficiency and sizing. Long-term wetlands studies are further needed in order to understand the wetlands systems more and to extrapolate and implement the laboratory scale results to the field scale, followed by the calibration and evaluation of the numerical models. Therefore, the objective of this study is to investigate the role of nutrient loading and design of horizontal flow treatment wetland in long-term controlled conditions.

3 Materials and Methods

Engineered treatment wetland experiments and microcosm experiment was performed to evaluate dissolved toluene's biodegradation rates in rhizospheric (wetland dominated area) and vadose zone respectively.

3.1 Experimental Investigation

Four laboratory-scale horizontal flow treatment wetlands (HFTW) of dimensions 0.6 m L × 0.3 m W × 0.3 m H were fabricated with plexiglass for wetland experiment and controlled. The design of wetland consists—a central chamber filled with experimental media and two sidewalls of length 5 cm to maintain the uniform flow through the root zone. The constructed HFTW were homogeneously filled with the single main layer of gravel with an average particle size of 10 mm and depth of 15 cm. Ponding of 15 cm, the headspace of about 15 cm was provided in the HFCWs by monitoring the water recirculation using Peristaltic pumps. All wetland setup was uniformly planted with *Canna generalis* with a plant density of 7 plants/ft^2. Both wetlands were placed in climatically controlled conditions with temperature 21° ± 3 °C and a light intensity of 85–100 μmol photons m^{-2} s^{-1} for an average exposure time of 16 h d^{-1} maintained throughout the experiments. Figure 1 shows the systematic diagram of the experimental setup employed in the current study. The porosity of the experimental media used in the HFTW, hydraulic residence time (HRT), and effective volume of the treatment setups were estimated based on the dimensions of the HFTW.

Contemporaneously, a homogeneous 0.1 M toluene solution was prepared by dissolving Toluene (Mark, 99.999%) to Millipore water up to its solubility limits ($\simeq$490 mg L^{-1}) while stirring with a magnetic stirrer at 400 rpm for 24 h. The prepared solution was stored at 10 °C. Similarly, the nutrient solution with concentration of 10, 30 and 30 mg L^{-1} of mineral nitrogen (NH_4Cl) were prepared during three times in experimental periods. After seven days of the plantation, the groundwater containing dissolved 250 ppm toluene was applied to the all HFTW to allow the acclimatization of the rhizospheric microbes with the pollutant for approximately two days. Thereafter, the effluent samples were collected and analyzed for toluene concentration periodically. After 2 days, These HFTWs were applied to the variable nutrient loading on every alternate day by adding 10, 30 and 30 mg L^{-1} of mineral nitrogen (NH_4Cl) to HFTW 1, HFTW 2 and HFTW3, respectively, for the complete duration of the experiment, i.e., two months. Another replica of three HFTWs was kept unplanted and was used as a reference for quantifying and comparing the root zone impact on the degradation rates. Figure 2 shows the systematic diagram and an actual photo of the experimental setup.

Rhizospheric water samples were collected daily from the outlet of HFTW using attached needle syringe and were stored without headspace at 4 °C until analysis. The concentration of toluene was analyzed using the selective ion method (SIM) of Gas Chromatography (Agilent 7890B) Mass Spectrophotometer (Agilent 5977A). The 2μL water samples were taken in vial (Agilent 5183–2067) using gold standard attached needle syringes (Agilent 9301–0713). A chrompack capillary column (30 m length, 0.25 I. D, 0.0.25 μm film) was used for the analysis. Helium was employed as the carrier gas at a flow rate of 25 mL/min. The temperature of the injector port, detector port, and oven were kept at 161 °C, 100 °C, and 150 °C, respectively.

3.2 CW2D-HYDRUS Simulation Run

Different degradation kinetic models can be used to determine the rate of degradation like Constant or zero-order, Linear or first-order, Monod or Michaelis–Menten, Logistic, Logarithmic. In zero-order kinetics, the rate of depletion of the contaminant is taken as a constant irrespective of the contaminant concentration in soil water at a particular time. For a completely mixed batch system where dilution and advection are not the factors that influence contaminant concentrations, the degradation rate of hydrocarbon is often described by a first-order decay regime with respect to the contaminant concentration:

$$\frac{dC}{dt} = -K_1 C \tag{1}$$

where K_1 is first order decay coefficient and C is contaminant concentration.

It can be integrated to yield:

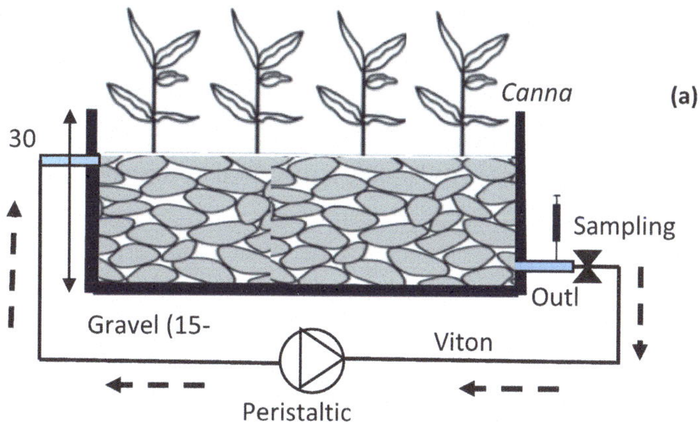

Fig. 2 **a** Schematic representation, **b** actual photo, of the horizontal flow engineered treatment wetland configuration used in this study

$$\frac{c}{c_o} = e^{-K_1 t} \quad (2)$$

where c_o initial concentration and t is is time.

On rearranging (3) we get

$$t = \ln\frac{\frac{C}{C_0}}{K_1} \tag{3}$$

The first-order or linear kinetics model assumes the rate of contaminant degradation proportional to the LNAPL concentration at a particular time. [8]. Solute transport in a variably-saturated soil profile is described by the advection–dispersion equation (ADE), Eq. 1, as

$$\frac{\partial(n_e C + \rho_b S_D)}{\partial t} = -\frac{\partial}{\partial z}\left(qC - n_e D^w \frac{\partial C}{\partial z}\right) - S_b \tag{4}$$

where C [ML^{-3}] is the contaminant concentration in the soil solution, n_e [L^{3}L^{-3}] and ρ_b[ML^{-3}] are the effective porosity and bulk density of the porous media, respectively, S_D [MM^{-1}] is contaminant adsorbed to the porous media and S_b [ML^{-3}S^{-1}] is the sink term for the degraded solute amount by microbial biomass.

The CW2D module considers 12 components and 9 processes. The components include dissolved oxygen; organic matter (three fractions of different degradability, i.e., readily and slowly biodegradable, and inert); ammonium, nitrite, nitrate, and nitrogen gas; inorganic phosphorus; and heterotrophic and two species of autotrophic microorganisms. Organic nitrogen and organic phosphorus are modelled as nutrient contents of the organic matter.

For simulating the HFT wetlands with aerobic conditions, the CW2D module was activated within the HYDRUS graphical user interface by selecting the 'Wetland Module' in the 'Solute Transport' window. The mass and length units were set according to the chosen units of all other parameters (Table 2). The transport domain with length 60 cm, width 15 cm and depth 15 cm (only one main layer of gravels as per the experimental set up) was created and discretized with elemental size 10 mm, to mimic the gravel size used in experimental setup. This resulted in a structured two/three-dimensional finite element mesh consisting of 4587 nodes within a layer. Ammonium oxidation and nitrate oxidation along with hydrolysis, aerobic growth and denitrification (anoxic growth) were modelled by assuming heterotrophic bacteria in experimental domain. The ponding boundary condition was implemented

Table 2 Solute transport and reaction parameters investigated in the experimental domain which was used to simulate numerical domain of HFTW in this study

Parameter	Value
Bulk density, g cm^{-3}	1.65
Longitudinal dispersivity, cm	8.5
Diffusion coefficient in water, cm^2 h^{-1}	0.03096
Diffusion coefficient in gas, cm^2 h^{-1}	313.2
Flow velocity, cm h^{-1}	0.625
Adsorption coefficient (K_d), h^{-1}	0.1
Henry's coefficient (K_h), dimensionless	0.26
Gas constant (R), atm m^3 mol^{-1} K^{-1}	8.20575e−5

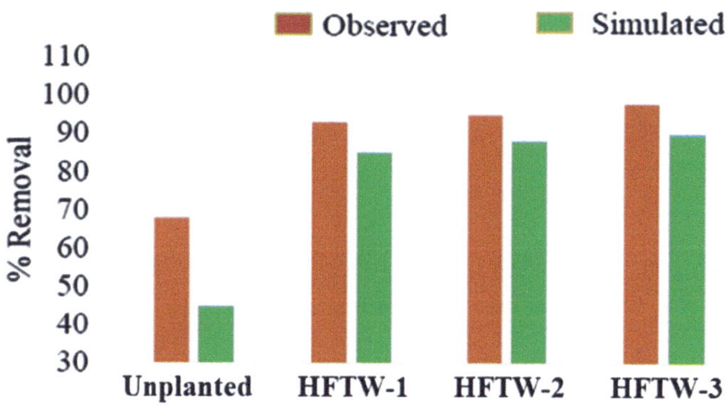

Fig. 3 Removal efficiency of toluene from planted HFTWs and unplanted HFTW (Observed in red color and simulated in green color)

as a modification of the upper atmospheric BC in the model and a constant pressure head BC (constant head of −2 cm) was applied to the bottom of the system. No flux boundary was taken at lower and side faces of the tank setups. The outlet point of HFT wetland was considered as a free drainage element. Surrounding greenhouse conditions were taken as the top atmospheric boundary condition (Fig. 2).

4 Results and Discussion

The influent toluene (HFTW1–HFTW3) concentrations varied between 0.01 and 250 ppm. The removal efficiencies was 68% for unplanted HFTW and ranged from 75 to 100% for planted HFTW. The mean toluene removal efficiencies from the laboratory set up and simulation domain were 58.3 (±52.0) %, 89.4 (±12.9) %, and 93.3 (±11.5) %, >99%, for HFTW1, HFTW2 and HFTW3, respectively. The HFTW2 and HFTW 3 showed highest performance with 93.3% and >99% toluene removal from experimental and simulation domains respectively. Figure 4 shows percentage removal of toluene with time in experimental and numerical domain with time. Where HFTW1, HFTW2, and HFTW3 represents nutrient loading of 10, 30 and 30 mg L^{-1}. High removal rate clearly indicates the positive impact of nutrient application of toluene degradation in HTFW-3 and 2.

The simulated removal rates of toluene showed a good agreement with the experimental values for all the systems. The observed breakthrough curves show a slow removal rate proportional to the simulated first order kinetics during the initial days. The removal rate, after the toluene concentration becomes less than 80 ppm, follows a transition from first-order to second-order kinetics sporadically. The percentage removal of toluene was near 98% at 56th day of experiments and was >99% at the end of experiment.

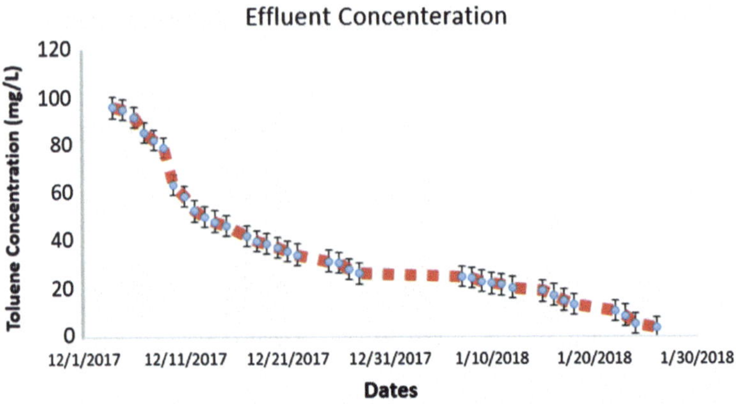

Fig. 4 Observed toluene concentration in effluent samples measured during experimental duration

Nutrients were added to enhance the degradation ability of the HFTWs. Previous studies have reported the significance of the addition of nitrogen and phosphorus as well as the supply of oxygen for the biodegradation of petroleum hydrocarbons Yadav et al. [14], Omari et al. Similarly, Basu et al. [2] reported high performance of HFTWs in case of nutrient supply by applying wastewater. In this study, it was observed that there was variability in the effect of nutrient on the performance of the three HFTWs for the different petroleum contaminants considered. The study domain representing HFTW2&3 shows complete degradation of toluene due to comparatively high nutrient dose as compared to HFTW1. The high removal efficiency may be resulted by biostimulation due to sufficient electron acceptor or aerobic growth of heterotrophic bacteria as modelled in CW2D numerical domain. Furthermore, applied concentration of nitrate accelerate the bio-kinetic reactions in duplex-HFTWs. Incorporating Monod type biokinetic processes to predict biochemical transformation and degradation of these compounds along with nutrient amendment in subsurface flow HFTWs is well justified as simulated and observed concentrations are well matched. This study will help in implementing the bio-remediation strategies in realistic agrarian field conditions where the nutrients' loading followed by their leaching to the vadose zone are major causes of groundwater pollution.

5 Conclusions

The wetlands are the self-sustaining green aquatic ecosystems and are preferred over conventional physio-chemical treatment methods. The biological processes of periphyton (microorganism community of wetland ecosystem) along with physical removal of pollutants by plant root biomass are responsible for pollutant removal in Horizontal Flow Treatment Wetlands (HFTW). Indeed the viability of any treatment wetland depends on factors like type of plants, porous media, and characteristics of polluted water along with the prevailing environmental conditions. Presence of plants to the treatment wetland system drastically improved the decontamination efficiency from 68% to >90% in the current case study which fortifies the crucial role of plants on bioremediation.

This efficiency can further be enhanced by addition of the nutrients which will increase the microbial metabolic activities. In the current case study, biodegradation rate of toluene was found to be augmented by addition of the nutrient in the planted wetland. The high removal efficiency of toluene is attributed to biostimulation of the periphyton in the root zone. This study will help in implementing the bioremediation strategies in realistic agrarian field conditions where the nutrients' loading followed by their leaching to the vadose zone are major causes of groundwater pollution. Further studies are required to quantify the nutrient and pollutant uptake by plant root biomass to their subsequent translocation to shoot biomass and to incorporate evapotranspiration losses.

6 Recommendations

Management and remediation of hydrocarbon polluted site is major challenge for policymakers, geoscientists to produce safe drinking water demands. Hence, an effort has been made to develop an effective and low-cost plant-based bioremediation processes by optimizing the nutrient application in a hydrocarbon polluted domain. The findings of this study help to make recommendations for management and remediation of hydrocarbon polluted site as-

(a) Bioremediation processes must be assisted with the nutrient loadings for getting an enhanced efficiency for the decontamination of hydrocarbon pollutants. More studies can be performed using variable dimensions of the horizontal flow constructed wetlands along with the different environmental conditions of temperature, sunlight and plantation, in order to establish a more robustness and closeness of fit between the observed and simulated bio-kinetics of the removal mechanisms.

(b) *Canna generalis* with plant density of 7 plants/ft^2 can play a significant role in the elimination of hydrocarbon compounds from soil–water systems without any further chemical amendments. Further studies are required to quantify the nutrient and pollutant uptake by the plant root biomass to their subsequent translocation to shoot biomass and to incorporate evapotranspiration losses.

(c) In initial days of hydrocarbon spills, biodegradation rates can be enhanced significantly by applying appropriate nutrient loading and oxygen supply at polluted sites. However, further works are needed to optimize site conditions and water quality parameters for a better implementation of bioremediation techniques. Machine learning techniques may help to optimize such operational parameters along with complex subsurface variables.

(d) Lastly, it is important to investigate bio-clogging and its impact of future removal effective of the constructed wetland.

References

1. Ballesteros F, Vuong TH, Secondes MF, Tuan PD (2016) Removal efficiencies of constructed wetland and efficacy of plant on treating benzene. Sustain Environ Res 26(2):93–96
2. Basu S, Yadav BK, Mathur S (2015) Enhanced bioremediation of BTEX contaminated groundwater in pot-scale wetlands. Environ Sci Pollut Res 22:20041
3. Cottin N, Merlin G (2008) Removal of PAHs from laboratory columns simulating the humus upper layer of vertical flow constructed wetlands. Chemosphere 73(5):711–716
4. Eke PE, Scholz M (2008) Benzene removal with vertical-flow constructed treatment wetlands. J Chem Technol Biotechnol 83:55–63
5. Gupta PK, Yadav B, Yadav BK (2019) Assessment of LNAPL in subsurface under fluctuating groundwater table using 2D sand tank experiments. ASCE J Environ Eng 145(9)
6. Gupta PK, Yadav BK (2020) Three-dimensional laboratory experiments on fate and transport of LNAPL under varying groundwater flow conditions. J Environ Eng 146(4):04020010
7. Kim S, Choi DH, Sim DS, Oh Y (2005) Evaluation of bioremediation effectiveness on crude oil-contaminated sand. Chemosphere 59:845–852
8. Knight RL, Kadlec RH, Ohlendorf HM (1999) The use of treatment wetlands for petroleum industry effluents. Environ Sci Technol 33(7):973–980
9. Langergraber G, Šimůnek J (2011) HYDRUS wetland module manual, version 2. Hydrus software series 4. Department of Environmental Sciences, University of California Riverside, Riverside, CA, USA, p 46
10. Mustapha HI, Gupta PK, Yadav BK, van Bruggen JJA, Lens PNL (2018) Performance evaluation of duplex constructed wetlands for the treatment of diesel contaminated wastewater. Chemosphere 205:166–177
11. Šimůnek J, van Genuchten MTh, Šejna M (2016) Recent developments and applications of the HYDRUS computer software packages. Vadose Zone J 15(7):25. https://doi.org/10.2136/vzj2016.04.0033
12. Tang X, Eke PE, Scholz M, Huang S (2008) Processes impacting on benzene removal in vertical-flow constructed wetlands. Biores Technol 100:227–234
13. Vaajasaari K, Joutti A, Schultz E, Selonen S, Westerholm H (2002) Comparisons of terrestrial and aquatic bioassays for oil-contaminated soil toxicity. J Soils Sediments 2:194
14. Yadav BK, Hassanizadeh SM (2011) An overview of biodegradation of LNAPLs in coastal (semi)-arid environment. Water Air Soil Pollut 220:225
15. Yadav BK, Ansari FA, Basu S, Mathur A (2013) Remediation of LNAPL contaminated groundwater using plant-assisted biostimulation and bioaugmentation methods. Water Air Soil Pollut 225:1793–1801

Municipal Wastewater—A Remedy for Water Stress in India

S. Swarnalatha, N. Vinayagamoorthy, and G. Sekaran

Abstract Water is considered to be one of the essential renewable resources for the survival of living beings. Due to continuous rise in population growth, growing urbanization, accelerated industrial growth along with increase in agricultural production, the demand for water is increasing substantially in India. The water supply in India is inadequate for the domestic, industrial and agricultural demand; therefore, an efficient water management is the hour of need. Proper municipal wastewater treatment and recycling of treated wastewater are the potential sources for addressing the growing water demand issues in rapidly growing urban centers. This chapter briefly illustrates the challenges in water supply in India, quantity of municipal wastewater generation and treatment capacity in India, characterization of municipal wastewater, conventional treatment options followed for municipal wastewater and their limitations in addressing the removal of emerging contaminants in municipal wastewater. The chapter also includes the advanced secondary biological treatment techniques such as Fluidized bed reactor (FBR), Membrane bioreactor (MBR) and low cost immobilized cell reactor technologies such as Fluidized immobilized carbon catalytic oxidation (FICCO) and Chemoautotrophic activated carbon oxidation (CAACO). The tertiary treatments such as tertiary clarification, disinfection methods, pressure sand filter and activated carbon filter are also discussed in this chapter. This chapter also includes the challenges for the decision makers to satisfy the demands for complete recycle of treated wastewater or recharge the aquifers to increase the water table.

S. Swarnalatha (✉)
Environmental Science and Engineering Division, CSIR – Central Leather Research Institute (CSIR – CLRI), Adyar, Chennai 600020, Tamil Nadu, India
e-mail: swarnalatha@clri.res.in

N. Vinayagamoorthy
Department of Environmental Health Engineering, Faculty of Public Health, Sri Ramachandra Institute of Higher Education & Research (Deemed to be University), Porur, Chennai 600116, Tamil Nadu, India

G. Sekaran (✉)
Environmental Science and Engineering Division, SRMIST Ramapuram campus, Ramapuram, Chennai 600089, Tamil Nadu, India

© Springer Nature Switzerland AG 2022

S. Yadav et al. (eds.), *Wastewater Assessment, Treatment, Reuse and Development in India*, Earth and Environmental Sciences Library,
https://doi.org/10.1007/978-3-030-95786-5_10

Keywords Municipal wastewater · Conventional wastewater treatment · Advance secondary biological treatment · Fluidized immobilized carbon catalytic oxidation · Chemoautotrophic activated carbon oxidation · Tertiary treatment

1 Introduction

Water is an essential commodity for the human metabolic activities and animals, vegetative growth of plants. Water is needed for human consumption, production of agriculture products, forestation, development of grasslands, and cultivation of commercial tree crops, generation of hydro-electric and thermo electric energy, production of the non-agricultural products and maintenance of the ecological balance and biodiversity.

Global use of water was increased by a factor of 6 over the past 100 years and continues to grow at about 1% rate per year [1]. The water use will continue to increase, which is attributable to population growth, economic development, urbanization, industrial growth and agricultural growth. The major growth in water demand is expected to occur in developing and emerging economies. By 2050, world population is expected to increase in between 9.4 to 10.2 billion. Global water demand has been estimated to be 4,600 km^3 per year and it has been projected to increase at the rate of 20–30% per year by 2050 [2, 3]. Global water cycle intensifies due to global warming such that wetter regions generally become wetter and drier regions becoming even drier [4].

Water crisis seems to be emerging as a serious threat to human security because of (a) depletion of ground water table, (b) the deterioration of the environmental quality and its assimilating capacity and (c) the contamination of surface water and ground water sources by domestic and industrial activities.

2 Challenges to Indian Water Supply System

The total available good water in India is about 4000 billion cubic meters per annum (BCM). Out of this, potable water resources are only 1137 billion cubic meters; 690 billion cubic meters are contributed from surface water and 447 billion cubic meters is contributed by ground water [5]. While the availability of good water is most crucial, it is equally important to provide potable water for irrigation to ensure food security for the increasing population and to provide employment to weaker sections in the country. The water requirement for various sectors was assessed by the "National Commission on Integrated Water Resources Development (NCIWRD)" based on the assumption that irrigation efficiency will increase to 60% from the current level of 35–40%.

Figure 1 presents the projected highest water demand by different sectors in

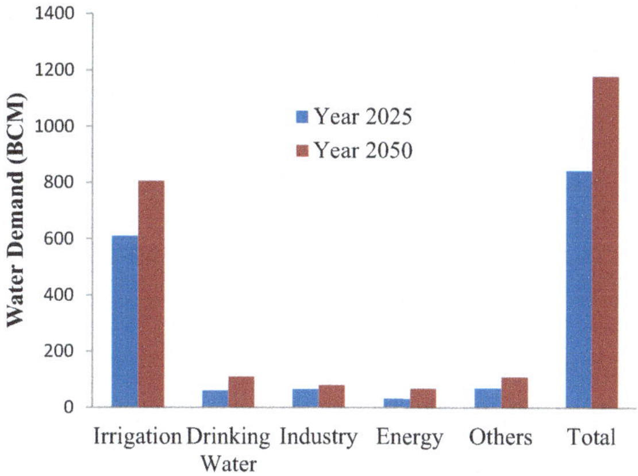

Fig. 1 Projected highest water demand per annum in India

India as per the National Commission for Integrated Water Resource Development (NCIWRD) (MOSPI 2015).

High demand for water from various sectors such as irrigation, drinking water, industry, energy and others is expected to rise to 843 BCM in the year 2025 and further to 1180 BCM in the year 2050 [6]. Irrigation sector is the largest water consumer, which accounts for 85% of the water demand followed by domestic use (6%), energy development (3%), and industries (6%). The water demand in domestic, energy and industrial sector will grow by 2.6 times, 3.7 times, and 2.2 times respectively [7].

Though the water demand by industrial sector (including energy demand) at present constitutes only about 8% of the total water demand and its water share is rising rapidly and the demand is expected to increase to about 13% of the total projected water use by the year 2050.

In India, per capita annual water availability is decreased from 1820 m^3 in the year 2001 to 1544 m^3 in the year 2011 and 1441 m^3 in the year 2015 [5, 8]. If India is estimated to touch 1.6 billion populations by 2051, the per capita water availability is expected to decrease to 1174 m^3 per year [8]. 'Falkenmark indicator' or 'water stress index' is the most commonly used measure of water scarcity. According to it, if the per capita annual water availability in a country is below 1,700 m^3, that country is experiencing water stress; if it is below 1,000 m^3 it is experiencing water scarcity; and below 500 m^3 is the indication of absolute water scarcity in that country [5, 9].

As indicated in Niti Aayog's report, India's water demand is estimated to be doubled in year 2030 which indicates implication of severe water scarcity and eventually will lead to 6% loss in the country's GDP. About 600 million population are facing high to extreme water shortages and about 2,00,000 population in India die every year due to lack of access to safe water [10].

Thus, many parts of India are expected to confront with severe water scarcity issues in the near future. The water supply in India will be a great challenge, and hence, the hour of the need is to address the complexity of this problem effectively before it gets converted into a security-threat.

3 Municipal Wastewater Generation and Its Treatment Capacity

Municipal wastewater may be defined as "the liquid waste originating from a community and it may be composed of domestic wastewater and/or industrial discharges [11]. Domestic sewage includes human excreta, urine and the associated sludge (collectively known as black water), and wastewater generated through bathing and kitchen (collectively known as grey water).

Total sewage generation in India is 61754 Million Litre Per Day (MLD), while the available sewage treatment capacity is only for 22,963 MLD. The combined status of wastewater generation and its treatment capacity in class I cities and class II towns with respect to population growth is presented in Fig. 2. The combined sewage generation of Class I cities and Class II towns in the year 2008 and 2017 is about 38,000 MLD and 75,000 MLD respectively [11–13]. However, the installed sewage treatment capacity is for only 12,000 MLD in the year 2008 and 26,000 MLD in the year 2017. There remains a gap of about 65% between sewage generation and

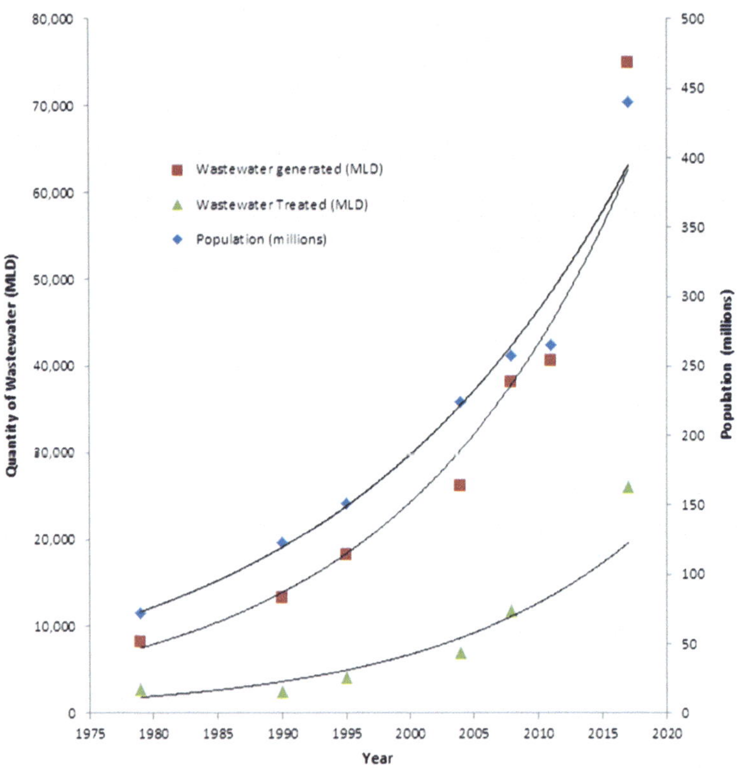

Fig. 2 Trend of wastewater generation and treatment capacity in Class I Cities and Class II towns

installed sewage treatment capacity. Even in some of the cities the existing treatment capacity remains poorly managed [14].

About 15,644 MLD of sewage is generated from the 35 metropolitan cities (having population more than 10 Lakhs) in India. But, the existing treatment capacity is only for 8040 MLD. The total capacity gap is 9031 MLD and, planned treatment capacity is 1549 MLD.

As the population is at positive growth rate, demand and shortage of water become unmanageable. Based upon the projected urban population growth in India, the wastewater generation is expected to be around 121 L/capita/day by the year 2051 [12]. The wastewater generation will be to be around 132,000 MLD based on the projected population for the year 2051 [15].

Thus, it becomes indispensable for the country to ensure municipal wastewater treatment by 100% and to enable the reuse of water may be one of the remedies for water stress in India. By reusing water, the country can be significantly benefited by increasing its effectiveness by exploiting all available water sources.

4 Characteristics of Municipal Wastewater

The treatment design for municipal wastewater depends on the pollution load (product of pollution strength and volume of the wastewater generated). Some of the important pollutants of municipal wastewater are given below.

4.1 Oxygen-Demanding Substances

Dissolved molecular oxygen is an important parameter in assessing the water quality to support the aquatic life. Many pollutants place a demand on the dissolved oxygen for their stability. This is termed as biochemical oxygen demand (BOD), and is used as one of the key parameter to know the efficiency of a sewage treatment plant. If the treatment efficiency is low and the treated water contains a significant concentration of high content of pollutants consequently, they demand more oxygen for stabilization and make that water to have less oxygen to support aquatic life [16]. The pollutants such as ammonia, sulphide, sulphite, nitrite and organic compounds are "oxygen-demanding" substances. These substances are usually degraded or converted into other compounds by microorganism (aerobic bacteria) only in the presence of sufficient oxygen in the wastewater.

4.2 Nutrients

The essential chief nutrients present in sewage are carbon, nitrogen, phosphorus, sulphide and heavy metal ions (micro nutrients essential for microbial growth). Removal of nutrients cannot be achieved by conventional secondary biological treatment processes, but it converts their organic forms into mineral forms which may be utilized by plants. But, the release of high amounts of these nutrients causes nutrient enrichment resulting in excessive growth of algae. The algal bloom screens sunlight and also depletes dissolved oxygen in the water bodies during night time causing eutrophication.

4.3 Suspended Solids

Solids that are retained by a 0.45-micron filter are known as total suspended solids (TSS). It can be further classified into organic (volatile) and inorganic (fixed) fractions. If the suspended solids present in sewage are discharged without proper elimination, it leads to the formation of huge amount of sludge and subsequently creates anaerobic conditions.

4.4 Emerging Contaminants

Emerging contaminants are natural or synthetically occurring substances not commonly monitored in the environment, having known or suspected undesirable effects on humans and the ecosystem. This group includes compounds such as pharmaceutical and personal care products (PPCPs), pesticides, and hormones that have adverse effects on human and wildlife endocrine systems [17]. The term "PPCPs" broadly refers to any product related to health care or medical purposes for humans and/or animals. The excreted PPCPs may either retain their original concentrations and structures or be mobilized and converted into other active (or inactive) compounds during their lifespan in aquatic matrices [18].

4.5 Pathogens (Bacterial Parameter)

Bacterial parameters, such as fecal coliform (FC) is an important indicator of fecal pollution. It becomes mandatory to control the pathogens in treated municipal wastewater meant for human access. Coliform bacteria are found in the intestinal tract of human beings. Every person discharges about 100–400 billion coliform bacteria per day. A group of coliform bacteria includes genera *Escherichia* and *Aerobacter*.

5 Conventional Technologies for Municipal Wastewater Treatment

A typical municipal wastewater treatment plant comprises of preliminary, primary, and secondary units. An "advanced" or "tertiary" treatment step may be required at specific locations or depending upon the usage of treated water to protect human health or quality of life. Conventional municipal wastewater treatment with tertiary treatment is considered to include screening, grit chamber, equalization, primary clarifier, biological treatment, secondary clarifier and tertiary filter. Schematic flow diagram/unit operations for a typical STP having the objective of water recovery are shown in Fig. 3.

5.1 *Preliminary Treatment*

Preliminary treatment consists of screen (coarse, medium or fine screens) to remove large size floating objects. Generally screens are positioned in a channel or chamber and inclined towards the direction of flow of the wastewater. The minimum width of the screen channel is usually 0.6 m. The sewage flow velocity in this screen channel should be the maximum of 1 m/s and minimum of not less than 0.3 m/s. The size of the screens is generally 10 mm × 50 mm and coarse screens are spaced at 75 mm, medium screens are spaced at 20–50 mm and finer screens are spaced at 20 mm [19]. The floating objects of large size present in the wastewater are to be trapped on the upstream surface of the screen, and thus, it can be cleaned either manually (in coarse screens) or mechanically (in fine screens). After screening, the wastewater is allowed to flow into a grit chamber to settle denser particles of size 0.15 mm and specific gravity 2.40 such as grit, cinders, sand, and small stones at the bottom of the tank. Generally, the grit chambers are designed for 3 × dry weather flow (DWF) and hydraulic retention time of 3 min. The per capita variation of grit accumulation per

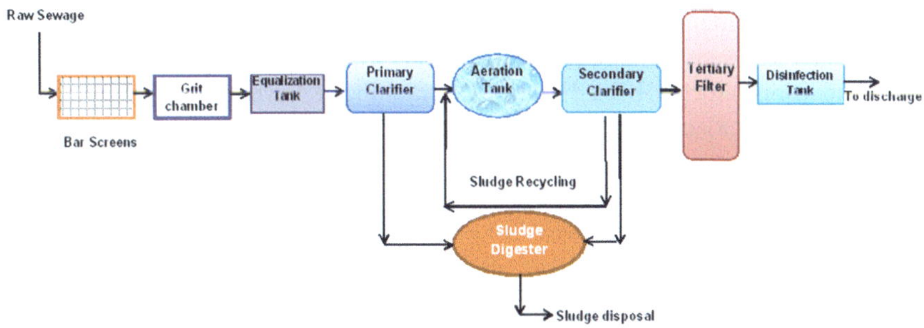

Fig. 3 Treatment scheme/unit operations of a conventional STP for municipal wastewater

day may be 5–15 g in India, volume of grit is in the range of 0.15–0.17 m^3/1000 m^3 [19].

5.2 Primary Treatment

The primary wastewater treatment separates settleable solids in the primary settling tank (primary clarifier) through gravity sedimentation and the suspended organic solids settle at the bottom of the sedimentation tank as primary sludge. Dissolved air flotation (DAF), the latest process for removing suspended solids, consists of injecting air in which oxygen is dissolved in the wastewater under pressure to achieve clarification and then air is released at atmospheric pressure in the flotation tank or basin [20, 21].

5.3 Secondary Biological Treatment

The secondary treatment process stabilizes most of the organic matter in wastewater as CO_2, water and biomass. It can be achieved either by attached growth processes or by suspended growth processes. The organic/bio organic compounds are oxidized to carbon dioxide and other end products by aerobic/anaerobic bacteria, and the balance provides the resources and energy needed to sustain the bacterial diversity. Biologically flocculated microorganisms form settleable particles in the process and this excess biomass is separated in secondary clarifier and is known as "secondary biological sludge".

Various commonly used treatment technologies for municipal wastewater in India include activated sludge process (ASP) and up-flow anaerobic sludge blanket (UASB) process [12]. Majority of the treatment plants are based on primary settling followed by activated sludge process technology or UASB followed by ASP and Polishing Pond technology.

The other prominent biological treatment technologies are Moving Bed Biofilm Reactor (MBBR), Sequencing Batch Reactor (SBR), Fluidized Bed Reactor (FBR), Membrane Bioreactor (MBR), Fluidized Immobilized Carbon Catalytic Reactor (FICCO) and Chemoautotrophic Activated Carbon Oxidation Reactor (CAACO).

5.3.1 Activated Sludge Process

In the aeration tank containing inoculated bacteria, wastewater is vigorously agitated with air for several hours and microorganisms are acclimatized to the wastewater in the form of slurry (activated sludge) such that, it allows the microorganisms to break down the organic matter. The excess biomass is removed in secondary clarifier by

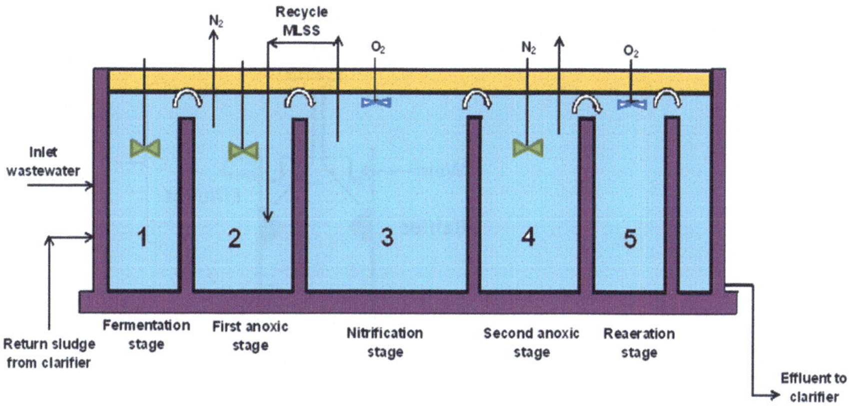

Fig. 4 Different stages in Bardenpho process

settling and the effluent is discharged or treated further. Some of the biomass is again returned to the aeration tank for mixing with incoming wastewater.

In the activated sludge process large proportion of the biomass is recycled which results in a large number of microorganisms that oxidize organic matter in a relatively short time [22]. The suspended solids concentration present in the aeration tank is referred to as the mixed liquor suspended solids (MLSS). Mixed-liquor volatile suspended solid (MLVSS) is the organic part of the MLSS, which is the non-microbial organic matter as well as dead and living microorganisms and cell debris [20, 21].

In India, many of the water supply and sewerage boards apply ASP for the treatment of municipal wastewater. For e.g., Chennai Metropolitan Water Supply and Sewerage Board (CMWSSB) provide water supply and sewage treatment (727 MLD) by ASP to the metropolitan city of Chennai (India) and areas around it [23, 24]. Amongst the common sewage treatment plants (CSTP) in Chennai, the Kodungaiyur CSTP supplies about 36 MLD secondary treated sewage to the nearby Industries in Manali town at the rate of Rs. 8.75 per kilo litre [23, 24]. The secondary treated wastewater received by the industries is further treated in their own tertiary treatment plants/RO plants to treat further for their reuse applications such as cooling tower make ups and toilet flush outs.

The Bardenpho process is an advanced modification of the activated sludge process, which results in nutrient removal such as nitrogen and phosphorus through microbial processes that occur in a biological reactor through multistage such as fermentation, first anoxic, nitrification, second anoxic and oxic (Fig. 4). It removes high levels of BOD, suspended solids, nitrogen and phosphorus [20, 21].

5.3.2 Up-Flow Anaerobic Sludge Blanket (UASB) Process

The Up-flow Anaerobic Sludge Blanket (UASB) reactor was developed by Lettinga with his coworkers in the late 1970s [25]. Among the various anaerobic wastewater

Fig. 5 Basic design of UASB reactor

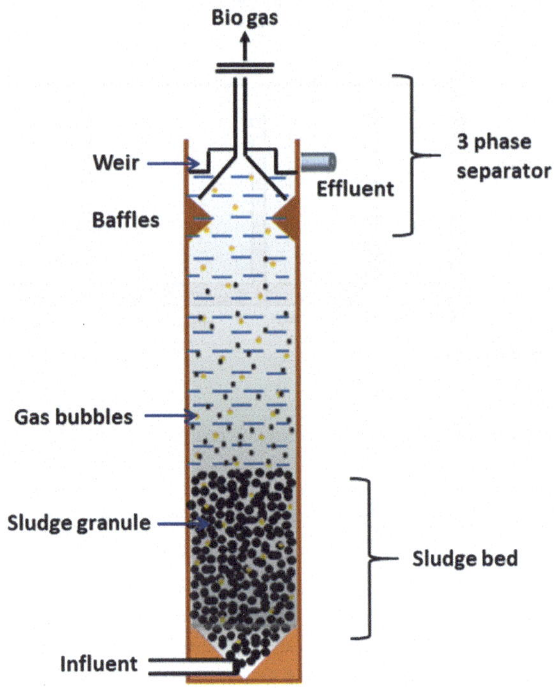

treatment technologies, UASB system had achieved a considerable success. It acts as a high-rate reactor system and retains high amount of biomass containing sludge, granules or aggregates of microorganisms. Figure 5 shows the typical UASB reactor design.

The basic conditions to be met in UASB are efficient retention of viable anaerobic (methanogenic) sludge and a sufficient contact time between incoming wastewater and the retained sludge. One of the main features of the UASB is their ability to form a blanket of granular type of anaerobic sludge, which has a high methanogenic activity and good settleability, so that the reactors can be operated at high volumetric and high COD loading rates [26, 27].

Sewage is introduced into the reactor through the bottom of the reactor in the upward flow mode through a biologically activated sludge blanket, which is usually in the form of granular aggregates. The internal mixing is caused by the gases (carbon dioxide and methane) produced under anaerobic conditions, which supports the formation and maintenance of granular aggregates. A gas–liquid–solid separator (GLSS) is provided at the top of the reactor for the effective separation of gas, liquid, and biomass granules. The gas entrapped biomass granules strike the bottom of degassing baffles in GLSS and return back into the sludge blanket and thereby the gas is released to the gas dome provided at the top of the reactor and the treated water is collected at hopper provided at the top of the reactor [28].

Bio-chemical activities in UASB Digesters comprise of three phases, namely hydrolysis, acidogenesis or acetogenesis and methanogenesis [12].

The advantages of UASB includes low hydraulic retention time (8–10 h), no pretreatment requirement, no filling media, no mixer or aerators be provided in the reactor and minimum maintenance. But, the most difficult problem with the UASB system is corrosion and thus it requires corrosion resistant constructional and plumbing materials.

5.3.3 Sequential Batch Reactor (SBR)

The Sequential Batch Reactor (SBR) system for wastewater treatment consists of a set of tanks which is operated on a fill and draw activated sludge system. Each tank in the SBR is filled with wastewater during a period of time and then operated as a batch reactor. Then, the mixed liquor is allowed to settle and the supernatant is withdrawn from the tank. The SBR experiences five cycles in sequence namely Fill, React, Settle, Draw and Idle (Fig. 6). According to aeration and mixing procedures, there are several types of fill and react periods. The unit operations employed for the SBR and conventional activated sludge processes are almost the same. Difference between the SBR and conventional activated sludge processes is that using a timed control sequence SBR performs equalization, secondary biological treatment and clarification in a single tank. But, these unit processes takes place in separate tanks in a conventional activated sludge system. The other advantages of SBR over conventional activated sludge systems are operating flexibility; minimum requirement of footprint and less capital cost. The disadvantages of SBR over conventional activated sludge system includes requirement of high level of maintenance and sophistication and potential plugging of aeration devices [29].

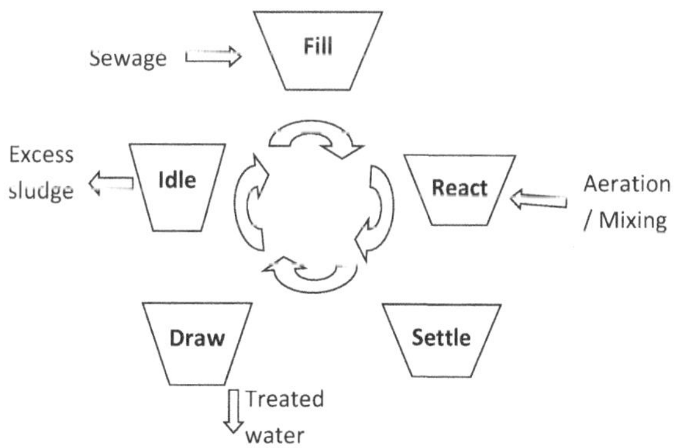

Fig. 6 Treatment cycles in SBR

5.3.4 Moving Bed Biofilm Reactor

Moving Bed Biofilm Bioreactor (MBBR) is a combination of both suspended growth (ASP) and attached growth (biofilter processes). The concept of MBBR process is to use the whole tank volume for biomass growth. It uses simple floating media carriers for attached growth of biofilms. The advantages of MBBR include carrier media in MBBR remain in suspension within the reactors due to perforations or screen arrangement at the discharge. And also the carrier matrix is designed in such a way to have a high specific surface area to accommodate biofilm growth which further eliminates the sludge recirculation arrangement to achieve the required biomass concentrations. The additional advantage of the MBBR is the quantity of sludge production in MBBR is lower than the typical activated sludge process and thus, cost of sludge disposal in MBBR is relatively lower than the conventional activated sludge systems. The movement of biofilm carrier in MBBR is caused by aeration in aerobic reactor and mechanical stirring in anoxic/aerobic reactor [30]. Figure 7 shows the aerobic and anaerobic MBBR in which movement is caused by diffused aeration and vertical shaft respectively.

5.4 Challenges in the Removal of Emerging Contaminants (EC) in STPs Using Conventional Technologies

Emerging contaminants (ECs) are generally present in sewage at concentrations in the order of parts-per-trillion (ng/L) to parts-per-billion (μg/L) [31], but they are either present as individual molecules or as complexes such that they become toxic or inhibitory effects to activated sludge microorganisms [32, 33]. Hence, the removal efficiency of EC using conventional technologies in STPs is low [34, 35]. Thus, STPs

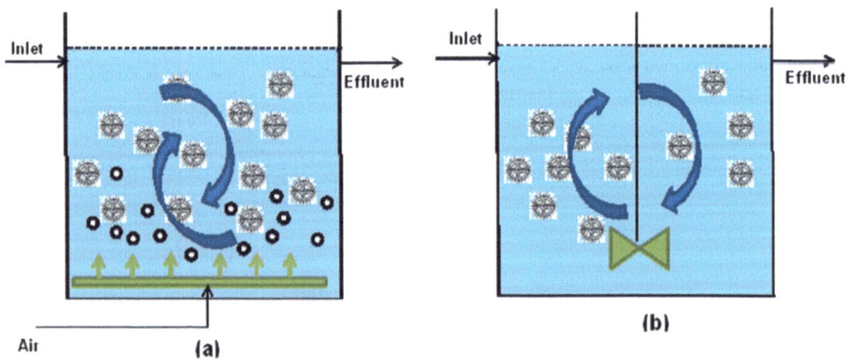

Fig.7 **a** Oxic/aerobic MBBR using diffused aeration and **b** anoxic/anaerobic MBBR using vertical shaft stirring arrangement

becomes a primary source of emerging contaminants in the aquatic environment [36, 37].

The fate of ECs in STPs includes retention in sludge, biodegradation or escaping into water bodies depends upon their chemical configuration and their secondary metabolites. Mostly the ECs are entrapped in the sludge/biomass of the STPs employing conventional technologies. Subedi et al. [38] studied twenty-nine pharmaceuticals and six metabolites in two sewage treatment plants (STPs) in South India located at Udupi and Mangalore. STP located at Udupi (STP_U) employed conventional activated sludge as secondary biological treatment followed by clarification and disinfection by chlorine, whereas STP located at Mangalore (STP_M) employed additional Up-flow Anaerobic Sludge Blanket (UASB) reactor along with the treatment scheme mentioned in STP_U. Subedi et al. [38] reported that the removal of recalcitrant PPCPs were not observed in STP_U, while it was removed in STP_M. But, in both the STPs, most of the removed PPCPs were ended up in sludge.

Some of the challenges involved in removing ECs using conventional technologies are low efficiencies in removal of pharmaceuticals and β-blockers; regeneration and disposal issues of sludge containing ECs; processing of ECs sludge can increase the total cost by 50–60% and ECs cannot be degraded properly. Thus, the new advanced technologies are in demand for the sewage treatment plants [39].

6 Advanced Secondary Biological Treatment Technologies

6.1 Fluidized Bed Reactor

Fluidized bed reactor (FBR) is one of the well demonstrated technologies for the treatment of municipal wastewater. It works on the principle of fluidization of carrier matrices (of bacteria immobilized or bacteria adhered) by the injected air and wastewater. Fluidization is a process which involves flow of a fluid through a bed of solid matrices with velocity such that to fluidise the carrier matrices. Figure 8 represents the basic design of FBR. The solid phase matrices can be sand, granular activated carbon, or polymeric materials. The advantages of fluidization include excellent contact between organics in wastewater and carrier matrices embedded with bacteria, uniform distribution of temperature, high mass transfer rate [40] and low sludge generation. FBR in wastewater treatment involves two phases (solid–Liquid) and three phase flow system (Solid–Liquid–Gas). Solid–Liquid fluidization hydro dynamics works only through upward flow of a fluid through a bed of solid particles to fluidize the carrier matrices. In Solid–Liquid–Gas fluidization, both the liquid stream and air bubbles fluidize the carrier matrices. The fluidized-bed bioreactor is an attached-growth process. The microbes adhered to the carrier matrices in the fluidization tank and form surface microbial film. Fluidization in the reactor is either caused by the wastewater and/or by the air stream. FBR can be operated either as batch, semi-continuous, and continuous loading. FBR doesn't require back wash,

Fig. 8 Basic design of FBR

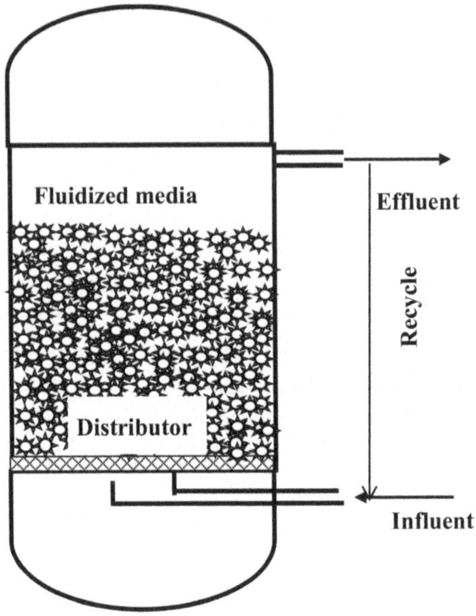

possess low head loss and high biofilm surface area. Thus, it can be used under continuous mode with ease [41].

FBR can be classified based on the direction of the flow of the fluid. Fluidization is attained through the upward fluid flow as it uses solid particles that are denser than the fluid phase in conventional FBR. However, in inverse fluidized bed reactor (IFBR) the density of the solid particles are lower than the fluid phase and fluidization is by the downward fluid flow. In IFBR, uncontrolled growth of biomass does not affect the hydrodynamics of bioparticles. The IFBR possess superior hydrodynamic characteristics than conventional FBR, but the downward flow of IFBR requires higher superficial fluid velocity [42].

6.2 Membrane Bioreactor

Membrane bioreactor (MBR) is the combination of biological treatment with membrane separation such as microfiltration or ultrafiltration. MBR process consists of membrane separation unit either located off line bioreactor or in line of the bioreactor (Fig. 9). The porosity of the membranes generally ranges from 0.035 microns to 0.4 microns. The membrane maintains high biomass concentration within the bioreactor as well as retains biomass. MBR is generally maintained as aerobic bioreactor but it can also be operated under anaerobic conditions or with alternating

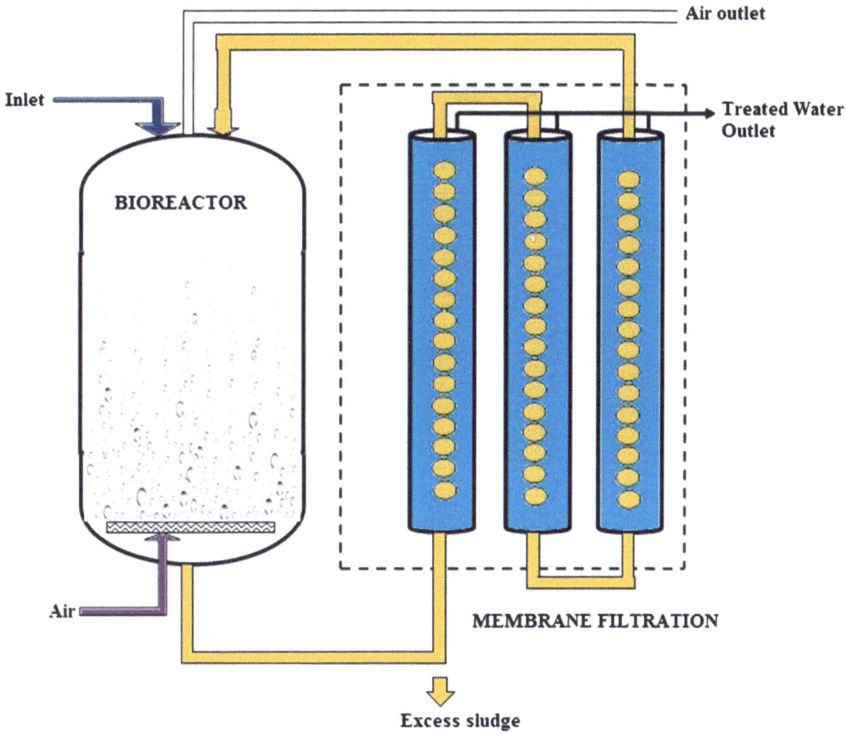

Fig. 9 Basic design of MBR

phases under aerobic/anaerobic conditions to facilitate the elimination of Nitrogenous Oxygen Demand through microbial nitrification and denitrification processes simultaneously with COD elimination.

The advantages of MBR include (i) less carbon footprint than the conventional activated sludge process (ii) better effluent quality (iii) effectively displaces conventional treatment plant process steps such as primary treatment, secondary activated sludge process and disinfection and (iv) less generation of sludge. But, the major disadvantage of MBR is high operating costs and frequent membrane fouling [20, 21].

6.3 *Fluidized Immobilized Carbon Catalytic Oxidation (FICCO) Reactor*

Fluidized Immobilized Carbon Catalytic Oxidation (FICCO) reactor [43] is a fluidized bed reactor. It is the patented technology of Council of Scientific and Industrial Research (CSIR) and it consists of three zones (Fig. 10). The first zone is known

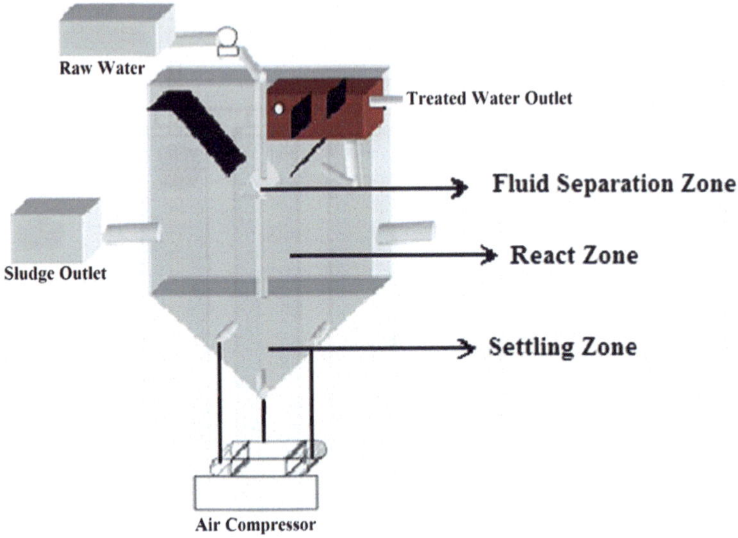

Fig. 10 Cross sectional view of FICCO reactor

as "React zone" comprised of the immobilized nanoporous activated carbon (NPAC) matrix which is fluidized by air and wastewater at upflow velocity of 5 m/min. The dissolved organics adsorb onto the carbon matrix and diffuse into the immobilized microbes and metabolized. The metabolized products are diffused back into the bulk medium. The oxidation of organics is facilitated by hydroxyl radicals generated from molecular oxygen [acquired from air]. The second zone is the fluid separation zone. The unspent oxygen and nitrogen in air are separated using a triangular septum provided at the optimum height of the reactor [44]. The separated air is collected through the perforated chamber provided at the top of the reactor. The third zone is the settling zone. The treated wastewater enters through the aperture and is allowed to settle in the inclined baffled zone. The angle of the inclined plate is based on the zone settling velocity of the NPAC dispersed in the wastewater. The settling tendency of the NPAC is enhanced by providing the functional groups onto the NPAC particles. The NPAC media used in FICCO reactor has high surface area. The attached suspended solids are sloughed off from the media on exceeding a critical thickness. The sloughed suspended solids slide back into the reactor through the aperture provided at the top of the reactor. The sludge accumulated in the reactor can be withdrawn through sludge withdrawal pipe line provided in the reactor.

Some of the advantages of FICCO system are low foot print requirement, low operation cost and minimum sludge production compared to conventional system.

Figure 11 depicts FICCO based wastewater treatment plant (1500 m^3/day) at Ghana, West Africa and STP (300 m^3/day) at an educational Institute in Chennai, India.

(a)

(b)

Fig. 11 FICCO based (**a**) wastewater treatment plant (1500 m^3/day) at Ghana, West Africa (**b**) STP (300 m^3/day) at an Educational Institute, Chennai, India

6.4 Chemoautotrophic Activated Carbon Oxidation (CAACO) Reactor

Chemoautotrophic Activated Carbon Oxidation (CAACO) reactor is also CSIR patented technology comprised of *Bacillus sp.* immobilized NPAC packed bed reactor (Fig. 12) for the oxidation of dissolved organics in wastewater [43, 45].

The air required for the oxidation of organics in wastewater is provided through packed bed reactor at two levels through perforated pipelines [46]. The wastewater to be treated is applied from the top of the reactor in the downward direction. The treated wastewater is collected from the bottom of the reactor [47]. The air required for the oxidation of organics is decided on the COD load in the wastewater. The pressure of air is decided on the head loss that would encounter during the oxidation of organics in wastewater.

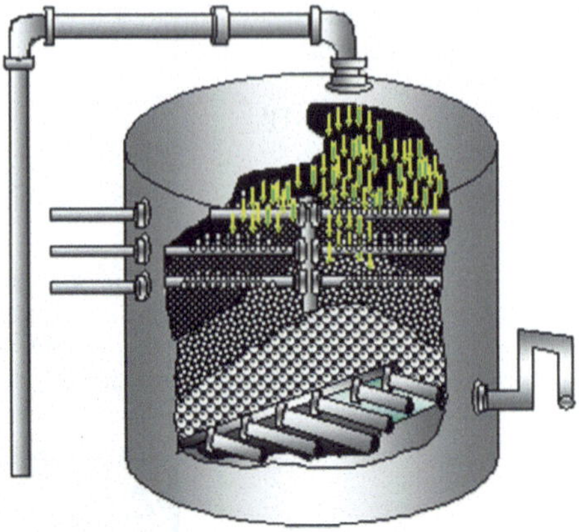

Fig. 12 Cross sectional view of CAACO Reactor

The following mechanism (Eqs. 1–6) for the oxidation of organics in CAACO reactor was proposed [46].

$$-\mathrm{NPAC}\left(e^{-}\right)+\mathrm{O}_2 \rightarrow -\mathrm{NPAC}\left(\mathrm{O}_2^{\bullet}\right)_{\mathrm{ads}} \tag{1}$$

$$-\mathrm{NPAC}\left(\mathrm{O}_2^{\bullet}\right)_{\mathrm{ads}}+\mathrm{H}^{+}_{(\mathrm{aq})} \rightarrow -\mathrm{NPAC}\left(\mathrm{HO}_2^{\bullet}\right)_{\mathrm{ads}} \tag{2}$$

$$-\mathrm{NPAC}\left(\mathrm{HO}_2^{\bullet}\right)_{\mathrm{ads}}+\mathrm{H}^{+}_{(\mathrm{aq})} \rightarrow -\mathrm{NPAC}(2\mathrm{OH}^{\bullet})_{\mathrm{ads}} \tag{3}$$

$$\begin{aligned}-\mathrm{NPAC}(h^{+})+(\mathrm{Organic}) &\rightarrow -\mathrm{NPAC}(h^{+})(\mathrm{Organic})_{\mathrm{ads}}\\ &\rightarrow -\mathrm{NPAC}^{*}(h^{+})(\mathrm{Cleavedorganic})_{\mathrm{ads}}\end{aligned} \tag{4}$$

$$\mathrm{OH}^{\bullet}{}_{\mathrm{ads}}+-\mathrm{NPAC}^{*}\left(h^{+}\right)(\mathrm{Cleavedorganic})_{\mathrm{ads}} \rightarrow -\mathrm{NPAC}(\mathrm{Organic})^{\bullet+}{}_{\mathrm{ads}}+\left(\mathrm{OH}^{-}\right)_{\mathrm{aq}} \tag{5}$$

$$\mathrm{OH}^{\bullet}{}_{\mathrm{ads}}\mathrm{or}\mathrm{HO}^{\bullet}_{2\mathrm{ads}}+-\mathrm{NPAC}(\mathrm{Organic})^{\bullet+} \rightarrow \mathrm{H}_2\mathrm{O}+\mathrm{CO}_2+-\mathrm{NPAC}^{*}(h^{+})+e^{-} \tag{6}$$

The detail of the CAACO technology is available in MSME Technology Facilitation Centre [48]. Merits of CAACO technology based municipal wastewater treatment system are (i) less land requirement (0.3 m^2/m^3), (ii) less residence time (1–4 h), (iii) very minimum sludge production (0.01 kg/Kg of COD destructed).

Figure 13 depicts some of the CAACO based treatment plants in India.

Fig. 13 CAACO based (**a**) treatment plant (750 m^3/day) in a textile industry, Chennai (**b**) STP (50 m^3/day) in a software park, Bangalore

7 Tertiary Treatment

Tertiary treatment of secondary biological treated wastewater involves a series of additional steps to eliminate residual organics, turbidity, nitrogen, phosphorus, and pathogens. Most processes involve some type of physico-chemical treatment such as chemical precipitation; filtrations such as pressure sand filter, activated carbon filter, membrane separation and application of disinfectants. Tertiary treatment of wastewater is employed for the refinement of the secondary biological treated wastewater for applications in water intensive areas such as in agricultural sector, tourism and development, constructional industry, surface transport sector and human consumption.

7.1 Chemical Precipitation

Chemical precipitation such as coagulation and flocculation (which is also used under primary treatment) is carried out using coagulating and flocculating agents to remove suspended and colloidal solids. The chemicals such as alum, lime, or iron salts are added to the wastewater to remove phosphorus by 95% [16]. The disadvantages of tertiary clarification by applying chemical flocculants and coagulants are (i) increases the TDS content of secondary biological treated wastewater (ii) substantially increases the cost of treatment and (iii) generation of tertiary chemical sludge causing disposal problem [49].

7.2 Pressure Sand Filter

The pressure sand filter (PSF) can typically handle up to 50 mg/L of suspended solids in an economical manner. PSF consists of multiple layers of coarse silica aggregate, medium size silica aggregate and fine size silica aggregate followed by sand as the top layer to screen out turbidity with minimum pressure drop.

7.3 Activated Carbon Filter

Activated carbon filter uses activated carbon of high surface area (>1000 m^2/g) as filtration media to remove residual biorefractory organic compounds, soluble microbial pollutants, turbidity and colour through adsorption. An activated carbon filter improves/refines effluent quality by improving the parameters such as BOD, COD, clarity (turbidity), color and odor.

7.4 Membrane Separation Processes

The membrane separation processes use membranes with various pore sizes and high pressure to transport the water through the membranes. The permeate meets the discharging standards of varied applications and the reject stream is disposed after treatment. Depending upon the separation mechanism and pore size of the membranes, it can be classified into microfiltration, ultrafiltration, nanofiltration, and reverse osmosis.

The pore size of microfiltration varies in the range of 0.1–10 μm; nominal pore size of ultrafiltration is 0.002–0.1 μm and pore size of nanofiltration membrane is lesser than 0.002 μm. Microfiltration separation has two flow categories, one is dead-end type in which the flow of the inlet water is perpendicular to the membrane surface (there is no reject stream) and another one is cross-flow filtration, where the inlet water flows tangentially across the membrane surface. In dead-end type, solute adsorption and deposition on the surface of the membrane pores takes place while permeate drags all the solids in the inlet water to the membrane surface. The main advantage of dead-end filtration is simple operation and high product recovery; however, the filter cartridge cannot be backwashed or cleaned. The cross-flow operates in a turbulent regime by shear forces and continuously removes particles from the membrane surface by interception, hence the fouling is reduced. Cross-flow microfiltration is effective in the removal of suspended solids, bacteria and possesses the capacity to reduce algal loading from 400,000 algae cells/ml to less than 50 algae cells/ml [50]. Microfiltration operates at pressure lesser than 2 bars.

Ultrafiltration is used for removing high molecular weight organic compounds such as colloids, proteins, polysaccharides, oils and large size microorganisms. The permeate from ultrafiler is characterized by very low turbidity, having less than 0.1 NTU [50]. Nanofiltration removes the mono, di- and oligo- saccharides and polyvalent ions [51]. The ultrafiltration and nanofiltration operates at the pressure of 1–10 bars and 5–35 bars respectively. Reverse osmosis (RO) operates using a semi permeable membrane having pore size lesser than 0.002 μm, mainly to remove ions by applying pressure to overcome osmotic pressure. Reverse osmosis can remove dissolved solids, glucose, amino acids etc. It operates at the pressure of 15–150 bar [51].

7.5 *Disinfectants*

Disinfectant is a physical or chemical agent that destroys pathogenic or other harmful microorganisms, but does not necessarily kill all microorganisms. Though there are different types of disinfectants, most commonly used disinfection methods are chlorination, ozonation and ultraviolet disinfection [20, 21].

7.5.1 Chlorination

When chlorine gas is added to water, it acts as a strong oxidizing agent, forms a mixture of hypochlorous acid (HOCl) and hydrochloric acids as presented in Eq. 7

$$Cl_2 + H_2O \rightarrow HOCl + HCl \tag{7}$$

Hypochlorous acid dissociates as H^+OCl^-. Chlorine and its compounds are the most commonly used disinfectants for treating drinking water and wastewater. Chlorine in the form of HOCl or ^-OCl is known as 'free available chlorine'. Free chlorine is quite efficient in inactivating pathogenic microorganisms.

HOCl combines with ammonia and organic compounds to form 'combined chlorine'. The reactions of chlorine with ammonia or nitrogen-containing organic substances are of great importance in water disinfection. It results in the formation of monochloramine (Eq. 8), dichloramine (Eq. 9) and trichloramine (Eq. 10).

$$NH_3 + HOCl \rightarrow NH_2Cl + H_2O \tag{8}$$

$$NH_2Cl + HOCl \rightarrow NHCl_2 + H_2O \tag{9}$$

$$NHCl_2 + HOCl \rightarrow NCl_3 + H_2O \tag{10}$$

Chlorine, being a strong oxidizing agent, reacts with any organic molecule including proteins, lipids, carbohydrates and nucleic acids to disrupt their structure. Bacterial inactivation by chlorine may be due to altered permeability of the outer cellular membrane which results in leakage of critical cell components; interference with cell-associated membrane functions (e.g., phosphorylation of high-energy compounds); impairment of enzyme and protein function as a result of irreversible binding of the sulfhydryl groups and denaturation of nucleic acid.

7.5.2 Ozonation

Ozone (O_3) is a powerful oxidizing agent, produced by passing an electric discharge through a stream of air or oxygen. Though, ozone is more expensive than chlorination it does not produce carcinogenic byproducts such as trihalomethanes or other chlorinated byproducts. Ozone can break down complex organic compounds into simpler compounds which serve as substrates for bacterial growth, but no residual ozone is left in the treated water. Generally, ozonation may be followed by chlorination to prevent regrowth of microorganisms. The effectiveness of ozonation is not influenced by pH and ammonia.

7.5.3 Ultraviolet (UV) Disinfection

Ultraviolet radiation damages microbial DNA or RNA at a wavelength 260 nm. It causes thymine dimerization, which blocks nucleic acid replication and effectively inactivates microorganisms. Initially, UV exposure damages the genome in viruses, followed by structural damage to the virus protein coat. Ultraviolet disinfection does not produce carcinogenic or toxic byproducts or taste and odor problems, thus, the usage of ultraviolet disinfection of wastewater gains increased popularity.

Disadvantages of ultraviolet disinfection are higher costs than halogens, no disinfectant residual, difficulty in determining the UV dose, maintenance of UV lamps, and possible photo reactivation of some enteric bacteria [22]. However, commercial application of UV treatment for municipal wastewater treatment can be increased by making advancements such as providing more efficient, lower cost, lamps and more reliable equipment.

8 Reuse of Treated Sewage Water

Treated sewage water can be potentially used in many areas such as industrial usage, agricultural purpose, domestic applications and aqua culture etc. Many industries such as thermal power plants, fertilizer and chemical manufacturing industries require large quantity of water for the integrated activities. Treated water from STP can be an alternate source to ground water sources for the above mentioned industrial usages as there is a limitation to draw water from ground water sources for the high tech industries. Commercial malls, theatres and office complexes require large quantity of water for maintaining air-conditioning and cooling, for which treated water from STP is the better option. Properly treated and disinfected wastewater from STP may be a boon to the farmers for agriculture to sustain their activity. Even in households, treated sewage water can be very well used for the activities such as gardening, flushing and car washing after proper disinfection.

Bangalore Water Supply and Sewerage Board (BWSSB) treats around 721 MLD of sewage [52] upto tertiary level treatment. Treated sewage is presently supplied to a number of industries.

9 Conclusion

Global water demand is estimated to be about 4,600 km^3 per year and projected to increase by 20–30% per year by 2050. Water demand in India is estimated to be doubled in 2030 implying severe water scarcity for hundreds of millions of people with an eventual loss of 6% in the country's GDP. Based on the projected urban population growth in India by the year 2051, the wastewater generation is expected to be around 121 L/capita/day and the wastewater generation will be around 132,000 MLD. As the water supply in India is going to be a great challenge, there is a necessity to generate water from wastewater especially municipal wastewater by recycle and reuse. Thus, strategies should be devised to harvest and use the treated wastewater by 100% for the various applications.

Many treatment technologies are available to treat municipal wastewater efficiently. However, various issues limit the efficiency of conventional treatment technology and management of sewage treatment plants in India. Important criteria to be considered for selecting wastewater treatment technologies or constructing a sewage treatment plant are as follows:

(i) addressing the removal of various pollutants efficiently
(ii) having the flexibility in the treatment processes/scheme to incorporate future expansions and improve the efficiency
(iii) should be simple in operation, maintenance and control
(iv) selection of footprint area based on the land availability e.g., if the land availability in the area is less, the treatment schemes to include technologies such as UASB, FBR, MBR, FICCO and CAACO

(v) using minimum sludge production technologies to reduce the operation and maintenance cost incurred due to the disposal of sludge
(vi) avoiding odour nuisance problems

Most of the treatment plants face improper operation and maintenance because of lack of proper power supply, lack of skilled manpower, peak load and feed water interruptions due to blockage of sewer lines. If the appropriate treatment schemes are selected along with proper addressing of the above O&M problems, good quality water can be harvested from municipal wastewater which could be considered as a great remedy for water stress in India.

10 Recommendations

The following suggestions may be considered for the proper management of Municipal wastewater discharged in India to reduce the water stress.

- The molecular level characterization of Municipal Wastewater must be mandatory to select the suitable technology for the treatment and reuse of wastewater.
- Only coarse screens must be used in screening operation and the usage of fine screens must be discouraged, so that fine substances also can enter into the treatment process and get treated.
- The treatment technologies must be selected either to avoid sludge generation or to keep it as minimum.
- The selected technologies used for the treatment of wastewater must ensure the degradation of organics in wastewater including volatile suspended solids.
- The COD of the municipal wastewater after secondary biological treatment and tertiary treatment must be less than 50 mg/L and 20 mg/L respectively to reuse it effectively after treatments.
- The selected technology must have provision not to discharge any form of rejects from the wastewater to the environment.
- Continuous monitoring of the performance of the installed technologies should be carried out to ensure the proper treatment.

Acknowledgements The authors place the record of appreciation and acknowledgement for CSIR—Central Leather Research Institute, SRMIST, Ramapuram campus and Sri Ramachandra Institute of Higher Education & Research for the publication of this book chapter.

References

1. Wada Y, Flörke M, Hanasaki N, Eisner S, Fischer G, Tramberend S, Satoh Y, Van Vliet M, Yillia P, Ringler C, Wiberg D (2016) Modeling global water use for the 21st century: water Futures and Solutions (WFaS) initiative and its approaches. Geosci Model Dev 9:175–222

2. Burek P, Satoh Y, Fischer G, Kahil MT, Scherzer A, Tramberend S, Nava LF, Wada Y, Eisner S, Flörke M, Hanasaki N (2016) Water futures and solution—fast track initiative (Final report). IIASA Working Paper. IIASA, Laxenburg, Austria: WP-16-006
3. UNESCO (2018) Nature-based solutions for water. The United Nations World Water Development Report 2018. http://www.unesco.org/new/en/natural-sciences/environment/water/wwap/wwdr/2018-nature-based-solutions/. Accessed 21 July 2019
4. IPCC (2015) Climate change 2014, synthesis report. https://www.ipcc.ch/site/assets/uploads/2018/05/SYR_AR5_FINAL_full_wcover.pdf. Accessed 21 July 2019
5. NITI Aayog, GoI (2018) Strategy for New India @ 75. https://niti.gov.in/writereaddata/files/Strategy_for_New_India.pdf. Accessed 21 July 2019
6. CWC, GoI (2015) Water and related statistics. http://cwc.gov.in/sites/default/files/water-related-statistics-2015.pdf. Accessed 21 July 2019
7. Asthana V, Shukla AC (2014) Water security in India: hope, despair, and the challenges of human development. Bloomsbury Publishing Inc., NY
8. MOSPI, GoI (2018) EnviStats India 2018 (supplement of Environmental Accounts). http://www.mospi.gov.in/publication/envistats-india-2018. Accessed 21 July 2019
9. Chris W (2012) Understanding water scarcity: definitions and measurements. Water Security. http://www.globalwaterforum.org/2012/05/07/understanding-water-scarcity-definitions-and-measurements/. Accessed 21 July 2019
10. NITI Aayog, GoI (2018) Composite water management index (A tool for water management). http://pib.nic.in/newsite/PrintRelease.aspx?relid=179962. Accessed 21 July 2019
11. CPCB, MoEF&CC, GoI (2009) Status of water supply, wastewater generation and treatment in class-I cities & class-II towns of India. http://www.indiaenvironmentportal.org.in/files/status%20of%20water%20supply.pdf. Accessed 21 July 2019
12. India Infrastructure Report (2011) Water: policy and performance for, sustainable development. Oxford University Press, New Delhi, p 341
13. India Infrastructure Research (2018) Sewage treatment market in India 2018. http://www.indiainfrastructure.com/reportpdf/report-sewage-treatment-market-in-india-june2018.pdf. Accessed 13 Aug 2019
14. MoEF&CC, GoI (2019) National status of waste water generation & treatment. http://www.sulabhenvis.nic.in/Database/STST_wastewater_2090.aspx. Accessed 21 July 2019
15. Venkatesh KR, Rajendran M, Murugappan A (2009) A correlation study on physico-chemical characteristics of domestic sewage. Nat Environ Pollut Technol 8(1):141–145
16. USEPA (2004) Primer for municipal wastewater treatment systems. EPA 832-R-04-001. https://www3.epa.gov/npdes/pubs/primer.pdf. Accessed 21 July 2019
17. Rodriguez-Narvaez OM, Peralta-Hernandez JM, Goonetilleke A, Bandala ER (2017) Treatment technologies for emerging contaminants in water: a review. Chem Eng J 323:361–380
18. Yang Y, Ok YS, Kim KH, Kwon EE, Tsang YF (2017) Occurrences and removal of pharmaceuticals and personal care products (PPCPs) in drinking water and water/sewage treatment plants: a review. Sci Total Environ 596:303–320
19. Arceivala SJ (2003) Wastewater treatment for pollution control, 2nd edn. Tata McGraw-Hill publication, New Delhi
20. Pepper IL, Gerba CP, Gentry TJ (2015) Municipal wastewater treatment. Environmental microbiology. Academic Press, MA, USA, pp 157–175
21. Pepper IL, Gerba CP, Gentry TJ (2015) Disinfection. Environmental microbiology. Academic Press, MA, pp 645–662
22. Bitton G (2011) Wastewater microbiology, 4th edn. Wiley-Blackwell, New Jersey
23. CMWSSB (n.d) Sewerage system. https://chennaimetrowater.tn.gov.in/seweragesystem.html. Accessed 21 July 2019
24. CMWSSB (n.d) Sewage treatment plant. https://chennaimetrowater.tn.gov.in/pdf/stp.pdf. Accessed 21 July 2019
25. Metcalf E, Eddy I (2003) Wastewater engineering treatment and reuse, 4th edn. McGraw-Hill Education, New York

26. Lettinga G, Hobma SW, Hulshoff Pol LW, De Zeeuw W, De Jong P, Grin P, Roersma R (1983) Design operation and economy of anaerobic treatment. Water Sci Technol 15(8–9):177–195
27. Aiyuk S, Forrez I, van Haandel A, Verstraete W (2006) Anaerobic and complementary treatment of domestic sewage in regions with hot climates—a review. Biores Technol 97(17):2225–2241
28. Daud MK, Rizvi H, Akram MF, Ali S, Rizwan M, Nafees M, Jin ZS (2018) Review of upflow anaerobic sludge blanket reactor technology: effect of different parameters and developments for domestic wastewater treatment. J Chem. https://doi.org/10.1155/2018/1596319
29. USEPA (1999) Wastewater technology fact sheet—sequencing batch reactors. https://www3.epa.gov/npdes/pubs/sbr_new.pdf. Accessed 21 July 2019
30. Barwal A, Chaudhary R (2014) To study the performance of biocarriers in moving bed biofilm reactor (MBBR) technology and kinetics of biofilm for retrofitting the existing aerobic treatment systems: a review. Rev Environ Sci Bio/Technol 13(3):285–299
31. Dai G, Wang B, Huang J, Dong R, Deng S, Yu G (2015) Occurrence and source apportionment of pharmaceuticals and personal care products in the Beiyun River of Beijing, China. Chemosphere 119:1033–1039
32. Thomaidi VS, Stasinakis AS, Borova VL, Thomaidis NS (2015) Is there a risk for the aquatic environment due to the existence of emerging organic contaminants in treated domestic wastewater? Greece as a case-study. J Hazard Mater 283:740–747
33. Thomaidi VS, Stasinakis AS, Borova VL, Thomaidis NS (2016) Assessing the risk associated with the presence of emerging organic contaminants in sludge-amended soil: a country-level analysis. Sci Total Environ 548:280–288
34. Behera SK, Kim HW, Oh JE, Park HS (2011) Occurrence and removal of antibiotics, hormones and several other pharmaceuticals in wastewater treatment plants of the largest industrial city of Korea. Sci Total Environ 409(20):4351–4360
35. Tsang YF (2015) Environmental protection and pollution management in China. The entrepreneurial rise in Southeast Asia. Palgrave Macmillan, New York, pp 309–334
36. Focazio MJ, Kolpin DW, Barnes KK, Furlong ET, Meyer MT, Zaugg SD, Meyer MT, Barber LB (2008) A national reconnaissance for pharmaceuticals and other organic wastewater contaminants in the United States—II) Untreated drinking water sources. Sci Total Environ 402(2–3):201–216
37. Padhye LP, Yao H, Kung'u FT, Huang CH (2014) Year-long evaluation on the occurrence and fate of pharmaceuticals, personal care products, and endocrine disrupting chemicals in an urban drinking water treatment plant. Water Res 51:266–276
38. Subedi B, Balakrishna K, Joshua DI, Kannan K (2017) Mass loading and removal of pharmaceuticals and personal care products including psychoactives, antihypertensives, and antibiotics in two sewage treatment plants in southern India. Chemosphere 167:429–437
39. Ahmed MB, Zhou JL, Ngo HH, Guo W, Thomaidis NS, Xu J (2017) Progress in the biological and chemical treatment technologies for emerging contaminant removal from wastewater: a critical review. J Hazard Mater 323:274–298
40. Tisa F, Raman AAA, Daud WMAW (2013) Basic design of a fluidized bed reactor for wastewater treatment using fenton oxidation. Int J Innov Manag Technol 5(2):93–98
41. Nelson MJ, Nakhla G, Zhu J (2017) Fluidized-bed bioreactor applications for biological wastewater treatment: a review of research and developments. Engineering 3(3):330–342
42. Bello MM, Raman AAA, Purushothaman M (2017) Applications of fluidized bed reactors in wastewater treatment–a review of the major design and operational parameters. J Clean Prod 141:1492–1514
43. Saranya P, Selvi PK, Sekaran G (2019) Integrated thermophilic enzyme-immobilized reactor and high-rate biological reactors for treatment of palm oil-containing wastewater without sludge production. Bioprocess Biosyst Eng 42(6):1053–1064
44. Mannacharaju M, Natarajan P, Villalan AK, Jothieswari M, Somasundaram S, Ganesan S (2018) An innovative approach to minimize excess sludge production in sewage treatment using integrated bioreactors. J Environ Sci 67:67–77
45. Arceivala SJ, Asolekar SR (2007) Wastewater treatment for pollution control and reuse, 3rd edn. Tata McGraw-Hill publication, New Delhi

46. Sekaran G, Ramani K, Kumar AG, Ravindran B, Kennedy LJ, Gnanamani A (2007) Oxidative destabilization of dissolved organics and E. coli in domestic wastewater through immobilized cell reactor system. J Environ Manag 84(2):123–133
47. Sekaran G, Karthikeyan S, Nagalakshmi C, Mandal AB (2013) Integrated Bacillus sp. immobilized cell reactor and Synechocystis sp. algal reactor for the treatment of tannery wastewater. Environ Sci Pollut Res 20(1):281–291
48. MSME, GoI (n.d) Chemo autotrophic activated carbon oxidation (CAACO) for treatment of wastewater. https://www.msmetfc.in/focus-caaco. Accessed 21 July 2019
49. Maharaja P, Mahesh M, Chitra C, Kalaivani D, Srividya R, Swarnalatha S, Sekaran G (2017) Sequential oxic-anoxic bio reactor for the treatment of tannery saline wastewater using halophilic and filamentous bacteria. J Water Process Eng 18:47–57
50. Singh R (2006) Hybrid membrane systems for water purification: technology, systems design and operations. Elsevier BV, Amsterdam
51. Mona AA (2018) Nanofiltration systems and applications in wastewater treatment: review article. Ain Shams Eng J 9(4):3077–3092
52. BWSSB (n.d) Overview of the sewerage system. https://www.bwssb.gov.in/com_content?page=3&info_for=4. Accessed 21 July 2019

Biosorptive Removal of a Herbicide, Glyphosate Using Waste Activated Sludge of Tannery Industry

Lata Ramrakhiani, Sourja Ghosh, Swachchha Majumdar, and S. N. Roy

Abstract Glyphosate [N-(phosphonomethyl)-glycine] has been applied as non-selective, broad-spectrum herbicide against all types of weeds. Extensive applications of this herbicide have originated water pollution, toxicity and carcinogenicity to human health. Hence, remediation of such pesticides is required to prevent its adverse impacts on the environment. The tannery industry waste activated sludge was used as a source of low cost complex biosorbent. In present study, the metal laden spent biosorbent produced after removal of Ni(II), Co(II), Zn(II) and Cd(II) ions in multi metal system were reused as biosorbent for removal of glyphosate from aqueous solution. The produced spent biosorbent showed >92% herbicide removal efficiency at original pH of herbicide, with 22–24 h of contact time. Feasibility study for actual effluent treatment showed 89% glyphosate removal efficiency in simulated municipal wastewater. Zeta potential, BET, FESEM, FTIR and XPS techniques were applied for characterization of biosorbent before and after glyphosate adsorption.

Keywords Herbicide · Glyphosate · Waste derived biosorbent · Biosorption

1 Introduction

Herbicides are substances or mixture of substances that are normally applied for prevention of the agricultural crops from growth of unwanted plants including long grasses, broad-leaved weeds, sedges and woody plants. Once herbicides and/or pesticides are commenced into the ecosystem, whether by manufacture, transportation,

L. Ramrakhiani · S. Ghosh (✉) · S. Majumdar
Water Technology Division, CSIR-Central Glass and Ceramic Research Institute, 196, Raja S.C. Mullick Road, Kolkata 700 032, India
e-mail: sourja@cgcri.res.in

S. N. Roy
Ceramic Membrane Division, CSIR-Central Glass & Ceramic Research Institute, Kolkata 700032, India

Adamas University, Kolkata, India

© Springer Nature Switzerland AG 2022

S. Yadav et al. (eds.), *Wastewater Assessment, Treatment, Reuse and Development in India*, Earth and Environmental Sciences Library,
https://doi.org/10.1007/978-3-030-95786-5_11

application, discarding, dumping or a spill, it can be persuaded by several practices. These processes determined the fate of the herbicide or pesticide by affecting its persistence and movement in the surrounding. Many urbanized countries have prohibited several herbicides and pesticides owing to their potential hazard to human and environment.

Glyphosate [N-(phosphonomethyl)-glycine] is the most widely used an organophosphorus herbicide. Glyphosate have been applied for several years in silvicultural, agricultural and urban environment as effectual, successful, non-selective, broad-spectrum and systemic herbicide against all types of weeds. It has been simply used as direct spray to plant foliage and enters the plant via diffusion. Commercial preparations of glyphosate ("Roundup") contain three components the isopropylamine salt (IPA salt) of glyphosate, polyethoxylated tallow amine (POEA) (a surfactant) and water. The surfactant present in the commercial preparations promotes the penetration of glyphosate across the plant cuticle. Glyphosate ($HOCOCH_2$-$NHCH_2PO(OH)_2$) have three functional groups (–COOH, –NH and –H_2PO_3) in its structure that can make complexes with various metal ions [1]. Large amount of agricultural applications of glyphosate can affect soil metal behaviours. Studies on co-sorption of glyphosate and zinc on red soil and wushan soil and suggested that the complexation of glyphosate with metal cations can affect physiology of glyphosate and plant metabolism depending on type of metal cation present in soil [2]. It also affects glyphosate distribution, bioavailability and degradation in soil.

Continuous applications of such components, accumulation in soil and leaching to ground and surface water led to toxicity problems and serious environmental pollution [3, 4]. Glyphosate is reportedly present in soil, water, fruits, crops and even human serum [5]. The occurrence of herbicide glyphosate in the aqueous environment like municipal and agricultural wastewater has directed the need to develop techniques for its remediation from water and wastewater [6]. The US Environmental Protection Agency [7] and WHO [8] set the maximum contaminant level (MCL) for glyphosate is 700 $\mu g\, L^{-1}$ in drinking water. Existing water treatment techniques such as chemical coagulation, filtration using sand, micro/ultra- filtration, reverse osmosis, adsorption onto activated carbon, ozonation, photocatalytic and/or biological degradation can be undertaken to deal their removal problems. Some of these techniques are specific to particular contaminants, involve large infrastructure investments and are become expensive that limits its application in large scale. As a result, there is upward need of the alternative low-cost techniques.

Biosorption process is an efficient, eco-friendly, cost effective and attractive option for removal of synthetic organic matter, as well as, toxic metals from aqueous solution. Furthermore, the process occurs via physicochemical interactions such as adsorption, ion exchange, electrostatic attraction, binding or complex formation between ions and functional groups present on the biosorbent surface. Biosorbents reported for the metal and organics removals mainly come under the categories like algae, fungi, bacteria, plant/ animal parts, industrial waste activated sludge, agricultural wastes and other polysaccharide matters. Recently, the researches on adsorbents for the remediation studies have been signified on waste derived materials due to its complex characteristics, cost effective, easy accessibility, and good adsorption properties. Use of different materials such as waste industrial residues like alum sludge [9], clay based substances [10], and activated carbon [11, 12] has been used for adsorptive elimination of glyphosate.

In our prior research, the dried sludge from tannery industry was effectively used for elimination of toxic heavy metal ions such as, nickel, cobalt, zinc and cadmium from aqueous environment [13, 14]. In the current study the exhausted biosorbent after adsorption of various metals was collected and further recycled for glyphosate herbicide removal. It had been observed that application of the dried activated tannery sludge resulted in only 26.18% removal of glyphosate whereas, the removal efficiencies were doubled with metal loaded biosorbent which could be further enhanced by increasing the metal loading. The prospective of the exhaused biosorbent for glyphosate removal motivated us to exploit this kind of toxic sludge as adsorbents for herbicide remediation in wastewater. This investigation presents a new and inventive move towards the reuse of dried sludge of tannery industry for remediation of toxic metal bearing industrial effluent and further reuse of the sorbent material in elimination of emerging contaminants such as glyphosate from contaminated surface water/ wastewater. The study necessitated for selection of suitable, effective and low cost complex biosorbent, the process optimization, feasibility in real wastewater and identification of removal mechanism.

The detailed objectives were to establish experimental parameters to obtain the optimum adsorption efficiency of the metal loaded biosorbent using simulated solution of glyphosate. The biosorbent surface was characterized initially and after glyphosate adsorption by various instrumental techniques such as, field emission scanning electron microscope (FESEM), Fourier Transform Infrared Spectroscopy (FTIR), X-ray photoelectron spectroscopy (XPS), zeta potential and surface area measurement (BET) for elucidation of biosorption mechanism. As a result, the biosorbent was further used in the treatment of simulated municipal wastewater.

2 Materials and Methods

2.1 Materials

The herbicide glyphosate is popularly known as "Roundup" (glyphosate 41% soluble liquid; Monsanto India Ltd., Mumbai, India) was used for present study. Nickel chloride, cobalt sulphate, cadmium nitrate, zinc sulphate, ninhydrin, sodium molybdate, and other reagents used for wastewater characterization were of analytical grade.

2.2 Biosorbent Preparation

Waste activated tannery industry sludge was collected in zip-lock bags from central leather complex in the city of Kolkata, India. After collection, samples were washed thoroughly, dried, powdered and sieved through a 150-mesh sieve and stored in an air tight container. The prepared biosorbent was studied and reported for biosorption of nickel, cobalt, zinc and cadmium ions in mixed metal system [14]. The metal loaded exhausted adsorbent was applied biosorbent for glyphosate removal and used as control.

2.3 Glyphosate Determination

Stock solutions (500 mg/L) of glyphosate and suitable aliquots were quantified using spectrophotometric method reported by Bhaskara and Nagaraja [15]. Briefly, 1.0 mL each of 5% sodium molybdate and 5% ninhydrin was added to 1.0 mL of glyphosate bearing sample and set into the water-bath at 90–95 °C for 10–12 min. Then the samples were allowed to cool and the capacity was make-up to 10 mL with deionized water. The absorbance of the resulted purple color complex was measured at 570 nm against the corresponding reagent blank using UV–Vis spectrophotometer model [Cary 50 bio, Varian, Australia].

2.4 Sorption Experiments

The batch biosorption experiments were carried out for glyphosate removal at initial concentration of 25 mg/L in room temperature by shaking the solution at 350 rpm for 24 h using metal laden spent dried activated tannery sludge as biosorbent (control biosorbent). The pH of the original glyphosate solution varies from 4.9 to 5.4 which was used as the working pH. The biosorption process was optimized for the equilibrium state using variation in parameters such as pH (3–9), dose of biosorbent (2–14 g/L), initial solution concentrations (5, 15, 25, 50, 75 and 100 mg/L) and contact time (1–28 h).

The difference in amount of glyphosate before and after biosorption were expressed as % adsorption (Eq. 1)

$$Biosorption(\%) = \frac{C_i - C_e}{C_i} \times 100 \quad (1)$$

The glyphosate amount retained in the adsorbent phase was obtained using Eq. 2.

$$q_e = \frac{V(C_0 - C_e)}{m} \quad (2)$$

where q_e is the glyphosate uptake (mg/g), C_i is the initial glyphosate concentration (mg/L), C_e is the concentration after biosorption, m is the biosorbent amount (g) and V is the volume of glyphosate solution (L).

2.5 Biosorbent Surface Characterization

Surface chemistry of the biosorbent before and after glyphosate biosorption was examined using zeta-potential measurements [Zetasizer, Malvern Instruments, UK], BET (Braunauer-Emmett-Teller) surface area [Quantachrome Autosorb, USA], field emission scanning electron microscopy [FESEM, Zeiss, Germany], Fourier transform infrared spectra [FTIR, Perkin Elmer, USA] and X-ray photoelectron spectroscopy [PHI 5000 XPS-analyzer, Versaprobe-II, USA].

2.6 Biosorption Study of Glyphosate in Simulated Municipal Wastewater

The adsorption efficiency was evaluated for glyphosate herbicide removal from practical wastewater. Municipal wastewater was collected from one of the sewerage pumping station of Kolkata region. The experimental setup was designed with simulated wastewater, prepared by adding 15 mg/L of glyphosate to the collected wastewater. Removal of glyphosate in simulated municipal wastewater was performed without adjusting the pH and concentration was determined by spectrophotometric method. The wastewater was analyzed for various physicochemical parameters by standard methods [16].

3 Results and Discussion

3.1 Experimental Variables of Biosorption Process Optimization

Biosorptive remediation of glyphosate herbicide were carried out using generated metal laden biosorbent after removal of multi metals in aqueous phase with waste dried activated tannery sludge as a purpose of material recycle. The biosorbent exhibited surface pH of 6.55 organic matter of around 25% and having about 75% of ash content. Other physico-chemical properties of the biosorbent include the bulk-density 0.44 g/cm^3, porosity 83.4% and average particle size 150 μm, the zeta potential value in the range of −13.7 to −18.5 mV and mobility was found as −1.071 to −1.448 μm cm/Vs.

Metal laden sludge showed 85–92% biosorption capacities for glyphosate. The maximum biosorption of glyphosate were found at original initial pH of glyphosate solution which is varying from 5.13–5.18. Moreover, studies also showed no significant variation in adsorption efficiency at different pH. The rate of biosorption increased as solid/liquid ratio was increased, but changed slightly above 10 g/L of biomass dosage which was selected as optimum solid/liquid ratio for further studies. The biosorption of glyphosate followed gradual removal and the equilibrium phases were attained at 24 h of contact time. Similar observations were reported for high glyphosate removal efficiencies with woody biochar [10] biopolymer membranes [17], as well as with MgAl-layered double hydroxide [18].

3.2 Surface Characterization of Biosorbent

The metal laden biosorbent generated after bioremediation of Ni(II), Co(II), Zn(II) and Cd(II) ions was used as control biosorbent for removal of glyphosate herbicide [14]. The biosorbent samples before and after glyphosate adsorption were characterized using various analytical techniques such as Zeta potential, BET surface area, FESEM, FTIR and XPS techniques [19].

3.3 Zeta Potential

The zeta potential determinations were performed at different pH 3, 5, 7, 9, 11 using 0.01 M NaCl solutions of the control and glyphosate loaded biosorbent to find out the net charge on biosorbent surface and the point of zero charge (pH_{ZPC}), ZPC is described as the pH at where the total surface charges of material become zero. Zeta potential values for both the biosorbent samples were negative confirming presence of no charge on biosorbent surface at different pH (Fig. 1). These results indicated that the bound metal ions at biosorbent surface along with the various functional groups can be involved in glyphosate removal.

3.4 BET-Surface Area

Surface area of biosorbent before and after glyphosate loading were measured by multipoint Braunauer-Emmett-Teller (BET) method and recorded as 20.08 m^2g^{-1} and 18.82 m^2g^{-1}, respectively (Fig. 2a, b). Reduction in surface area of glyphosate loaded biosorbent demonstrated that the binding sites were partially filled by glyphosate molecule and part of the glyphosate could possibly bind with the surface bound metal ions on the control biosorbent.

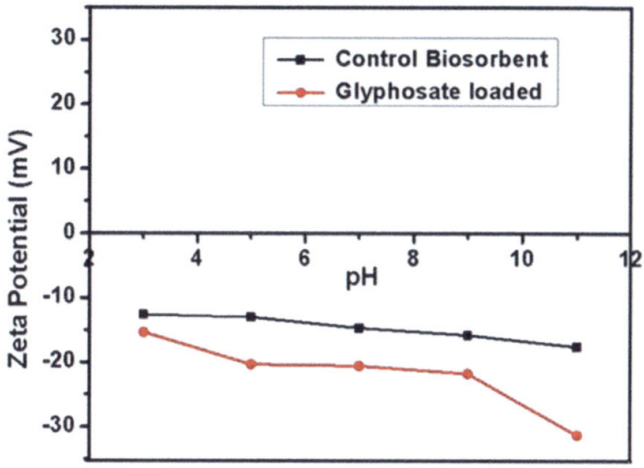

Fig. 1 Zeta potential of the biosorbent (before and after glyphosate adsorption) as a function of pH

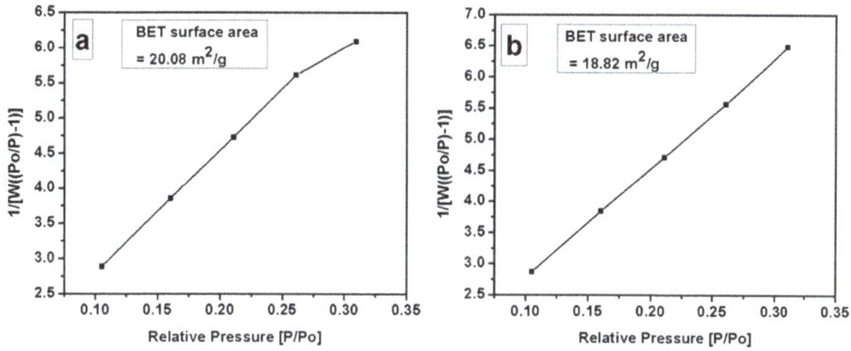

Fig. 2 BET plots of **a** control biosorbent and **b** glyphosate loaded biosorbent

3.5 FESEM Analysis

FESEM technique offered significant role to monitor alterations in surface characteristics and texture the biosorbent surface. A clear agglomeration was found in the topography of the glyphosate loaded biosorbent when compared to the FESEM micrographs of control biosorbent (Fig. 3a, b). This might be due to the surface adherence plus formation of complex between glyphosate molecule and multi metals bound on the biosorbent surface.

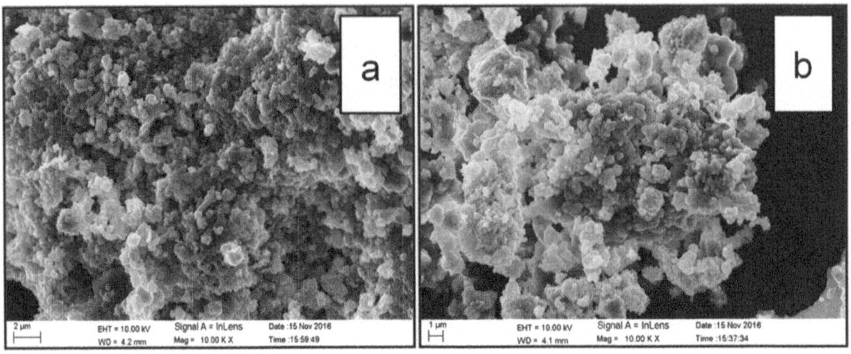

Fig. 3 FESEM micrograph **a** control biosorbent and **b** glyphosate loaded biosorbent

3.6 FT-IR Spectroscopy

FT-IR spectra of control were compared with that of the glyphosate loaded biosorbent to observe the changes in various transmittance peaks (Fig. 4). The peaks located in the control biosorbent spectra peaks corresponding to 3402.53 cm^{-1} frequencies match up to O–H stretching due to hydroxyl groups, carboxyl groups contributed peaks formation at 2957.83, 2916.24, 2846.65, 2506.73, and 1798.09 cm^{-1}, peak at 1617.33 cm^{-1} owing to amide groups and at 1422.97 cm^{-1} from carbonyl group are observed. In 900–1200 cm^{-1} region no peaks were observed for control biosorbent corresponding to phosphate groups.

Glyphosate loaded biosorbent exhibited an important difference in its FTIR spectrum those reappeared peaks at 1083.05 and 1027.06 cm^{-1} due to the phosphate groups. The result indicated that phosphate group of glyphosate molecule was mainly

Fig. 4 FTIR spectra of **a** control biosorbent and **b** glyphosate loaded biosorbent

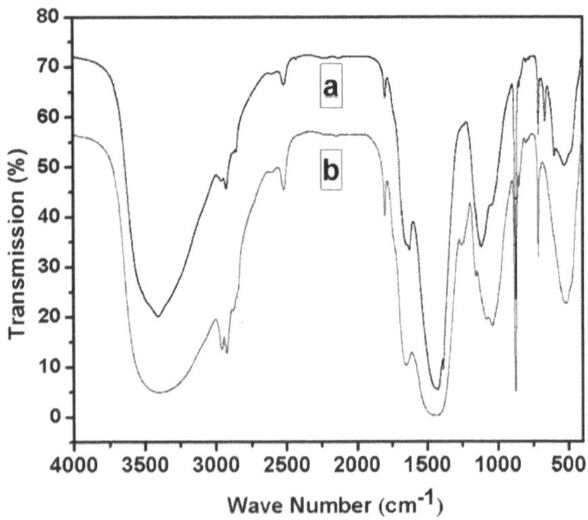

involved for the complex formation with surface bound metal ions of the control biosorbent. Moreover, changes in peak frequencies of amine, carboxyl, hydroxyl, carbonyl, and sulphide groups were also reported that confirmed their role in surface adsorption of glyphosate [20, 21].

3.7 XPS Spectroscopy

X-ray photoelectron spectra of the biosorbent samples before and after glyphosate biosorption was analyzed and observed from the shifts in binding energies of surface bound metal ions and the various functional groups [22, 23].

XPS wide scan spectra were represented in Fig. 5a, b). The XPS spectrum of control biosorbent exhibited photoelectron peaks at 852.61 eV, 780.89 eV, 1021.64 eV, 403.64 eV, 398.84 eV, 285.53 eV, and 530.81 eV, which are recognized as the Ni(2p), Co(2p), Zn(2p), Cd(3d), N(1s), C(1s) and O(1s), respectively. The strong peaks observed in glyphosate adsorbed biosorbent were at 854.35 eV (Ni2p), 781.58 eV (Co2p), 1022.12 eV (Zn2p), 404.86 eV (Cd3d), 399.95 eV (N1s), 284.43 eV (C1s) and 530.65 eV (O1s).

Significant difference were observed for Ni2p, Co2p, Zn2p, Cd3d peaks of biosorbent sample after glyphosate sorption exhibiting shifts in the binding energies (Table 1) [24]. This can be interpreted as direct interaction of metals with glyphosate. The XPS results were found in good agreement with the FTIR analysis.

The P2p peaks showed binding energy of 132.94 eV in control biosorbent and 133.04 eV in glyphosate loaded biosorbent is attributed to the phosphate moiety (Fig. 5a, b) and increase in binding energies could be due to the participation of phosphate group in the sorption process [25].

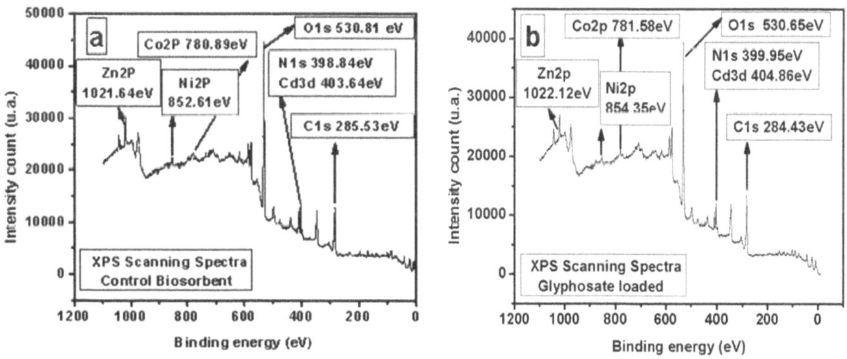

Fig. 5 Wide scan XPS spectra **a** control Biosorbent; **b** Glyphosate loaded biosorbent

Table 1 Results and assignments of XPS spectral peak

Element	Binding energy (eV)		Atomic concentration (%)		Assignments
	Control biosorbent	Glyphosate loaded	Control biosorbent	Glyphosate loaded	
C 1s	285.53	284.43	34.37	44.41	C–C, C–O, C–N, C=O, O–C–O, O–H, O–C = O
O 1s	530.81	530.65	39.42	35.28	O–H, C=O, –C–O, Phosphates, SiO_2
N 1s Cd 3d	398.84	399.95	9.18	6.90	– NH_2, –NH–, Nitride, N, – NH_2, –NH–, organic matrix
	403.64	404.86			CdO, CdO_2, CdS, $Cd(OH)_2$, $CdCO_3$, Nitrites,
Zn 2p	1021.64	1022.12	9.71	6.89	Zn in Phosphide, ZnS, $ZnCr_2O_4$, ZnO, $Zn_4Si_2O_7(OH)_2.2H_2O$
Co 2p	780.89	781.58	1.15	0.90	Co, CoO, Co_2O_3, CoOOH, $CoSO_4$, $Co(NH_3)_3Cl_3$
Ni 2p	852.61	854.35	1.23	0.81	Ni as NiO, Ni_2O_3, NiS, $Ni(OH)_2$, $Ni(NO_3)_2$, $Ni(dimethylglyoxim)_2$ NiSi, $NiSi_2$ and Silicides
P 2p	132.94	133.04	1.04	1.26	Phosphate, Pyrophosphate P_4O_{10}
Ca 2p	347.11	347.06	3.91	3.56	Ca in $CaCO_3$, CaO, $CaCrO_4$, $Ca_3Si_3O_9$, $Ca(NO_3)_2$

3.8 Glyphosate Removal from Simulated Municipal Wastewater

The biosorption efficiency was evaluated in the simulated municipal wastewater. The municipal wastewater sample showed an initial pH value of 7.29 and the optimum conditions obtained from the synthetic solution study was used for simulated wastewater system without adjustment of pH. The biosorbent showed about 89% removal efficiency of glyphosate (Table 2). The results revealed that the multi metal laden spent tannery sludge could be recycled for effective treatment of the municipal wastewater bearing such type of herbicides.

4 Conclusion

Selective biosorption of glyphosate herbicide was investigated in synthetic system and in simulated municipal wastewater using metal laden spent biosorbent. The

Table 2 Glyphosate removal in real wastewater and physicochemical characterization of municipal wastewater as such, with glyphosate addition and after biosorptive treatment

Parameter	Sample detail		
	Municipal wastewater	Glyphosate simulated municipal wastewater	Biosorbent-treated glyphosate simulated municipal wastewater
pH	7.30	7.29	7.65
Conductivity (mS/cm)	816	813	920
TDS (mg/L)	917	881	581
Salinity (mg/L)	388	419	386
TOC (mg/L)	21.97	17.21	15.17
COD (mg/L)	520	500	100
BOD (mg/L)	187	174	48.6
Glyphosate (mg/L)	0.0	15	1.65

spent biosorbent was produced after mixed metal removal of nickel, cobalt, zinc and cadmium ions using dried activated waste tannery sludge and was recycled as suitable and efficient complex biosorbent for glyphosate remediation. The following conclusions were reached based on the study:

- Incorporation of the multi-metal content in biosorbent derived from tannery sludge resulted in efficiency enhancement of glyphosate removal from 26.18% to >92%.
- Various removal mechanisms were established in the glyphosate biosorption process such as physical adsorption, complexation with surface functional groups, Vander waal forces, chelation, and ligand formation.
- The remediation potential in real actual effluent treatment study showed 89% glyphosate removal in simulated municipal wastewater.

Acknowledgements The financial support received by LR from the Department of Science and Technology (DST), Government of India vide grant no. DST/WOS-B/2017/111-ETD dated 10.05.2018 is gratefully acknowledged. The authors sincerely acknowledge the support from the Council of Scientific & Industrial Research (CSIR), Govt. of India.

References

1. Rigau J, Saitua H (2015) Optimization and modeling of glyphosate removal by nanofiltration at a pilot scale, using response surface methodology. World J Environ Eng 3(4):126–132
2. Wang YJ, Zhou DM, Sun RJ, Cang L, Hao XZ (2006) Co-sorption of zinc and glyphosate on two soils with different characteristics. J Hazard Mater A137:76–82
3. Rubí-Juarez H, Cotillas S, Sáez C, Canizares P, Barrera-Diáz C, Rodrigo MA (2016) Removal of herbicide glyphosate by conductive-diamond electrochemical oxidation. Appl Catal B Environ 188:305–312

4. Borggaard OK, Gimsing AL (2008) Fate of glyphosate in soil and the possibility of leaching to ground and surface waters: a review. Pest Manag Sci 64:441–456
5. Yue Y, Zhang Y, Zhoum L, Qin J, Chen X (2008) In vitro study on the binding of herbicide glyphosate to human serum albumin by optical spectroscopy and molecular modeling. J Photochem Photobiol B Biol 90:26–32
6. William AB, Karen CR, Michael JF, Sue S, Robert XB (2009) The occurrence of glyphosate, atrazine, and other pesticides in vernal pools and adjacent streams in Washington, DC, Maryland, Iowa, and Wyoming, 2005–2006. Environ Monit Assess 155:281–307
7. US EPA (2006) Ultraviolet disinfection guidance manual for the final long term 2 enhanced surface water treatment rule, EPA 815-R-06-007, US EPA, Washington, DC
8. WHO (2005) Glyphosate and AMPA in drinking-water. Background Document for Development of WHO Guidelines for Drinking water Quality, World Health Organization, WHO/SDE/WSH/03.04/97, Geneva
9. Hu Y, Zhao Y, Sorohan B (2011) Removal of glyphosate from aqueous environment by adsorption using water industrial residual. Desalination 271:150–156
10. Gimsing AL, Szilas C, Borggaard OK (2007) Sorption of glyphosate and phosphate by variable-charge tropical soils from Tanzania. Geoderma 138:127–132
11. Nourouzi MM, Chuah TG, Choong TSY (2010) Adsorption of glyphosate onto activated carbon derived from waste newspaper. Desalinat Water Treat 24:321–326
12. Mayakaduwa SS, Kumarathilaka P, Herath I, Ahmad M, Al-Wabel M, Ok YS, Usman A, Abduljabbar A, Vithanage M (2016) Equilibrium and kinetic mechanisms of woody biochar on aqueous glyphosate removal. Chemosphere 144:2516–2521
13. Ramrakhiani L, Majumdar S, Sarkar S, Ghosh S (2016) Heavy metal biosorption in multi component system on dried activated sludge: investigation of adsorption mechanism by surface characterization. Mater Today Proc 3:3538–3552
14. Ramrakhiani L, Halder A, Majumder A, Mandal AK, Majumdar S, Ghosh S (2017) Industrial waste derived biosorbent for toxic metal remediation: mechanism studies and spent biosorbent management. J Chem Eng 308:1048–1064
15. Bhaskara BL, Nagaraja P (2006) Direct sensitive spectrophotometric determination of glyphosate by using ninhydrin as a chromogenic reagent in formulations and environmental water samples. Helv Chim Acta 89:2686–2693
16. American Public Health Association (1989) American Water Works Association, Water Environment Federation, Standard Methods for the examination of Water and Wastewater, 17th edn. American Public Health Association, Washington, DC
17. Carneiro RTA, Taketa TB, Gomes Neto RJ, Oliveira JL, Campos EVR, de Moraes MA, de Silva CMG, Beppu MM, Fraceto LF (2015) Removal of glyphosate herbicide from water using biopolymer membranes. J Environ Manag 15:1353–1360
18. Li F, Wang Y, Yang Q, Evans DG, Forano C, Duana X (2005) Study on adsorption of glyphosate (N-phosphonomethyl glycine) pesticide on MgAl-layered double hydroxides in aqueous solution. J Hazard Mater B125:89–95
19. Ramrakhiani L, Ghosh S, Majumdar S (2016) Surface modification of naturally available biomass for enhancement of heavy metal removal efficiency, upscaling prospects, and management aspects of spent biosorbents: a review. Appl Biochem Biotechnol 180:41–78
20. Shoval S, Yariv S (1981) Infrared study of the fine structures of glyphosate and Roundup. Agrochimica 25:377–385
21. Ramstedt M, Norgren C, Shchukarev A, Sjöberg S, Persson P (2005) Co-adsorption of cadmium(II) and glyphosate at the water–manganite (γ-MnOOH) interface. J Colloid Interface Sci 285:493–501
22. Xu X, Xia L, Huang Q, Gu JD, Chen W (2012) Biosorption of cadmium by a metal-resistant filamentous fungus isolated from chicken manure compost. Environ Technol 33:1661–1670
23. Liu H, Yang F, Zheng Y, Kang J, Qua J, Chen JP (2011) Improvement of metal adsorption onto chitosan/Sargassum sp. composite sorbent by innovative ion-imprint technology. Water Res 45:145–154

24. Moulder JF, Stickle WF, Sobol PE, Bomben KD (1995) Handbook of x-ray photoelectron spectroscopy. In: Chastain J, King RC Jr (ed) A reference book of standard spectra for identification and interpretation of XPS data. ULVAC-PHJ-Inc., Japan
25. Yamaguchi NU, Bergamasco R, Hamoudi S (2016) Magnetic $MnFe_2O_4$-graphene hybrid composite for efficient removal of glyphosate from water. Chem Eng J 295:391–402

Wastewater Treatment Processes with Special Reference to Activated Sludge Process in Indian Conditions for Water Use Sustainability

Ram Karan Singh, Vineet Tirth, and Mansvee Singh

Abstract Wastewater treatment is one of the most important sanitation issue required for maintaining healthy ecosystem, various industrial, commercial and domestic activities produce huge quantities of wastewater which if not treated will be directly released into the water bodies. This would not only cause algal bloom and eutrophication of the rivers and lakes but will also cause serious water-borne diseases like typhoid, jaundice, cholera, etc. Three main steps in wastewater treatment are primary treatment, secondary treatment and tertiary treatment. Activated sludge process is an important type of secondary treatment, which aims to clean the wastewater by separating it into effluent and sludge. The effluent can then be discharged into any water body while the sludge can be used as manure or incinerated as per the requirement. This chapter presents the wastewater treatment technologies in general and activated sludge process in particular in Indian conditions for sustainable water use; it also takes a case study of Okhla Waste Water Treatment Plant located in Delhi India and draws important recommendations for future in context to activated sludge process.

Keywords Wastewater treatment · Primary wastewater treatment · Secondary wastewater treatment · Tertiary wastewater treatment · Aerobic digester · Anaerobic digestion · Water use sustainability

R. K. Singh (✉)
Department of Civil and Environmental Engineering, The NorthCap University, Sector-23A, Gurgaon 122017, Haryana, India
e-mail: ramkaransingh@ncuindia.edu

V. Tirth
Department of Mechanical Engineering, College of Engineering, King Khalid, University, Abha 61411, Kingdom of Saudi Arabia
e-mail: vtirth@kku.edu.sa

M. Singh
Department of Applied Psychology, The University of Massachutes Amherst, Amherst, USA
e-mail: mansvees@mtu.edu

© Springer Nature Switzerland AG 2022

S. Yadav et al. (eds.), *Wastewater Assessment, Treatment, Reuse and Development in India*, Earth and Environmental Sciences Library,
https://doi.org/10.1007/978-3-030-95786-5_12

1 Introduction

The pressure on fresh water sources is consistently increasing with increasing human population and development. Pollution and climate change also pose a serious threat. Wastewater reuse is a conservation strategy often implemented as a last resort due in part to a psychological resistance to the idea of reusing waste water. Emotional decision making and how it can override more pragmatic concerns as well as strategies to overcome it are relevant to conservation and sustainability in general and water sustainability in particular [1–3]. As an illustration of the complexity that can go into emotional decision making the negative affect/disgust that is contained with resistance to wastewater reuse is mediated by pathogen disgust sensitivity which in its turn can be a function of education and gender. One way in which emotional decision making can be countered is by innovative framing strategies that result in the desired behavior outcomes that are agreeable to a sustainable model of water use [4–6]. An experiment found that if the purified effluent from wastewater treatment facilities was referred to as "recycled water" instead of "treated wastewater" there was a greater willingness among both farmers and consumers to use it as well as its derivatives as well as pay for it. Naturally the name change itself is not the complete solution but it definitely plays an important role. In conclusion there are psychological and social factors that can be detrimental to sustainability and this need to be taken into serious consideration and be appropriately and ethically mitigated by communicational strategies like framing. The environmental deterioration is becoming a serious issue and one of the major contributors to this is wastewater and hence the importance of wastewater treatment to protect natural water bodies from being contaminated [7–11]. Wastewater is produced as a result of industrial processes, commercial processes and day-to-day human activities, although the amount of wastewater generated varies from place to place and time to time. Wastes in water includes organic compounds, inorganic nutrients such as Carbon, Nitrogen, heavy metals, radioactive elements, toxic organic and inorganic compounds, pathogens and any other compound which can impair the quality of water of receiving streams and lakes, making them toxic to the flora and fauna and also for the human consumption [12–15].

Wastewater treatment is a process which involves treatment of wastewater before being released into the natural environment so that after disposal it does not pose any threat to human or animal health. Broadly classified, wastewater contains about 70% organic and 30% inorganic materials [16]. Microbes use the organic portion of the wastewater as food while the inorganic portion passes on through the entire treatment process unaffected. The sludge generated can be discharged in landfill, incinerated or can be used as manure in fields depending on the concentration of contaminants it harbors. Wastewater treatment consists of a combination of physical, chemical and biological processes to remove physical, chemical and biological contaminants. The entire treatment process is divided into preliminary treatment, primary treatment, secondary treatment and tertiary or advanced treatment which may be followed by disinfection in some countries. Depending on the quality of final effluent, it can be discharged into a water body, can be used for irrigation, or groundwater recharge.

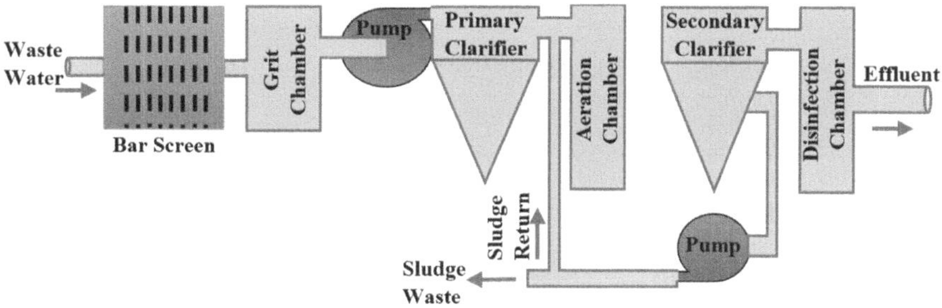

Fig. 1 A typical waste water treatment process diagram

The following Fig. 1 shows the diagram of typical wastewater treatment process [17–19].

Preliminary treatment is the first step in wastewater treatment. Its main objective is to remove coarse solids and other large materials often found in raw wastewater. This process is very important as it enhances the operation, maintenance and efficiency of further treatment processes. The process typically includes coarse screening and grit removal. In the preliminary wastewater treatment process the screens are was for the removal of large floating impurities such as rags which may damage or may interfere with the other subsequent treatment units next unit known as grit removes inorganic solids that would settle in channels and may interfere with the treatment process. The quality of wastewater improvement is very little in the process of preliminary treatment units [5, 16, 20–25].

Primary treatment is intended to remove larger floating particles and suspended solids. The main purpose of the primary treatment is to produce a generally homogeneous liquid capable of being treated biologically and a sludge that can be processed separately. Two major processes are: sedimentation involving removal of organic and inorganic solids which can settle down and skimming which is removal of materials that float over the water surface. The typical materials that are removed as a result of primary treatment are fats, oils, grease, sand, gravel, rocks, etc. Primary treatment helps in reducing around 25–50% of the incoming biological oxygen demand (BOD_5), 50–70% of the total suspended solids (TSS) and 65% of the oil and grease [5]. The constituents that are removed are contained in primary sludge.

Secondary treatment is the further treatment of the effluent from primary treatment to remove residual organics and suspended solids. The biodegradable dissolved and colloidal organic matter is removed using aerobic biological treatment process which is performed in the presence of oxygen by aerobic microorganisms. These microbes metabolize the organic matter in the wastewater producing inorganic end products, principally carbon dioxide, ammonia and water. There are various types of biological processes like activated sludge process, trickling filter, oxidation lagoon or rotating biological contractor [26–29]. The principle involved behind these processes is that the microorganisms biologically flocculate to form particles which can settle down, and following biological treatment, excess biomass is separated in sedimentation tanks as a concentrated suspension called "secondary sludge" to produce clarified

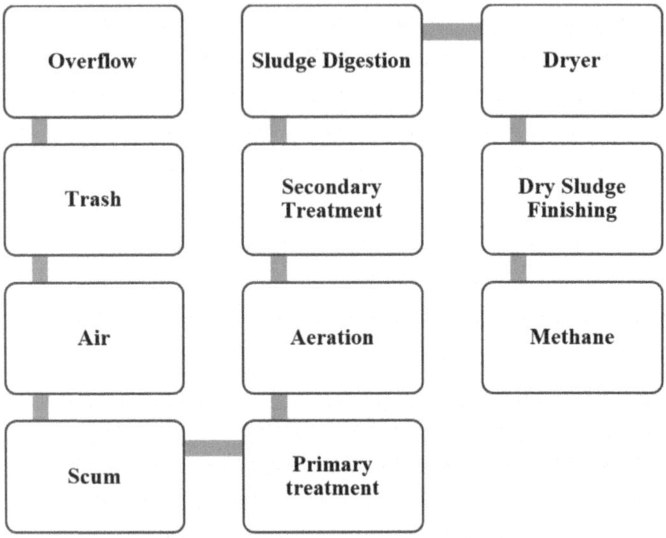

Fig. 2 Flow diagram for a wastewater treatment plant

secondary effluent. This process removes around 85 to 95% of BOD and TSS in the secondary effluent [30, 31].

The advanced wastewater treatment process also known as tertiary treatment is used for the municipal wastewater treatment plants when the higher quality effluent is desired. After secondary treatment the amount of BOD contained is around 30 mg/L or less but at time this level of sewage treatment is not sufficient to protect the aquatic environment. Also the high concentration of Nitrogen and Phosphorous are not reduced by secondary treatment which may cause algal blooms and eutrophication. Therefore tertiary treatment becomes important as it can reduce 99% of the pollutants from raw sewage and can produce an effluent of almost drinking water quality. But the cost of this process is significantly high and sometimes may double the cost of entire treatment process [32, 33].

Sometimes water is also disinfected before it is released to make it free of pathogens. It is a process that deactivates virtually all recognized pathogenic microorganisms, but not necessarily all microbial life. Most common method of disinfection is by using Chlorine but the use of Ozone gas and UV rays is also becoming common. Other ways of disinfection can be by heat treatment or chemical treatment (Fig. 2).

2 Activated Sludge Process

In the activated sludge process organic waste are stabilized with the help of activated Micro-organisms under in presence of oxygen (aerobic conditions). The component of treatment process as depicted in Fig. 3 starts with screening followed by grit

Fig. 3 A typical diagram of activated sludge process

chamber, primary sedimentation, and aeration, settling tank and disinfections where it is mixed with aerobic micro-organism mostly. Bacteria and protozoa that degrade organic matter into carbon dioxide, water, new cells, and other end products. The bacteria involved in activated sludge systems are primarily 15 g-negative species, including flock formers and non-flock formers, and aerobes and facultative anaerobes, carbon oxidizers, nitrogen oxidizers. The protozoa, for their part, include flagellates, ciliates and amoebas. An aerobic environment is maintained with the help of aeration, which also serves to keep the contents of the reactor (or mixed liquor) completely mixed. With retention time, the mixed liquor passes into the secondary clarifier, where the sludge is allowed to settle and a clarified effluent is produced for discharge. The process recycles a portion of the settled sludge back to the aeration basin to maintain the required activated sludge concentration (see Fig. 3). The process also intentionally wastes a portion of the settled sludge to maintain the required solids retention time (SRT) for effective organic removal.

Sludge is the residual semi-solid material left from industrial, wastewater treatment processes. Since hundred per cent solids in the wastewater may not be digested by the bacterial culture and in the process these solids eventually settle out as the sludge in the clarifier known as the settling chamber. For stabilization of this sludge two biological processes can be used: aerobic or anaerobic digestions, depending on various factors like time available, money, policy issues, etc.

The process where oxygen is required is known as the aerobic treatment process. In the aerobic process the microbes transform the waste material into the non-polluting material. The dissolved and the solid pollutants are converted into the cell mass, gases, water and the hydrogen etc.Activated sludge process (ASP) is one of the most widely used methods of wastewater treatment process [18, 19]. The characteristics of the process depend on the characteristics of microorganisms and configuration of aeration tank. The quality of effluent discharged by this process depends on how biological sludge mass can be separated from the treated wastewater which is done by sedimentation of aggregating flocks by gravity sedimentation. The waste water and its treatment for the Delhi city wastewater [20]. The following Fig. 4 shows the typical diagram of Aerobic Treatment System.

The activated sludge process is aerobic treatment processes in which the non-settle able substances, along with colloidal and dissolved form are converted to biological flock or activated sludge. The process uses microorganisms to feed on organic contaminants in wastewater, producing a high-quality effluent. The basic principle behind all activated sludge processes is that as microorganisms grow, they

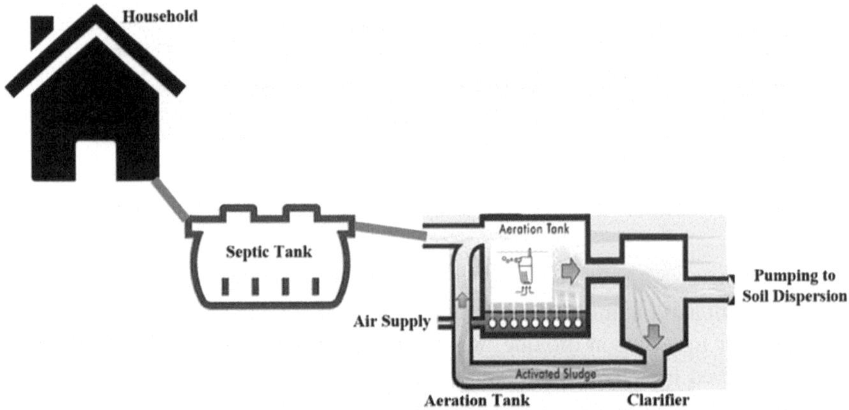

Fig. 4 Flow diagram of an aerobic treatment system

form flock that clump together. The BOD in the form of suspended solids is removed by physical enmeshment with biological flock. Colloidal organics are flocculated and absorbed with the biological flock. These flocks are allowed to settle to the bottom of the tank, leaving a relatively clear liquid free of organic material and suspended solids. The oxygen is supplied by dissolution from entrained air. Under aerobic conditions, aerobic bacteria rapidly consume organic matter as a source of energy and convert them into carbon dioxide. Ultimately the entire organic content of sludge gets used up by bacteria. In presence of plenty of nutrients, the bacteria divide and increase their population. After the entire sludge gets degraded, there is a lack of organic matter, bacteria die and are used as food by other bacteria. This stage of the process is known as endogenous respiration.

In the wastewater the biological conversion of organic materials takes place by the mixed culture of bacteria in aerobic conditions. The secondary clarifier gets the influent after long retaining period in the primary aeration tank. The end products of clarification process are discharged into open natural water bodies. The small part of sludge returned to primary aeration tank is called as returned activated sludge. The microbes get the basic food from the influent wastewater and the biodegradable organic material (carbonaceous), present in the sludge, is converted into new bacterial cells and other end products include CO_2, NO_3, SO_4 etc.

The bacteria multiply rapidly with sufficient food and oxygen and are able to oxidize the dissolved and carbonaceous organic matter into simple end products and additional biomass. The biological equation for bacterial respiration, using organic matter as substrate is given below also the following Fig. 5 shows the typical plan of.

Organic Matter + O_2 + Nut. → CO_2 + NH_3 + New Biomass + End Products

Fig. 5 A photograph of aeration tank [18]

3 Factors Affecting Activated Sludge Process

The characteristics of the influent that is going to the aeration tank: since the influent flow does not vary by more than 10% from day to day, a relatively stable loading is being applied to the tank. Toxic substances also retard the treatment process.

The environment that must be maintained in the environment basin: The requirements of aeration basin environment are the correct amount of oxygen, mixing and food to maintain the zoo of microorganisms. The food is supplied mainly in the form of dissolved and suspended solids in the wastewater itself. Dissolved oxygen level higher than 1.0 mg/L is desirable and it has to be monitored as it is affected by BOD level. Mixing of the contents in the tank must be properly done. Excessive mixing will cause flock shearing and slow mixing can make the tank septic. In order to achieve good treatment, the mass of microbes should be maintained at the correct level needed to consume all food that enters the system. Conventional systems run at F: M ratio between 0.25 to 1.00. Adequate temperature should also be maintained inside the tank, the higher temperatures speed up the biological process where as the lower temperature tends to slow down the biological process. The treatment process remains active by the heat created by aerobic process along with the heat from the electrical components. The cold weather can have the adverse effects on the performance of aerobic units the problems are sometime avoided by the insulating around the units.

Secondary clarifier conditions: proper inlet and outlet flow should be maintained in the clarifier to avoid any overflow and also maintain the efficient treatment speed.

Effective sludge removal from the bottom of clarifier should be properly done and should be timely scraped out.

3.1 Advantage of Aerobic Treatment Systems

- The capital cost of aerobic digester is lower and at the same time the digestion occurs much faster than the anaerobic process.
- The aerobic digester provides an alternative for sites not suited for conventional septic tank systems.
- The aerobic digestion process can result in the destruction of disease-causing microorganisms and parasites to a sufficient level.
- To allow the resulting digested solids to be safely applied to land used as a soil amendment material (with similar benefits to peat) or used for agriculture as a fertilizer provided that levels of toxic constituents are sufficiently low.
- Provide a higher level of treatment than a septic tank and help protect valuable water resources where septic tank systems are failing.

3.2 Disadvantages of Aerobic Treatment Systems

- Are more expensive to operate than a septic system digestion because of energy costs for aeration needed to add oxygen to the process.
- Requires electricity.
- Include mechanical parts that can fail or malfunction.
- Require more routine maintenance than conventional septic systems.
- Are subject to upsets under sudden heavy loads or when neglected (NSFC, 1996).

4 Evaluation of Activated Sludge Process Based Sewage Treatment Plant at Okhla, New Delhi

Okhla sewage treatment plant was the first treatment plant to be constructed in Delhi in 1938 which has a treatment capacity of 140 million gallons per day (MGD). This plant receives sewage from Sarita Vihar, Tigri, Deoli, and Tughlakabad village, areas along M.B. Road, Badarpur Road and Mathura Road. It's typical units involve 2 mechanical bar screen, 2 grit chambers, Parshall flume with ultrasound-type flow meter, 2 primary settling tanks with central drive mechanical sludge rakes, raw sludge pump house, aeration tank with 20 nos. surface aerator, 2 secondary settling tanks, return sludge pumping house, 6 digesters, 2 gas holders and 18 sludge drying beds. The plant has been designed for a daily flow of 3000 m^3/hr, peak hourly flow of 6067 m^3/hr. and lean flow of 1517 m^3/hr. This study involves the literature review of the dissertation "performance evaluation of ASP based 16 MGD sewage treatment plant at Okhla, Delhi" [26].

The raw sewage first enters bar screen in which large floating matters are screened. Grit Chamber then removes grit consisting of sand, gravel, cinders and other solid materials that have subsiding velocities or specific gravities substantially greater than those of the organic putrescible solids in wastewater. Wastewater then is passed on to the primary clarifier which enables settlement of suspended solids in raw sewage. For the oxidation of organics (BOD) by aerobic biological treatment, aerators are used in aeration tank to ensure that the biomass developed in the tank remains in suspension and is uniformly distributed throughout the tank for optimum stabilization of sewage. It also ensures that the maximum oxygen is transferred to liquid phase for utilization by micro-organisms developed in the tank. The aeration tank contents are settled in secondary clarifiers. Settled sludge from the secondary clarifier is conveyed to a return sludge pump for recycle to the aeration tanks. The sludge generated is digested by anaerobic method in the digesters for around 30 days. The digested sludge is applied on the sludge drying beds to form dry cake.

The sampling was done before and after each treatment unit in primary clarifiers performing primary treatment, aeration tanks and secondary clarifiers performing secondary treatment. For evaluating the performance of secondary digester, sampling was done for raw sludge and digested sludge. The sample was collected by the plant staffs from various locations like from channel feeding (raw sewage sample), from primary clarifier (primary effluent sample) and from outlet channel of aeration tank (aerated effluent sample). With standard methods for examination of water and wastewater lab tests were done. Physical tests were made on temperature, total solids, and suspended solids and dissolved solids. Chemical tests made involved pH, alkalinity, BOD5, COD, chloride, oxygen absorption etc. Effluent from each clarifier was tested for different measures. In particular the aeration tank was tested for pH, alkalinity, chloride, total solids, and suspended solids, dissolved solids, oxygen absorption, and dissolved oxygen. The performance of plant was studied during the seasons of both summer and winter in 2004 and 2005.

5 Result and Discussion

The main purpose of the study was to see the efficiency of OSTP and whether it achieves the required effluent standards or not. Because the waste load in water varies largely from summer to winter season, therefore the performance of the plant had to be measured for both the seasons. The following Table 1 gives the standard values of Suspended Solids, BOD and COD during the winter and summer seasons.

Average percentage removals were 93.3%, 93.7% and 9.7% for suspended solids, BOD and COD respectively. This shows that efficiency of aeration tanks and secondary clarifiers were excellent and these units are treating the sewage near to maximum theoretical units. This shows the efficiency of the plant is better during winter but the reason behind this can be that the amount of wastewater generated is

Table 1 Standard values of suspended solids, BOD and COD during the winter and summer seasons [26]

Parameter	Standards (PPM)	During winter (PPM)	During summer (PPM)
Raw sewage			
Suspended solid		208–370	250–504
BOD		150–200	153–230
COD		340–472	320–520
Final effluent			
Suspended Solid	<30	9–26	10–30
BOD	<20	5–15.3	6–20
COD	<50	28–52	23–52

much higher during summer. Average of 10 mg/L in winter and 14 mg/L of ammonia was also found in the final effluent which indicates that complete nitrification has not taken place. But the overall efficiency of the plant for removing the carbonaceous fraction was good.

6 Recommendations

Mechanical drying of sludge can be carried out which will decrease the amount of space required for conventional dewatering process and lessen the ground water contamination possibilities. OSTP does not have the tertiary level treatment process which can be considered there. Incinerators can also be installed which will help in reducing excess sludge and even help in disposal of screenings and grit. Energy consumption can be reduced by substituting aerators with fine bubble aeration system.

Acknowledgements The authors gratefully acknowledge resources and facilities given by the Civil and Environmental Engineering Department, The NorthCap University, Gurugram, India

References

1. Droste RL (1996) Theory and practice of water and wastewater treatment. John Wiley and Sons, New York
2. Faeth P (2000) Fertile ground: nutrient trading's potential to cost-effectively improve water quality. World Research Institute, Washington, D.C.
3. Gupta VK (2009) Performance evaluation of ASP based16 MGD sewage treatment plant Okhla Delhi

4. Hidayah EN, Pachwarya RB, Cahyonugroho OH, Ramanathan AL (2020) Characterization of molecular weight-based fluorescent organic matter and its removal in combination of constructed wetland with activated sludge process. Water Air Soil Pollut 231(2):1–12
5. Menegaki AN, Mellon RC, Vrentzou A, Koumakis G, Tsagarakis KP (2009) What's in a name: framing treated wastewater as recycled water increases willingness to use and willingness to pay. J Econ Psychol 30(3):285–292. https://doi.org/10.1016/j.joep.2008.08.007
6. Psychological and social factors associated with wastewater reuse emotional discomfort. J Environ Psychol 42:16–23 (2015). https://doi.org/10.1016/j.jenvp.2015.01.003
7. Zhang S, Han B, Gu J, Wang C, Wang P, Ma Y, Cao J, He Z (2015) Fate of antibiotic resistant cultivable heterotrophic bacteria and antibiotic resistance genes in wastewater treatment processes. Chemosphere 135:138–145
8. Singh RK (2012) Water quality assessment and management for Delhi stretch of River Yamuna. Lambert Academic Publishing, Germany
9. Tyagi VK, Subramaniyan S, Kazmi AA, Chopra AK (2008) Microbial community in conventional and extended aeration activated sludge plants in India. Ecol Ind 8(5):550–554
10. Metcalf and Eddy (1991) Inc., wastewater engineering: treatment disposal and reuse, 3rd edn. McGraw-Hill, New York
11. American Water Works Association and American Society of Civil Engineers (1990) Water treatment plan design. McGraw-Hill, New York
12. Asia IO, Enweani IB, Eguavoen IO (2006) Characterization and treatment of sludge from the petroleum industry. Afr J Biotech 5(5):461–466
13. Balkema A (1998) Sustainability criteria for the comparison of wastewater treatment technologies. In: Paper prepared for the eleventh European junior scientist meeting, the myth of cycles versus sustainable 4. Water and material flux management, Wildpark Eekholt, Germany, 12–15 February 1998
14. Fluid Knowledge (2000) Backa ground on water and wastewater industry. http://www.fluidknowledge.com/waterlist/html/waterindustry.html
15. Gabal Al Asfar Treatment Plant. http://www.bbv-ltd.com/Projects/WSTD/GlobslEIATP.htm
16. Yang Y, Ok YS, Kim KH, Kwon EE, Tsang YF (2017) Occurrences and removal of pharmaceuticals and personal care products (PPCPs) in drinking water and water/sewage treatment plants: a review. Sci Total Environ 596–597:303–320. https://doi.org/10.1016/j.scitotenv.2017.04.102
17. Garrett TM Instrumentation, control and automation progress in the United States in the last 24 years. Water Sci Technol 37:21–25
18. Google Image (2019) https://www.google.co.in/search?hl=en&authuser=0&tbm=isch&sxsrf=ALeKk026hqTRZINZmHnfRqbu_nRmEUd9Mw%3A1586248223922&source=hp&biw=1366&bih=657&eig. Accessed 21 January 2020
19. Liu DHF, Lipták BG (1999) Wastewater treatment. Lewis, Boca Raton, Florida
20. McGhee TJ (1991) Water supply and sewerage. McGraw-Hill, New York
21. Snidaro D, Zartarian F, Jorand F, Bottero J-Y, Block J-C, Manem J (1997) Characterization of activated sludge flocs structure. Water Sci Technol 36(4):313–320
22. Yuan QB, Guo MT, Wei WJ, Yang J (2016) Reductions of bacterial antibiotic resistance through five biological treatment processes treated municipal wastewater. Environ Sci Pollut Res 23(19):19495–19503
23. Shchegolkova NM, Krasnov GS, Belova AA, Dmitriev AA, Kharitonov SL, Klimina KM, Kudryavtseva AV (2016) Microbial community structure of activated sludge in treatment plants with different wastewater compositions. Front Microbiol 7:1–15. https://doi.org/10.3389/fmicb.2016.00090
24. Abdul-Malik QY (1996) Yemen's water resources and treated wastewater. Sana'a: ministry of agriculture and irrigation of Yemen. In: Adham S, Jacangelo J, Lainé JM (eds) Characteristics and costs of MF and UF plants. J AWWA
25. Bataineh F, Najjar M, Malkawi S (2002) Wastewater reuse. In: Paper presented at the water demand management forum, Amman, Jordan. Available at: Black and Veatch

26. Singh RK (2011) Wastewater management in national territory of India. Lambert Academic Publishing, Germany
27. Su JQ, Wei B, Ou-Yang WY, Huang FY, Zhao Y, Xu HJ, Zhu YG (2015) Antibiotic resistome and its association with bacterial communities during sewage sludge composting. Environ Sci Technol 49(12):7356–7363
28. Sulfikar HR, Noguchi M, Yamamoto-Ikemoto R, Watanabe T (2018) Effect of sedimentation and aeration on antibiotic resistance induction in the activated sludge process. J Water Environ Technol 16(2):94–105
29. Wester J, Timpano KR, Çek D, Lieberman D, Fieldstone SC, Broad K (2015) Psychological and social factors associated with wastewater reuse emotional discomfort. J Environ Psychol 42:16–23. https://doi.org/10.1016/j.jenvp.2015.01.003
30. Zhang J, Sui Q, Tong J, Zhong H, Wang Y, Chen M, Wei Y (2018) Soil types influence the fate of antibiotic-resistant bacteria and antibiotic resistance genes following the land application of sludge composts. Environ Int 118(34–43):22
31. Reddy RC, Rao RR, Andey SP (1999) (1999) evaluation of sewage treatment plant-a case study. Ind J Environ Health 41(4):346–355
32. Metcalf and Eddy, Inc. (2003) Wastewater engineering: treatment and reuse, 4th edn. McGraw-Hill, New York
33. Peng J, Wang X, Yin F, Xu G (2019) Characterizing the removal routes of seven pharmaceuticals in the activated sludge process. Sci Total Environ 650:2437–2445

Assessment of Water Resources in Development of Rajasthan

Supriya Singh, Pratibha, Vanshika Singh, and Sudesh Kumar

Abstract With increase in urbanization, the threat to water resources is also increasing. There is an increasing need to save the traditional and historic water resources from being depleted completely. Water being a natural resource cannot be produced manually or by any technological force at the time of need. In order to avoid the over development it is agreed to control the desirable groundwater. There is a need for proper legislation and regulation over the water resources; else, the day will not be far enough to lose the sight of even a drop of water. A central legislative system for development of control system is needed. Prepared in accordance with the due knowledge of the adequacy of available water.

Keywords Water resources · Depletion · Urbanization

1 Introduction

One of the most essential resources for mankind apart from air is water [1]. Over the years, water has become a critical scarce resource in several parts of the world. Though water is in the move in all forms in solid, liquid and gas through hydrological cycle, the water in the usable form is contributed by precipitation in the form of rain, snow and also frost [2].

Water being a natural resource can't be produced manually or by any technological force at the time of need [3]. The composition of water on earth is fixed. Circulation of water across the earth is continuous process, which is also influenced by solar energy [4]. The steps involved in the cyclic circulation of water are evaporation, condensation and precipitation [5].

The demand for water is increasing tremendously across the globe and it is not possible to meet the needs of Urban Water Supply. This is first century of urban revolution and over the past 35–40 years, numbers of urban residents have tripled. A plant assessment of water is demanded based on country location and region.

S. Singh · Pratibha · V. Singh · S. Kumar (✉)
Department of Chemistry, Banasthali Vidyapith, Rajasthan 304022, India

© Springer Nature Switzerland AG 2022

S. Yadav et al. (eds.), *Wastewater Assessment, Treatment, Reuse and Development in India*, Earth and Environmental Sciences Library,
https://doi.org/10.1007/978-3-030-95786-5_13

The resources of water are randomly distributed. Regardless of the great cutthroat demands of water on its application, no available water can be utilised. The average rainfall across the country is approximate 119.4 CM [6]. This data is very huge with respect to the size of country of a similar size. The amount is unevenly distributed and hence results in some places as floods while others as droughts. It is thoughtful to create water storage structures for the optimum utilisation of rainwater falling on ground. By creating such catchment areas, the utilizable water status can be raised.

1.1 Demand for Water in Rajasthan

Through extraction of groundwater about 91% of Rajasthan's water demands are met [7]. Rest of the demands are fulfilled through several reservoir dams, which are also reported to have less than 40% of availability. Approximate 24% population of Rajasthan results in the urban areas which consumes 9% of water [8]. Most of the hand pumps and tube wells these days are dry and defunct because 40% of more water is continuously drawn from them, which consequently is resulting in lowering of the water table. 19 districts are facing water crisis.

The water demand could be grouped under three broad headings namely [9]:

- Residential
- Irrigational
- Industrial.

The demand shares vary from place to place and country to country. In the developing countries, the majority of consumptions are in the agricultural sector. Lion share of total water is consumed in the field of agriculture.

In context of India, over the ages a relative share of water consumption is observed changing as people are moving forward urbanization. Water consumption in industrial and residential areas may grow at a faster rate [10]. A Priority should be decided for use of water on the basis of available water sources [11]. With the help of data available to the water demand at district level is carried out. Data on residential demand industrial agricultural along with the population data is collected from the respective sources. Estimation should be done sector wise. A detailed estimation procedure needs to be involved.

Policymakers are giving a lot of thought to the problem of residential water shortage in urban areas.

Shortages can be defined both in quantity as well as quantity the shortages are analyse with respect to two angles [12].

The water requirements in urban and rural areas has increased very drastically. On one hand in urban areas the purpose and demand for water is 140 L per day whereas a person residing in rural areas will not require more than 50 L of water on daily basis. The state of Rajasthan is in extreme need of heavy rains so as to the charges ground aquifers or urgent measures should be taken in order to in order for recharge of groundwater beside connecting its rivers with more districts [13].

2 Ground Water Resource Management

The idea of sustainable development come to action when planning for a project or development [14]. To last indefinitely we need to plan our resources sustainably [14]. The rain water falling on ground is absorbed by the soil, and it's accumulated there [15]. The water penetrates in the soil, but after certain level, the pores are completely filled and the water could no longer penetrate. The point at which all the pores are filled with water is called water table. A few centimetres above the water table, the soil is found wet.

The water flow rate is relatively slow. The average rate is 2 m per day. The factors affecting this are:

- Permeability
- Slope.

When the water table meets the surface, water emerges out as a spring or swamp. The categorisation depends on the amount of water gathered.

Groundwater can be classified into two broad categories[16]:

- **Stable Water**: The economically exploitable water
- **Static Water**: The water which is still present on ground and it is not possible to be used for every year

2.1 Groundwater System Monitoring

To monitor the response of aquifer in space and time three parameters are taken in account. These are [17]:

1. Water level change
2. Groundwater abstraction rate
3. Chemical quality changes

Abstraction of groundwater monitoring poses difficulty, if the pumping is done by private sources, which are unregistered in the pumping records. The value of pumping records only provides the gross records [18].

2.2 Monitoring of Water Level

The recharge and discharge regime is responded by groundwater level [19]. For effective management of ground water its essential to look around the annual recharge of groundwater level [20]. If the rainfall and runoff remains undisturbed then the water level remains in dynamic equilibrium [21]. Man may alter the recharge and discharge relationship by introducing various new techniques such as ground water pumps, surface water irrigation, etc. By pumping, the water table gets lower. The lost due to pumping may be overlooked if the discharge due to pumping is replaced by other sources of discharge such as surface water runoff and evaporation. It is essential to understand the relationship between various components of the recharge and discharge relationships. This would help in interpretation of data obtained from water level monitoring. The data analysis should be able to review the drawings of filling water levels, which attribute to water pumping.

2.3 Water Quality Monitoring

Water aquifers are capable of recharge but this recharge is very low because of high dilution factor of exploited water compared to accumulation of recycled salt. Quality changes can occur rapidly if the under or over lain respectively material of aquifer is saline or brackish groundwater [19].

Due to induced flow of groundwater from Indus basin, have resulted in deterioration of water quality [22]. This is dreadful to many aquifers. It is normal to monitor the area half yearly only unless it is declared critical. The wells are production facilities and as a result, network of water quality monitoring is independent of water level monitoring. Electrical conductivity enhances the monitoring data. Electrical conductivity indicates the change in water quality. Presentation of data is done in the form of maps [23].

2.4 Control of Groundwater Development

In order to avoid the over development it is agreed to control the desirable groundwater [24]. The control measures for groundwater are mainly focused for private sector because the public sector does not require a formula for its control. In the case of developing countries it is important to plan of an effective legislation and to focus on the practical problems for suitable ensure. The developing countries should define a fixed volume which is extracted from the well.

2.5 Control and Equity

The 1981 review from World Bank of Agricultural project, projects in India, the issues of farmers for ground water [25]. The review identifies that under any spacing or density criteria, the first investors automatically pre-empt a water right. Water ownership rests with the owner of the overlying land [26].

Development of groundwater Public intervention the alluvial aquifer systems are suitable [27]. There are some aquifers which have very low permeability these demand very large number of water points for the complete development of resources, this in turn makes it difficult for direct public intervention. The aquifers like hand dug wells should impose control on any development by a single entity or a small group of farmers have a much localised effect [28].

It is essential to develop rational management of aquifer. Identification of the issues and problems of the equitable control is much simpler than developing a solution.

2.6 *Options for Conjunctive Use*

The management of the total usable water resource, which is available for an irrigation project to achieve the objective of the project and to avoid the development of detrimental hydrological conditions, is known as conjunctive use. The definition given above gives recognition to both irrigation water and benefits of developing groundwater for consumptive use. The surface water irrigation system is helpful in rising the available ground water resource within the area due to increased recharge resulting from the population of distribution and non-field losses from the surface system [29]. The absence of exploitable aquifer system just below the irrigated surface is an example of exceptions.

Very detrimental effects on crop production have been observed from the shallow water levels [30]. On the other hand the high water table is proved to be beneficial for the areas where the predominant crop is rice or in a place where there is sufficient rainfall water and the surface water supply which would flush out the salt from the soil profile [30, 31]. It has been testified, the farmers require additional water supply, which is helping in development of private sector development of groundwater for conjunctive use.

2.7 *Hand Dug Wells*

Before the development of shallow Tube well technology, dug wells was the source of irrigation in the developing world [32]. This was the only source of water, which was suitable for the development of conjunctive use in surface water. Now these motorised centrifugal pumps are used to abstract the water from the Wells.

2.8 *Small Capacity Tube Wells*

The private sector started the construction of comparatively Shallow tube Wells through hand drilling techniques [33]. These tube wells are used mainly for irrigation in the developing countries. The shallow tube wells are only successful in the areas having appropriate water levels to the aquifer. These technology never last high levels of development [34].

2.9 *Large Capacity Tube Wells*

Construction of large capacity tube wells is restricted to the areas which have thick alluvial aquifer [35].

These wells are aimed to deliver water either directly through Canal system or through a collector channel, which then discharges to measure canal.

3 Water Resource Exploitation

Water is one of the most important and useful assets to us. The values from the hydrological and climatic data are the representation of this asset's ever-growing value from both National and international perspective. The long-term needs are fulfilled chiefly from surface water and through rainfalls. Groundwater also is one of the chief important resource. The important necessities for long term records relate to:

- Records of the rainfall
- River flows
- Discharge of sediment
- Equality of climate

The rainfall records establishes a relationship between rainfall and runoff. Once the relationship is established, it helps to develop the average discharge made on unrecorded rivers in the similar climatic condition.

The phenomenal use of water has been marked in the twentieth century. With the over expanding population the use of water has been multiplied six folds. The oil crisis seems too often talking about the water problems. The expected water availability of the state in 2015 has been dropped down to 439 M cube from 840 m cube in 2001. The water resources of Rajasthan are categorised under two broad headings, which are **surface water** and **groundwater**. Surface water is further classified as surface water within the Rajasthan boundaries and surface water, which are imported. Imported surface water is the waters, which are brought to the state with the help of several projects and interstate agreement. Rajasthan acquires 1.11 MAF (Million Acres Feet) from Ganga canal 1.41 MAF from Bhakhra Canal, 0.5 MAF from Narmada, 0.91 MAF from Yamuna, 8.6 MAF from ravi, 3 7 from Mahi water and lastly 1.6 MAF from Chambal Kota. The total allotted water in MAF comes out to be 14.5.

4 Water Pollution

The water crisis in twenty-first century is going to increasing its tremendous rate in the coming years. Increasing global warming will end up in melted glaciers and elevated temperatures. Consequences of these extreme climatic changes are extreme. There will be either flood or exclusive droughts. Poor sanitation affects the only one third of the world.

India is a deeply religious country with vast diversity. Many of its rituals are performed near water bodies, which consequently end up polluting them [36].

The increasing population builds a lot of pressure on the quality and quantity of water [37]. To understand the importance of safe drinking water, it's essential to understand the relation between human health and safe drinking water [38]. Abut 75 billion m^3 of ground water is extracted on annual basis. Pollution, climatic change have been reduced flow of surface water, which is causing over extraction of groundwater [39].

In the developing countries, rich variety of people lives in rural environment. Water often gets contaminated due to many natural and anthropogenic causes [40]. Rural and Suburban areas have poorly developed sanitation facilities [41]. The poor sanitation is one of the major reasons behind the poor health and mortality of infants in the developing countries. There are certain substances which show the presence of pollution at the point of the origin. The causes of pollution are wide ranging, they are fertilizers runoff, sewage waste, textile wastewater etc. [42]. Biomarkers of water pollution are the microbes which are enclosed association of water bodies [43]. They are evidence of water pollution as they are as the defect accumulation of metabolites of pollutants on other harmful substances.

Water is the life saver [44]. This is now getting a limited commodity. The social economic development could not be defined without availability of proper water resources. Water is the backbone of existence of all sorts of flora and fauna [45]. Quantity of water plays an important role to the existence of all living forms. The physical properties define water's purity and its usefulness for various purposes. The freshwater source arises from groundwater. Water available on most of the part is marine water, which needs to be purified with the help of certain tools and techniques.

Water pollution acquires concerns of many scientist and environmental activists across the world. With the rapid rise of pollution, increase the possibility of start of various harmful diseases which are lethal to the living forms. Data states that water pollution is responsible for death of more than 14,000 people on a daily basis [46]. This problem is seen rising at a tremendous rate in developing countries and the problem is creating an alarming situation.

It is found that various substances accumulate in water bodies and act as toxins. Namely, pesticides entering the marine life are imposing threat to the aquatic life. Although the pesticides are species specific but their presence has been found to be lethal to the other aquatic organisms. This could be explained by the following fact that although the pesticides are species specific but there are certain properties, which common to all the animal life are putting them in danger of death. This further enters the food chain through soil and plants, gradually getting fatal for human life also. Pesticides entering into the human system start acting as xenohormones. Xenohormones defined as those chemical entities, which mimic the action of endogenous hormones.

As per the study, the concentration of fluoride in the groundwater resources was found to be 4.78 mg per litre. Fluoride in drinking water causes tooth decay and joint pains called fluorosis. Groundwater of Rajasthan is also tested positive for the presence of Boron. Boron exists in the form of two isotopes, which are B10 and B11 respectively. The source of Boron in groundwater is leaching from the soils. The interaction of water from discard of sewage wastewater causes Boron existence in the water.

5 Water Supply: Approaches and Issues

Rajasthan has an area of 3.42 lakh square kilometres [47]. It's one of second largest state of country and constitutes for 10% of area with exceptionally low water resources of only 1% point the annual rainfall is low in the areas of erratic and western Rajasthan [48]. Rajasthan comprises of 221 urban towns, out of which 14 are categorised as class 1 cities. Cities having population of more than 100,000 are categorised as class 1 cities [49]. All the various schemes are in execution to meet the water demands of people around the town. However, due to expansion of urban limits, developing standards and migrating people from rural places to cities, is getting difficult to meet the demands of water requirements. Urban population increased by 29%out of total 221 towns, 104 of them do not even have 50% of the services as per the norms laid by the government. The dependency on groundwater is more than compared to the surface water. Consequently, depletion in the groundwater source has begun, and the water has become non-potable.

Steps have been taken to fulfil the water needs of towns point through Indira Gandhi Nehar, the Himalayan water has been brought to Western parts of Rajasthan [50]. Further classified efforts have been made to benefit the towns, which perennially are suffering from insufficient supply of water.

Towns such as Kota and Dholpur are benefited by water from Chambal River. Water supply from Yamuna River was meant to benefit towns of Alwar and Bharatpur.

A lot of thinking is given for the utilisation of Narmada river water.

In provision of basic civil amenities, usually the small towns are often neglected point accelerated Urban Water Supply program was implemented for towns less than the population of 20,000 as per 1991 census. Assistance weight by government was 50% of the cost of the project for augmentation of water supply.

World Bank also holds important position in the list of water financers. Due to shortage of funds, an inadequate maintenance in the assets has been observed, and as a result, their degradation has been observed at a faster rate than usual. This is chief reason for demand for new augmented projects. Provision funds are very important for the maintenance of existing system. Exercise to raise tariff and develop a suitable tariff structure is on in Rajasthan. From the scale of water, only 40% of the total expenditure on maintenance of Urban Water Supply is recovered. Subsidy evolve on water is huge it becomes it is a very important to control and prioritise Investments need for the supply of water with the objective to conserve the water.

When the people of a city demand for water augmentation, this is crucial to analyse effective demand.

Whether or not the people would be able to be part of the cost involved in the augmentation. Without the introduction of cost recovery it would be difficult to recover the cost from beneficiaries or, it would also become difficult to control the demand for one of the chief and most valuable yet scarce resource.

Talking about per capita consumption of water in cities, flushing toilet holds great importance [51].

A flushing system or pulling chain system of toilets, whenever a person goes for urination, proximately 3 gallons or 13 L of water is flushed down the drain [52]. The estimate is calculated for a day, nearly 65% water is wasted which constitutes 70–90% of normal water supply. Hence there is need to avoid cistern flushing and adopt valve flushing systems. This will save 20–30% of water if the demand and availability trends continue then the recycling and reuse of used water may become distinct [53]. Over exploitation already depleting water resources has brought degradation in the quality of groundwater in certain places. Demand for water by agriculture and industrial field results in holistic approach in the water management.

The policies which would be made in future should include.

- Water conservation to the maximum extent.
- Recycling and reuse of water.
- Recovery of cost from the beneficiaries.

6 Water Supply: Problems

The population of India has been observed increasing decade after decade. This trend of urbanization is similar to any other urbanization observed in rest of the developing countries. The water sector is capital-demanding [54]. The ground cause of such speedy urbanisation was better employment opportunities, education potential, and availability of better lifestyle [55]. The urbanization is directly elevating progress but simultaneously it is associated with downfall of all the available natural resources because this would result in increasing the gap between demand and supply of requisite services.

Most of the dwellers migrating to the cities live in slums. The slum dwellers are considered to be living in the so called informal settlements. Slum dwellers are more prone to diseases and their lives lack availability of basic amenities [56].

To increase the living condition, the foremost thing to keep in mind is to improve the sanitation facility, safe drinking water supply are must. However, the increasing population is causing to develop a fraction between the demand and supply of basic amenities.

The data from year 1990 states that although 84% of urban population was provided with safe drinking water facilities but the situation of rest rather most of the town is not even satisfactory [57].

Water distribution system is very poor. Most of the water sources have been already tapped. The sources from all the far distant places have so been started utilising [58].

The national water policy has granted the utmost priority to the drinking water supplies [59]. To make long term planning of used available water sources is a wise act to do. Due to erratic monsoon and increasing demands for water, the available groundwater and surface water sources are depleting. Water management plants need to be made wisely. By only judicious means, we can conserve the water resources and further make them available for different beneficial purposes. Though ground water is a renewable resource but the excessive demand won't let it replenish. Instead, it is also deteriorating the existing available water. It's important to ensure that a limit is set for accessing the water resource and control the rate of exploitation within the permissible limits [60].

A central legislative system for development of control system is needed. Prepared in accordance with the due knowledge of the adequacy of available water. Several schemes are prepared without proper knowledge of the availability of water, which results in schemes getting dysfunctional. Proper marking down the drinkable sources should be done before laying out the master plans of society.

During the planning period of dams and reservoir it would be wise to start reserving drinking water for minimum period of 20–30 years to meet the demands of their respective town [61]. When constructing water dams for the purpose of irrigation, it's always advice to make provisions for fresh drinking water [62]. The pre emptying water sources are consequence of improper planning schemes in determination of water supply sources for water needs [63].

6.1 Affordability

The poor sections of society have to suffer the hierarchy reduction of the water supply. The common water supply is for fixed hours and insufficient. To collect water from far distant sources not only involves loss of time but also involve transportation charges. The cheapest source of water supply is through pipeline. This would be cheaper and also reduce repeated loss of money in transportation [64].

6.2 Privatisation

To make water supply more effective, it is essential to privatise the water and sanitation facilities. Privatisation helps in maintaining a financial discipline and curbs the waste consumption. Innovative schemes need to be present to meet the demand and supply of the country [65]. Observing from an economic perspective water scarcity may lead to rise in prices [66].

6.3 Technological Options

The recent trends depict the emergence of adopting technologies, which are in accordance of meeting the low-cost demands and maintenance of the treatment of waste. These technologies are not only low in cost but also take care of treatment of waste and sewage water respectively [64].

Anaerobic system was adopted to treat the water and adoption of low cost sanitation technologies.

6.4 Application of Duckweed for Disposal of Municipal Waste

A lagoon system of duckweed is developed for treatment of sanitary waste water [67]. This process is unique due to its simplicity and high efficiency. Duckweed treatment plants are better than the traditional system of lagoons because not only are the successful in destruction of pathogens but also efficient in removal of nutrients from the wastewater stream. These are also successful in separation of algae [68].

Most of effluents of the world systems are rich in algae. Algae is a transporter of nutrient from the treatment system to the receiving water body.

The duckweed systems also hold a potential following the capital cost and high effluent quality. The duckweed harvest on a dry weight basis can exceed 100 kg per hectare per day that is 10 times the best street attend for Soya beans.

It is also necessary to lay out a category for selection of sources of for water treatment. This criterion is for drinking water purpose.

7 Case Study

7.1 Udaipur

Water is an essential element of life and it is the basis of sustenance of all the forms of living organisms. Water is a sovereign ruler for the human activities. Earth is surrounded by two third of water. Being a watery planet with 97% as seawater and rest 2% is present in the form of frozen glaciers and snows. Humans are left with only a limited amount of water, which is fit for drinking, and consumption purposes. The average annual rainfall of India is 300–650 mms.

Ahar and Kotra river systems are the tributaries of Berachi River [69]. These two act as the non-perennial surface water resource to the city. The river enters the city from the Northwest region and finally drains into the Uday Sagar, covering a distance of 30 km within the basin of Udaipur [70]. However, these days this is really a city drain, which is a source of all round water pollution.

The Kotra River is a significant tributary of the Ahar River [71]. This is done at the Badipal, which further forms the lake Pichola and is the source of water for the Pichola Lake. During the rainy Seasons if the Pichola Fateh Sagar Complex gets brim fullness, the excess of water is directed into the Ahar as the essential outlets.

7.1.1 Effect of Population Growth on Water Sources

Udaipur is an important city of Rajasthan, which is growing in both size and number of population. Area of Udaipur is increased from 61.10 to 200 km^2 [72]. With the increase in population, the consumption of various natural resources also increased uniformly. Urbanisation not only involved increased consumption of water for the drinking purposes but the lifestyles at city also added to the consumption of water. The per capita rise and the consumption of water increased from 5 g in 1952 to 35 g in 1990. Similar increase was also observed in the number of connections in the domestic areas as well as commercial areas [73]. A population growth analysis of the former eighty years (1921–2001) shows population growth from 34,798 in 1921 to 49,0142 in 2001, the year during which last census was held. Therefore, the recorded population has enlarged by 1308% over 1921. The effect of rise in population was directly observed over water resources [74]. Exploitation of water has been also made easy and the technology for exploitation has improved a lot. Water is now not only being used for drinking purpose is but commercialization has also in increased consumption of water. Water is also used in the recreation tourist it needs. The development of industries such as mineral of marble, cement industries respectively is putting a great load of pressure on the limited water resources.

7.1.2 Sources of Water Supply

The city of Jodhpur has predominantly two water sources; these are **surface water sources** and **underground water sources**. The water sources accounting for surface water are chiefly lakes. The city is beautified and guarded by manmade historical lakes, which are: *Udaisagar* in the Eastern side, *Fateh Sagar*, *Swaroop Sagar*, *Rang Sagar and Pichola Lake*. The Swaroop Sagar and the Ram Sagar Lake was constructed in 1845 AD. These are meant to feed water. The Fateh Sagar lake are linked to the Pichola through Rang Sagar Lake. The depth of Ram Sagar Lake is about seven metres but its width extends to 245 m [75].

In the west side of the city lies the Pichola which provides the charm to the city and is a great attraction to the tourist' soul. The lake acquires a triangular shape and

Table 1 Position of water in lakes with respect to annual rainfall [77]

S. no	Lakes and dams	Net catchment (km^2)	Gross capacity in lakh
1	Pichhola	131	137
2	Fateh Sagar	41	121
3	Badi	16	105
4	Dewas State	28	36
5	Fateh Sagar feeder	–	–
6	Total	216	399

it was renovated and enlarged in the 1559 AD. The depth of this lake is 9.15 m, with the catchment area of 131.3 km^2 [76].

The *Fateh Sagar Lake* is a *pear* shaped Lake situated in the Northern western part of the city and it covers an area of 4 km^2. The catchment area of this lake is 40.6 km^2 and its annual yield is lower than the Pichola Lake (Table 1).

The underground water sources are the water sources of groundwater occurs from beneath the surface of earth, which feeds Wells, baoris and tubewells.

7.1.3 Water Supply Management and Problems

The majority portion of water for the daily supply is extracted from the Pichola and the Fateh Sagar Lake and some amount is also accepted from the Baoris and tubewells. The filter needs applied in gulabbagh Fateh Sagar and Doodh Talai. The rainfall is adequate the lakes again get brim-full and the raw water is sufficient but the production capacity of filter units is limited. There are many industries located around the water bodies. Despite their distance, their waste treatment is not proper hence polluting the surface waters [78].

7.1.4 Variations in the Supply of Water

The variation in the water supply is due to the increasing population densities and the variations in the number of suppliers in the city areas. Number of people per tap connection is increasing rapidly and there are lot of factors, which are caused for this variation some are the timings and suppliers locality boys, economic and cultural levels of people residing in different localities. These variations give birth to a lot of problems which include that of clean drinking water supply sanitation problems and the decreasing quality of drinking water [79]. The city has attained a relatively Rapid growth in the past three decades. The supply lines are not sufficient to fulfil the growing needs of the city. The topography of the city is very uneven and the differences are high. Many colonies face the problem of low pressure which are due to the distribution system where pressure is inadequate and due to the underdeveloped distribution pipelines. The frequently leaking pipes leads to the contamination of water and further deteriorate the quality of drinking water. An urgent need of replacement of the ruptured lines is demanded to overcome the health hazards.

On the other hand the industrial effluents are polluting the groundwater sources point the pesticides industries a discarding the wastes in the nearby river without proper pre-treatment of waste water.

The human interference increased in the people and for the Sagar lake, the quality has been degraded over the centuries due to increased sewage disposal, washing, bathing, etc.

7.1.5 Proposals for Augmenting the Water

On looking into the monsoon trend and the increasing water pollution it is clear that the water supply to the city simply cannot rely on chiefly lakes. A few schemes what implemented for augmenting the supply at the earliest, these are as follows:

- Jaisamand project
- Mansi wakal project.

7.2 *Jaipur*

The capital city of Rajasthan-Jaipur also known as Pink City [80]. The rate of growth of population is 4.48% per annum according to the Census of 1991.

One of the ten cities among the ten major Indian cities is Jaipur which has a population of around 3.1 Million [81]. Jaipur is basically a tourist city which attracts its tourist through the local handicraft industry [82]. The floating population of Jaipur is around 3000 tourists per day [83]. Looking back to 1727 AD, only a few wells comprised of the potable drinking water. As the time moves ahead, the population increased and so did the demand of water per person-increased. In order to meet the demands of people it became essential for a well-developed water supply system. To meet the increasing demand of population it became extremely important to be aware to take care of the limited resource. Earlier when the technologies were under-developed the augmentation of water supply faced difficulty, but later after independence the status of the availability of technology improved and efficient efforts were made for the availability of water to the city dwellers. Due to absence of proper pumping devices in the history, some attempts went unsuccessful [84]. After the independence, the growth of population the subsequent migration added to the inadequacy of water.

The population of Jaipur is expected to be increased by 10 folds in the last 50 years [85]. The Ramgarh reservoir came out to be to be inadequate and hence the water is not drawn from Ramgarh Lake. The underground water exploitation rate is much more than the recharge rate, because of this the groundwater table sank down to excessively low level. The outer colonies of Jaipur are supplied water from the tube wells. Deterioration the quality of groundwater has been observed, with the increased presence of chloride fluoride and nitrates. The chemical parameters of water have been changed at a faster rate. Augmentation is needed to provide portable drinking water. The government boarings resulting in lowering of water table and drying up of wells in the nearby area. The Public Health Engineering Department (PHED), states that the 50% off metered connections are not functional in the city [86]. Around 462 million litres of water is supplied per day to the residents of the city. The exact estimation could not be made because of a large number of floating populations.

The water ponds, which are still not under the urbanisation impact, should be kept pristine, as they would act as a good source of the recharge of aquifers with clean drinking water. The points, which are under the influence of pollution, should be restored because this water would end up in contamination of the groundwater through the recharge.

7.3 Jodhpur

Jodhpur also known as Sun City is situated in the West area of Rajasthan. Jodhpur was founded in the year 1459 AD [87]. The rainfalls are not certain, climatic condition here is erratic and undependable monsoon, which causes stress on the common person. Repeatedly, drought situations keep on arising with problem of drinking water shortage. To meet the increasing demand of the city for better living standards continuous efforts are made by the engineers and the state government administrators. At the beginning of twentieth century the city accounted for 216 water sources which comprised of tanks step wells, Jhalaras, and some open dug Wells [88]. People in this city also practiced rainwater harvesting at the rooftop. These water-conserving structures are very economical and successful for proper management and conservation of water.

The city has both surface and groundwater bodies, such as Nadis, talabs, tanks, Wells, Baories and around 74 other non-functional water bodies. Venue of the water bodies around the city are approximately 500 years old. As a demand for water increase the number of water bodies also increase around the city. The history of these water bodies existed back to the time of the rulers. When the public water system came into existence in the development of these water bodies was halted. The increasing water crisis is demanding to protect the old water bodies which partially meet the demands of people. The canal system Jodhpur demonstrates High degree of Engineering. Protection and maintenance is needed to ensure the trapping of rainwater and further directly towards the other water bodies.

7.3.1 Surface Water Bodies

Surface water bodies like Nadi, step wells, tanks and lakes serve the human and cattle population. Development and design of these water bodies increased responding to the increasing needs of water due to poor rainfall. These water bodies are not only excellent example of architectural engineering design but they also represent the harmony among the community for the common desire for water for all in the society. When the public water system was introduced in the country in the water bodies felt the ignorance [89].

7.3.2 Ground Water Bodies

The basic ground water bodies of Jodhpur are wells *Jhalaras* and *Baoris* [89]. The Dispersion of water bodies around the city tells a lot about the population dispersion around. The chief aims of the water bodies are to ensure a regular water supply to the nearby area. Unlike the surface water sources, ground water bodies do not have a catchment area of their own nor are they connected to a channel or water course. Various shape size and structures are observed in case of ground water bodies.

Baoris

Jodhpur is known to be the city of Baoris [90]. These groundwater bodies are the community step wells. The purpose of providing steps is to reach to the water. These structures are usually designed in a beautiful way by providing rules and entrance to the bottom. The steps are usually at an angle of 45 degree from the main entrance. Since the evaporation of water is almost negligible, therefore the water holding capacity of these structures is much more. One of the oldest Bawri was created in 78,488 is around 100 years ago at Mandore. Potable water is still available in 20 of them.

Chand Baori

Chand baori is one of the hidden secrets of India. It was constructed by *King Chanda* in the ninth century. This baori is about 64 ft. deep and has 3400 steps, which further descends down 20 m deep to the bottom. Before the modern day public water delivery systems developed, this was the source of fresh drinking water to the surrounding population for a very long time.

7.3.3 Present Scenario

Now as urbanisation paced up only a few of these structures are in existence. Most of the catchment areas of the tanks have been destroyed and are now base of the multi storey buildings. Even the feeder canal maintenance is not done time to time. This can also now decide for city garbage disposal, and is now considered a canal for city sewage disposal. Around 65 surface water sources existed in the past, but presently hardly 10 could only function properly and that too after proper maintenance. Out of total 75 dug wells used, only 30 could have portable drinking water [91]. There were more than 30 step wells, but only 11 were reported to have potable water.

7.3.4 Present Status: Problems

The enormous rise in population the demand for water is also increasing, and migration of rural population is occurring. The overall progress of society has increased which results in increasing demands towards qualitative and quantitative supply of water. The census is evident of the rising population of the Jodhpur town. This as in turn resulted in reduction in the continuous supply of water, and it is now being limited to only a few hours per day [92].

The surface and groundwater resources face indiscriminate exploitation.

Huge amount of water is being lost due to leakage point the average leakage range of water is about 30–40% [93]. This indicates a lack in the maintenance system. The leakage results in 8 gallons of water loss per week. The service lines accounts for 77% of the loss and rest 16% of the loss occurs near the walls. Sleeping girls and porous pipelines are the most neglected areas.

Required material is not being made available for maintenance and which could help avoiding the water loss and that good compensate the water crisis to a large extent.

8 Precautions

To make water economic it is essential to follow certain steps.

- Development of technology is focusing on research and development
- Make each and every citizen aware of the importance of water and this should be done by seeking their corporation
- Proper pricing policies, law, and order.
- To strictly follow the concept of recycling of water.
- Effective implementation of the awareness programs, stating how to use water economically.
- Proper maintenance of all the pipeline, service lines and feeder lines, canal and the catchment areas should be executed time to time.

9 Conclusion

The rise in number of population values in Rajasthan is now just a normal phenomenon with no exception. The population rise in Rajasthan follows a rate of 3.08% per annum compared to the overall population. With the increasing number of inhabitants, the throat cut demand for the sources also increases which results in degradation of the surroundings and deterioration of the quality of life. The reason behind shortage of resources is not only increasing population but one of the major causes of this is improper and insufficient management and usage of them. It is extremely essential to study the various aspects of consumption and management of various natural resources predominantly water.

10 Recommendations

The authors want to highlight to the following recommendations:

- Avoid cistern flushing and adopt valve-flushing systems.
- Improve the sanitation facility, safe drinking water supply are must.
- During the planning period of dams and reservoir, it would be wise to start reserving drinking water for minimum period of 20–30 years to meet the demands of their respective town.
- To make water supply more effective, it is essential to privatise the water. Privatisation helps in maintaining a financial discipline and curbs the waste consumption.

References

1. Manahan SE (2010) Water chemistry: green science and technology of nature's most renewable resource. CRC Press
2. Manahan S (2017) Environmental chemistry. CRC Press
3. Hawken P, Lovins AB, Lovins LH (2013) Natural capitalism: the next industrial revolution. Routledge
4. Duffie JA, Beckman WA (2013) Solar engineering of thermal processes. John Wiley & Sons
5. Rockstrom J (2013) Balancing water for humans and nature: the new approach in ecohydrology. Routledge
6. Pai D, Sridhar L, Rajeevan M, Sreejith O, Satbhai N, Mukhopadhyay B (2014) Development of a new high spatial resolution (0.25 × 0.25) long period (1901–2010) daily gridded rainfall data set over India and its comparison with existing data sets over the region. Mausam 65(1):1–18
7. Glendenning C, Vervoort R (2011) Hydrological impacts of rainwater harvesting (RWH) in a case study catchment: the Arvari River, Rajasthan, India: Part 2. Catchment-scale impacts. Agric Water Manage 98(4):715–730
8. Groenfeldt D (2013) Water ethics: a values approach to solving the water crisis. Earthscan
9. Tanguay GA, Rajaonson J, Lefebvre J-F, Lanoie P (2010) Measuring the sustainability of cities: an analysis of the use of local indicators. Ecol Ind 10(2):407–418

10. Barlow M, Clarke T (2017) Blue gold: the battle against corporate theft of the world's water. Routledge
11. Faust SD, Aly OM (2018) Chemistry of water treatment. CRC Press
12. Gibbons DC (2013) The economic value of water. RFF Press
13. Moreau A (2011) Control strategy for domestic water heaters during peak periods and its impact on the demand for electricity. Energy Procedia 12:1074–1082
14. Singh AP, Bhakar P (2021) Development of groundwater sustainability index: a case study of western arid region of Rajasthan, India. Environ Dev Sustain 23(2):1844–1868. https://doi.org/10.1007/s10668-020-00654-9
15. Chapin FS, Matson PA, Mooney HA (2002) Terrestrial decomposition. Springer
16. Gharari S, Hrachowitz M, Fenicia F, Savenije H (2011) Hydrological landscape classification: investigating the performance of HAND based landscape classifications in a central European meso-scale catchment. Hydrol Earth Syst Sci 15(11):3275–3291
17. Vicente-Serrano SM, Beguería S, López-Moreno JI (2010) A multiscalar drought index sensitive to global warming: the standardized precipitation evapotranspiration index. J Clim 23(7):1696–1718
18. Pfeiffer L, Lin C-YC (2012) Groundwater pumping and spatial externalities in agriculture. J Environ Econ Manage 64(1):16–30
19. Green TR, Taniguchi M, Kooi H, Gurdak JJ, Allen DM, Hiscock KM, Treidel H, Aureli A (2011) Beneath the surface of global change: impacts of climate change on groundwater. J Hydrol 405(3–4):532–560
20. Chinnasamy P, Maheshwari B, Dillon P, Purohit R, Dashora Y, Soni P, Dashora R (2018) Estimation of specific yield using water table fluctuations and cropped area in a hardrock aquifer system of Rajasthan, India. Agric Water Manag 202:146–155. https://doi.org/10.1016/j.agwat.2018.02.016
21. Asbjornsen H, Goldsmith GR, Alvarado-Barrientos MS, Rebel K, Van Osch FP, Rietkerk M, Chen J, Gotsch S, Tobon C, Geissert DR (2011) Ecohydrological advances and applications in plant–water relations research: a review. J Plant Ecol 4(1–2):3–22
22. Qureshi AS, McCornick PG, Sarwar A, Sharma BR (2010) Challenges and prospects of sustainable groundwater management in the Indus Basin, Pakistan. Water Resour Manage 24(8):1551–1569
23. Lee H, Choi TK, Lee YB, Cho HR, Ghaffari R, Wang L, Choi HJ, Chung TD, Lu N, Hyeon T (2016) A graphene based electrochemical device with thermoresponsive microneedles for diabetes monitoring and therapy. Nat Nanotechnol 11(6):566
24. Pearce D, Barbier E, Markandya A (2013) Sustainable development: economics and environment in the Third World. Routledge
25. Reij C, Scoones I, Toulmin C (2013) Sustaining the soil: indigenous soil and water conservation in Africa. Routledge
26. Prasad K (2013) Ground water governance in India: gaps and the way out. Indian J Public Admin 59(2):228–246
27. Mukherjee A, Saha D, Harvey CF, Taylor RG, Ahmed KM, Bhanja SN (2015) Groundwater systems of the Indian sub-continent. J Hydrol Reg Stud 4:1–14
28. Payen J, Faures J, Vallée D (2012) Small reservoirs and water storage for smallholder farming. AG Water Solutions, Colombo, Sri Lanka
29. Taylor RG, Scanlon B, Döll P, Rodell M, Van Beek R, Wada Y, Longuevergne L, Leblanc M, Famiglietti JS, Edmunds M (2013) Ground water and climate change. Nat Clim Change 3(4):322
30. Anjum SA, Xie X-y, Wang L-c, Saleem MF, Man C, Lei W (2011) Morphological, physiological and biochemical responses of plants to drought stress. Afr J Agric Res 6(9):2026–2032
31. Siebert S, Döll P (2010) Quantifying blue and green virtual water contents in global crop production as well as potential production losses without irrigation. J Hydrol 384(3–4):198–217
32. Famiglietti JS (2014) The global groundwater crisis. Nat Clim Change 4(11):945
33. Chakraborti D, Das B, Murrill MT (2010) Examining India's groundwater quality management. ACS Publications

34. Machiwal D, Jha MK, Mal BC (2011) GIS-based assessment and characterization of groundwater quality in a hard-rock hilly terrain of Western India. Environ Monit Assess 174(1–4):645–663
35. Bhattacharya AK (2010) Artificial ground water recharge with a special reference to India. Int J Res Rev Appl Sci 4(2):214–221
36. Makwana S (2020) Impact assessment of anthropological activities on water quality of historical Gulab Sagar pond, Jodhpur, Rajasthan (India). Int J Res Appl Sci Eng Technol 8(10):178–185
37. Kneese AV, Bower BT (2013) Managing water quality: economics, technology, institutions. Rff Press
38. Forman R (2014) Land mosaics: the ecology of landscapes and regions (1995). Springer
39. Coyte RM, Singh A, Furst KE, Mitch WA, Vengosh A (2019) Co-occurrence of geogenic and anthropogenic contaminants in groundwater from Rajasthan, India. Sci Total Environ 688:1216–1227. https://doi.org/10.1016/j.scitotenv.2019.06.334
40. Wuana RA, Okieimen FE (2011) Heavy metals in contaminated soils: a review of sources, chemistry, risks and best available strategies for remediation. Isrn Ecol
41. Alirol E, Getaz L, Stoll B, Chappuis F, Loutan L (2011) Urbanisation and infectious diseases in a globalised world. Lancet Infect Dis 11(2):131–141. https://doi.org/10.1016/S1473-3099(10)70223-1
42. Förstner U, Wittmann GT (2012) Metal pollution in the aquatic environment. Springer Science & Business Media
43. Hussain MA, Ford R, Hill J (2010) Determination of fecal contamination indicator sterols in an Australian water supply system. Environ Monit Assess 165(1–4):147–157
44. Skinner BJ, Murck B (2011) The blue planet: an introduction to earth system science. SKI 504(3):7
45. Hand SC, Menze MA, Toner M, Boswell L, Moore D (2011) LEA proteins during water stress: not just for plants anymore. Annu Rev Physiol 73:115–134
46. Loehr R (2012) Pollution control for agriculture. Elsevier
47. Hussain I, Arif M, Hussain J (2012) Fluoride contamination in drinking water in rural habitations of Central Rajasthan, India. Environ Monit Assess 184(8):5151–5158
48. Udmale P, Ichikawa Y, Manandhar S, Ishidaira H, Kiem AS (2014) Farmers' perception of drought impacts, local adaptation and administrative mitigation measures in Maharashtra State, India. Int J Disaster Risk Reduction 10:250–269
49. Mishra P, Mishra P (2015) Urban development and environmental challenges in Rajasthan. Geogr Aspects 19
50. Hooda SM (2017) Rajasthan water assessment: potential for private sector interventions. World Bank
51. Black M (2016) The atlas of water: mapping the World's most critical resource. Univ of California Press
52. Dimpfl M (2011) The human dimensions of household wastewater. alternative onsite wastewater treatment systems and the Skaneateles lake watershed composting toilet project. State University of New York College of Environmental Science and Forestry
53. Canfield EL, Cheslock EP (2012) Detecting unintended flush toilet water flow. Google Patents
54. Rouse M, El Achi N (2019) A road map to sustainable urban water supply. Water Sci Policy Manage Glob Challenge 309–328
55. Varghese P (2016) Exploring other concepts of smart-cities within the urbanising Indian context. Procedia Technol 24:1858–1867
56. Agarwal S (2011) The state of urban health in India; comparing the poorest quartile to the rest of the urban population in selected states and cities. Environ Urban 23(1):13–28
57. Freeman III AM (2010) Water pollution policy. In: Public policies for environmental protection. Routledge, pp 179–224
58. Gleick PH, Palaniappan M (2010) Peak water limits to freshwater withdrawal and use. Proc Natl Acad Sci 107(25):11155–11162
59. Organization WH, Supply WUJW, Programme SM (2015) Progress on sanitation and drinking water: 2015 update and MDG assessment. World Health Organization

60. Chiras D (2011) The homeowner's guide to renewable energy: achieving energy independence through solar, wind, biomass, and hydropower. New Society Publishers
61. Shah M (2013) Water: towards a paradigm shift in the twelfth plan. Econ Political Weekly 48(3):40–52
62. Kuiper E (2013) Water resources development: planning, engineering and economics. Springer
63. Khastagir A, Jayasuriya N (2010) Optimal sizing of rain water tanks for domestic water conservation. J Hydrol 381(3–4):181–188
64. Hunter PR, MacDonald AM, Carter RC (2010) Water supply and health. PLoS Med 7(11):e1000361
65. Cairncross S, Bartram J, Cumming O, Brocklehurst C (2010) Hygiene, sanitation, and water: what needs to be done? PLoS Med 7(11):e1000365
66. González-Gómez F, García-Rubio MA (2018) Prices and ownership in the water urban supply: a critical review. Urban Water J 15(3):259–268. https://doi.org/10.1080/1573062X.2018.1436187
67. Evdokimova GA, Ivanova LA, Mozgova NP, Myazin VA, Fokina NV (2016) Floating bioplato for purification of waste quarry waters from mineral nitrogen compounds in the Arctic. J Environ Sci Health Part A 51(10):833–838
68. Crites RW, Middlebrooks EJ, Bastian RK (2014) Natural wastewater treatment systems. CRC Press
69. Machiwal D, Rangi N, Sharma A (2015) Integrated knowledge-and data-driven approaches for groundwater potential zoning using GIS and multi-criteria decision making techniques on hard-rock terrain of Ahar catchment, Rajasthan, India. Environ Earth Sci 73(4):1871–1892
70. Rathore NS (2011) A historical perspective of the development of rain water harvesting techniques in the Mewar region, Udaipur, Rajasthan, India. Int J Water Resour Arid Environ 1(4):285–294
71. Wahi T (2013) Irrigation state and society in pre colonial India
72. Tailor R, Meena G, Sharma L, Sharma F (2012) Constraints faced by the tribal farmers in dairy farming in Udaipur district. Rajasthan J Extension Educ 20:187–189
73. Nair S, George B, Malano HM, Arora M, Nawarathna B (2014) Water–energy–greenhouse gas nexus of urban water systems: review of concepts, state-of-art and methods. Resour Conserv Recycl 89:1–10
74. Rathore NS, Rathore D (2018) Shrinking of water resource in Udaipur, Rajasthan (India) basin using multi-temporal satellite data. Int J Water Resour Arid Environ 7(2):160–171
75. Patil PN, Sawant DV, Deshmukh R (2012) Physico-chemical parameters for testing of water-a review. Int J Environ Sci 3(3):1194
76. Riddhi S, Vipul S, Sudan SM, Kumar VB, Rachana M, Singh GK (2011) Studies on limnological characteristic, planktonic diversity and fishes (Species) in Lake Pichhola, Udaipur, Rajasthan (India). Univ J Environ Res Technol 1(3)
77. Mishra V, Sharma S, Sharma B, Upadhyay B, Choubey S (2012) Phytoplankton, primary productivity and certain physico-chemical parameters of Goverdhan Sagar lake of Udaipur, Rajasthan. Univ J Environ Res Technol 2(6):569–574
78. Mishra A, Nagda B (2020) Study of water quality in Udaipur Region, India. Asian J Water Environ Pollut 17(1):51–57
79. Rufener S, Mäusezahl D, Mosler H-J, Weingartner R (2010) Quality of drinking-water at source and point-of-consumption—drinking cup as a high potential recontamination risk: a field study in Bolivia. J Health, Popul Nutr 28(1):34
80. Jethoo A, Poonia M (2011) Water consumption pattern of Jaipur city (India). Int J Environ Sci Develop 2(2):1–4
81. Trehan M, Agarwal S, Sharma S (2012) Applicability of Bolton's analysis: a study on Jaipur population. Int J Clin Pediatr Dent 5(2):113
82. Singla M (2014) A case study on socio-cultural impacts of tourism in the city of Jaipur, Rajasthan: India. J Bus Manage Soc Sci Res 3(2):22–30
83. Kulshreshtha S, Sharma B, Sharma S (2013) The Ramsar sites of Rajasthan: ecology and conservation of Sambhar Salt Lake, Jaipur and Keoladeo National Park, Bharatpur. In: Faunal heritage of Rajasthan, India. Springer, pp 173–219

84. Olivier L (2011) The solid waste management system of Jaipur: an overview and analysis
85. Sam K (2014) Analysis of population growth Dy Barddhaman planning
86. Mathur P, Vijay A (2013) Non-revenue water reduction-a tool for achieving 24 × 7 water supply. IOSR J Mech Civil Eng (IOSR-JMCE). E-ISSN:2278–1684
87. Aggarwal K, Aggarwal SB, Bhavet D, Raju MS, Sharma M, Mahandru S, Sharma R, Garg S, Zutshi A, Natrajan V (2018) Chief patron
88. Soni HB, Thomas S (2013) Preliminary assessment of surface water quality of tropical pilgrimage wetland of Central Gujarat, India. Int J Environ 2(1):202–223
89. Alliance C (2014) Inclusive heritage-based city development program in India
90. Saxena D Water conservation: traditional rain water harvesting systems in Rajasthan.
91. Prasad J, Purohit D, Sharma S, Ameta N (2017) Rain water harvesting through tanka in Western Rajasthan
92. Amit D, Jethoo A, Poonia M (2012) Impact of drought on urban water supply: a case study of Jaipur city. Int J Eng Innov Technol 1:170–174
93. Puust R, Kapelan Z, Savic D, Koppel T (2010) A review of methods for leakage management in pipe networks. Urban Water J 7(1):25–45

Waste and Sustainability

Sustainable Construction Practices for Residential Buildings to Reduce the Water Footprint

Vineet Tirth, Ram Karan Singh, Amit Tirth, Saiful Islam, and Manisha Bhatkulkar

Abstract The sustainable construction practices primarily include three aspects namely; the conservation and optimization of resources, waste management, and reduction of energy and water footprint. A large number of studies have been directed towards the construction practices adopted in different stages of building construction and its life cycle assessment. The present study is dedicated to the environment-friendly construction practices by optimizing the use of water during several construction phases of a reinforced cement concrete framed multi-story residential building constructed on a land of area 199 m^2. The building was constructed in seventeen months in the Moradabad city of western Uttar Pradesh. The direct water footprint of the residential building has been estimated during different phases of construction. The methods have been suggested to reorder the construction stages in correlation with the ambient conditions to reduce water consumption, recycle the water used for curing and minimize the wastewater in washing and cleaning. The proposed water resource management strategy is expected to reduce the water footprint of building construction significantly, hence adding to sustainable construction practices.

V. Tirth (✉)
Mechanical Engineering Department, College of Engineering, King Khalid University, Abha 61421, Saudi Arabia
e-mail: vtirth@kku.edu.sa

R. K. Singh
The ICFAI University Dehradun, Dehradun 248001, Uttrakhand, India
e-mail: ramkaransingh@ncuindia.edu

A. Tirth
Department of Public Health Dentistry, Kothiwal Dental College and Research Center, Moradabad 244001, UP, India

S. Islam
Civil Engineering Department, College of Engineering, King Khalid University, Abha 61411, Saudi Arabia
e-mail: sfakrul@kku.edu.sa

M. Bhatkulkar
Department of Zoology, Jawaharlal Nehru Arts, Commerce and Science College, Wadi, Nagpur 440023, India

© Springer Nature Switzerland AG 2022

S. Yadav et al. (eds.), *Wastewater Assessment, Treatment, Reuse and Development in India*, Earth and Environmental Sciences Library,
https://doi.org/10.1007/978-3-030-95786-5_14

Keywords Residential building · Water footprint · Water resource management · Water conservation · Water recycling · Construction stages

1 Introduction

Water is indispensable for the persistence of life on the planet. It is a primary necessity of mankind, animals, and plants [1]. A classification of different types of water that exists on the planet earth is given in Fig. 1 [2–5]. The marine water is in the oceans and it is saline. Although 67% of the earth is marine water. It is unfit for consumption due to the salt content and other contaminations. The desalination of marine water is costly and is not economically feasible. For the survival of life on the land, freshwater is essential. The marine water level is increasing with each passing day due to the melting of glaciers owing to global warming and the termination of all rivers into the sea. Therefore, the primary source of fresh water suitable for mankind is Blue water [2, 6]. The natural resources of Blue water are rivers, streams, water bodies, snow, glaciers, underground water, and rainfall. The clean water usable for humans, termed as Potable water, is required to meet certain standards of purity. Green water refers to the moisture content of the soil, absorbed by precipitation. This water is used by the plants for survival by transpiration. Green water is a major contributor to the hydrologic cycle and a constituent of evapotranspiration flux. In the areas where there is limited or no rainfall, the Blue water is supplied by different means. During the monsoon season, the rivers are flooded with water, which enriches the underground water tables.

The rain water also peculates through the soil to make up the water table. The higher the water table, the more suitable is the land for habitation. Post-monsoon, the underground water, charged by the rainfall, rivers, ponds, and lakes remains the most

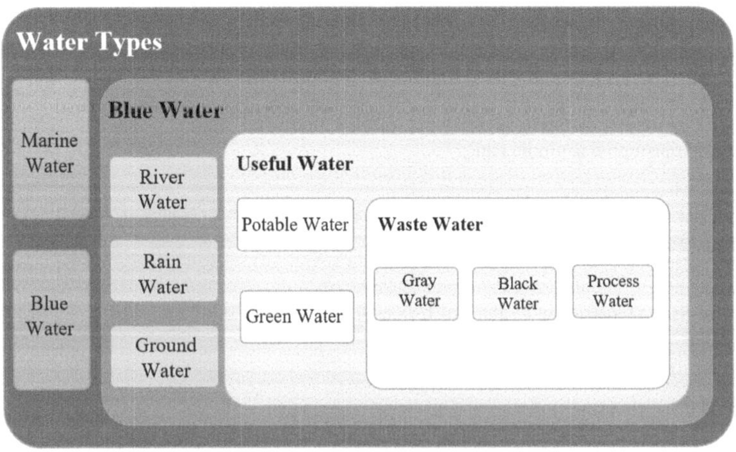

Fig. 1 Classification of water types

dependable water resource. Once humans, animals, and plants consume Blue water for drinking, cleaning, and other purposes, the water becomes wastewater, which is of three types. The Gray water is the wastewater contaminated by dust, dirt, soap and oily substances but it does not come in contact with fecal matter [3]. It is obtained from bathing, dishwashing, laundry, and agricultural runoff, which may include fertilizers and pesticides. Such water can be recovered and reused in different means including for agriculture [7]. It can be absorbed by the soil and is purified during its process of percolation to the water tables. The Black water is contaminated with fecal matter. It is the water obtained by flushing human or animal excreta. It has a high fecal coliform and numerous pathogens. The Black water should not come in contact with the Gray water [8]. There are some means to obtain fuel and manure from Black water. The water obtained from the industrial use and from the machines and equipment such as boilers, steam turbines, chillers, deep freezers, cooling towers, air conditioning plants, etc. is called Process water [2, 9]. Due to the exposure to high temperatures, dirt, chemicals, and hazardous materials, this water is not suitable for reuse unless it is filtered through a reliable purification process such as reverse osmosis. The toxins and heavy metals are the primary impurities that contaminate Process water [10, 11]. Hence, all types of waters can be recycled by suitable treatment. Recovery of wastewater is inevitable for the survival of mankind. A recent study [12] in Shenzhen, China revealed gross imbalance between the demand and supply of water. The shortage was approximately 81%. The building construction was found to be a primary sector responsible for wastewater generation in Europe [13]. A recent study focused on the barriers to sustainable construction [14] in Vietnam reported identified significant barriers. These are ignorance and incompetence of project managers, the hesitance to adopt new constructions practices and resistance to change, the unavailability of sustainable techniques and materials and low initiatives by the policy makers.

2 Review of Literature

Among human activities, building construction is a high water-intensive activity [15]. A brief review of the literature on water consumption during building construction is discussed in the following subsections.

2.1 Direct and Indirect Water Footprint in Building Construction

The different commodities and activities consume plenty of water to accomplish. The amount of water consumed by any activity, material, product or commodity to get completed is called its water footprint (WFP) [16]. Building construction is a human activity that consumes plenty of water. The water consumed during

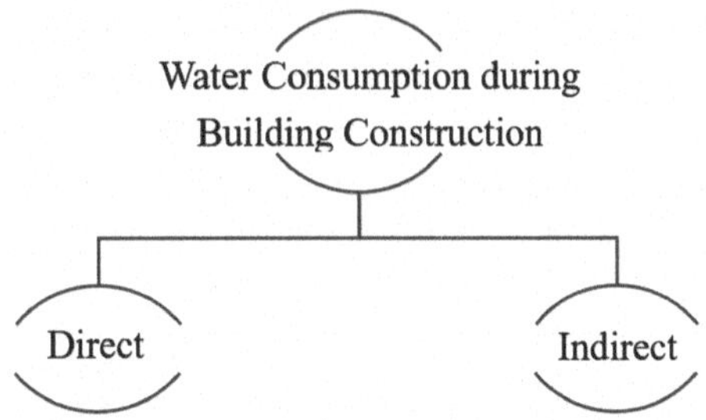

Fig. 2 Water consumption during construction activities

Table 1 The direct and indirect water footprint of building material [11]

Material	Direct water consumption
Bricks	0.71 kl/m^3
Cement	1 kl/Ton
Steel	200 kl/Ton
Aluminium	0.088 kl/Ton

building construction may be direct or indirect, as shown in Fig. 2. The direct water consumption (or WFP) is the amount of water consumed during the construction process for example, during mixing, binding, curing, hydration, and dilution, etc. [17, 18]. The indirect water consumption is the WFP during processing, packaging, and transportation of the building material to the construction site. A sum of direct and indirect WFP is the total WFP of the material or process [19–21].

Table 1 gives the direct WFP of the three main materials used in building construction. The fabrication of Clay Bricks, Cement, Steel, and Aluminium consumes plenty of water. Most high WFP material is steel, it consumes about 1000 L of water per Ton of steel produced, followed by Cement and Bricks.

3 Material and Methods

This study includes the assessment of the water footprint of a residential building in Northern India and pathways to reduce its water footprint by sustainable construction practices. The google map of the residential building understudy is given in Fig. 3. It is located at coordinates 28.865775, 78.753183 in the residential locality in Moradabad city in western Uttar Pradesh, India [22]. The Moradabad city is at the periphery of Western Uttar Pradesh in India. The environmental conditions of the location of the building are almost a transition between the tropical thorn and tropical dry deciduous. It can be categorized as humid sub-tropical. The Moradabad city is located at an altitude of 202 m from the sea level. The total annual rainfall was 976 mm and the average annual temperature was 24.5 °C in 2019. All four seasons are prominent in Moradabad. The precipitation is variable but most of the time, it is good. The summers are hot and humid and winters are quite cold [23, 24].

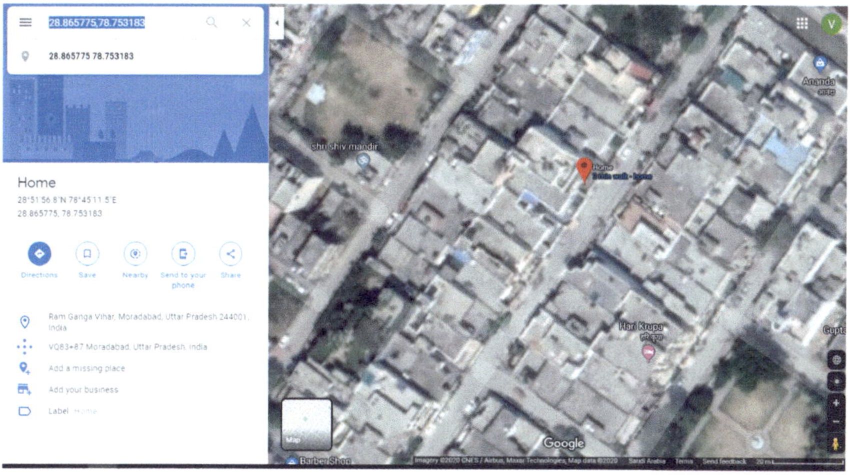

Fig. 3 Google map of the residential building under study [22]

The building is made on a piece of land with area 199 m^2. There are three floors in the building with a total covered area of 5018 ft^2.

The construction of the building was started in July 2011 and was completed in November 2012, in 17 months. The building was developed on a combined foundation of the pile and spread footing. All the construction was reinforced cement concrete framed. The direct water consumption during the construction of the building was carefully noted from day one until the building was completed. Indirect water consumption was not considered in the study. During the construction of the building, water from two sources was used. The water from the municipal supply as well as the underground water. The location of the building is about 1.2 km from the banks of Ramganga River, which is a major tributary of River Ganges. Due to the close vicinity of river banks, the groundwater table in the area is very high. For the construction purpose, water was obtained at a level of 45 feet through a four-inch bore well. Two water tanks of capacity 1000 and 500 L were installed at the construction site. The water consumption was measured on the complete filling of the water tanks. A makeshift toilet and a septic tank were made for the labor and caretakers who stayed at the site during the building construction. All types of wastewater were collected in a cement underground tank with an outlet to the municipal drain controlled by a valve. The wastewater was drained when extremely dirty or contaminated, into the municipal drain but before and after the drainage, the level of water was measured by the dipstick. The wastewater storage tank also had markings of water levels. All the wastewater collected due to the washing and cleaning of tools, equipment, curing was collected in this tank, including the wastewater generated due to the washing of hands and feet by the labor. This water was explicitly reused for hydration of Bricks, mixing with cement and curing.

The total water consumed, discarded, reused, and drained was monitored during the construction stages of the building and the results are discussed in the following section.

4 Results and Discussion

The amount of water consumed, wastewater generated, water recovered and reused, and water finally drained has been recorded during the construction of a residential building, which completed over a period of 17 months. The results are discussed in the following subsections.

4.1 *Stages of Building Construction*

The building construction practices vary from place to place according to the permitted norms and standards, the soil, geography, and requirements. The multi-story residential buildings in Northern India are generally RCC framed and Brick (clay) walls are for partition purposes. However, in single-story low rise houses, the Brick walls are also load-bearing. The present study includes RCC framed three-story residential building. The construction activities in the building have been divided into 16 stages, shown in Fig. 5. The water consumed during each stage has been shown by a symbol in Fig. 5. The bigger symbol indicates more water, the small symbol indicates less water and no symbol indicate no water was consumed during this stage of construction. The initial stage (1) was testing the soil to determine the strength and appropriate foundation. Thereafter, stage 2 was need analysis and discussions with the architect. Based on the requirements, several proposals were made and the best suitable one was selected by the client. The finalized proposal was submitted to the authority for approval. At stage 3 (preconstruction), the work time schedule was done and the contractors for excavation and foundation, RCC and masonry work, woodwork, flooring, and painting were finalized and contracts were signed. During this stage, the connectivity and warehousing facilities were developed on the plot. Moderate water was consumed at this stage. At stage 4, the resource generation was executed and orders for procurement of construction materials were placed based on the quality and price. No water was consumed at this stage. The map of the plot was approved by this stage. At stage 5, the plot was cleaned and dressed with a consumption of less or moderate water. The excavation work started at stage 6 and moderate water was consumed. The layout, markings were made and trenches for drainage, sewer and wastewater pipes were dug. Stage 7 included the laying of sewage and wastewater drainage pipes and the installation of main holes by the plumber. This stage consumed low to moderate water. At stage 8, a large quantity of water was consumed during the foundation. Structure up to the plinth level was completed at this stage. The utilities, temporary living facilities for the labor were constructed at stage 9, consuming plenty of water. The landfilling started at stage 10. Plenty of blue water was consumed at this stage for the settling of the landfill. Precautions were taken to prevent the evaporation of the water. A significant fraction of water used for the settling of the sand goes into the ground. The construction of the RCC framework started for floor level 1, at stage 11. It consumed a lot of water in binding, cleaning, hydration, etc. At this stage, the labor workforce increased and so the water consumed by human activities also increased. The RCC framework was followed by Brickwork and Roof Slab casting (stage 12), which consumed more water. After Roof casting, the concealed electric, plumbing, and sanitary lines were laid at stage 13, consuming less water. The plastering started at ground floor level and simultaneously, the RCC framework commenced on the first floor. More water was consumed at this stage. The construction stages repeated from 11 to 14 at each floor level. After the completion of the civil work (stage 14), the flooring work started (stage 15), which was water-intensive. Finally, at stage 16, the finishing work was completed, which included woodwork, bath fittings, electric fittings, painting and polishing. Hence all the 16 stages of building construction were completed in 17 months.

4.2 Schedule of Construction Stages

The construction stages are shown in Fig. 4 were completed in 17 months. The work-time schedule of the construction stages is given in Fig. 5. Here, M1, M2, ... M17 represents the number of the month. The construction stage (from Fig. 5) is represented in digits from 1 to 16. F1 indicates floor 1 and so on. The construction stage accomplished in the respective month is shown by shading.

4.3 Water Consumption in Different Construction Stages

The amount of water consumed during each stage of construction was indicated as a symbol (more, less or nil) in Fig. 4. The amount of water consumed in liters, in different construction stages is given in Table 2. The total water is drawn, water wasted, water collected in tanks and reused, and finally, the water drained is shown in respective columns of Table 2. The stages 12, 14, 8, 15, 6 and 10 are the most water-intensive construction stages. Stages 16, 9, 15, 11, 12 generated maximum wastewater, which had to be drained.

4.4 Water Resource Management and Optimization

A large quantity of water consumed during the construction may be recovered and reused if the runoff is prevented. In the building understudy, a separate tank for

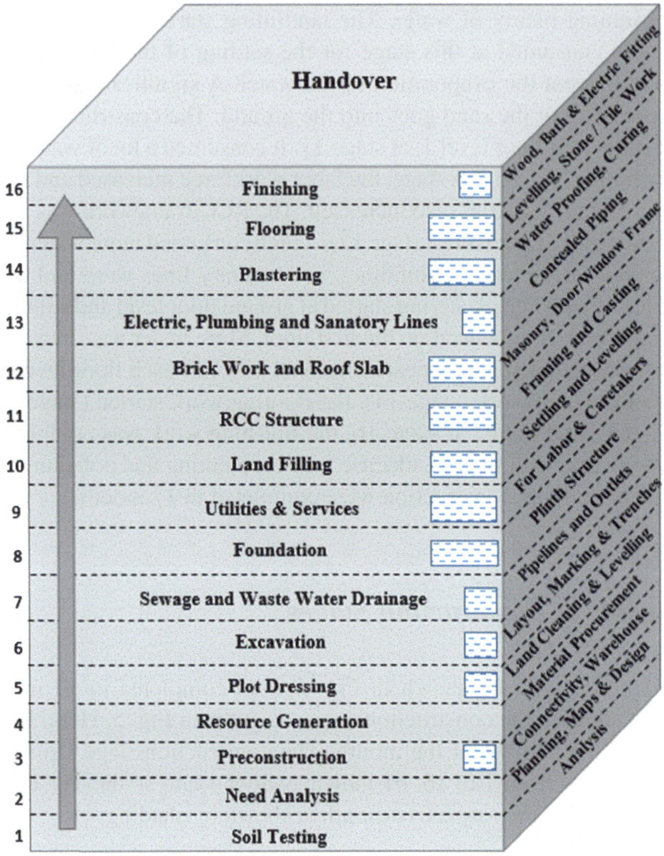

Fig. 4 Phases of construction of a residential building. The water consumed during each stage (more, less or nil) is shown by a symbol

collecting wastewater was made. Figure 6 shows the total amount of water consumed during the construction of the building, the wastewater generated, the wastewater reused, and the water finally discarded in drains. Total water consumed was 190,400 L, the water absorbed by building material was 120,725 L, and the water drained was 11,790 L.

The fraction of wastewater, wastewater reutilized, and wastewater drained is plotted in Fig. 7 as a percentage during different construction stages. About 95% of the wastewater generated in stage 7 (sewage and wastewater drainage) was reused. About 85% of the wastewater generated at stage 12 (brickwork and roof slab) was reused, reducing the drained water to a minimum. The utilization of the wastewater in such large fraction not only conserves the water resources, but it reduces the water footprint of the building and water pollution resulting in the drainage of this water. The maximum fraction of drained water was in stages 5 (plot dressing), 6 (excavation), and 15 (flooring).

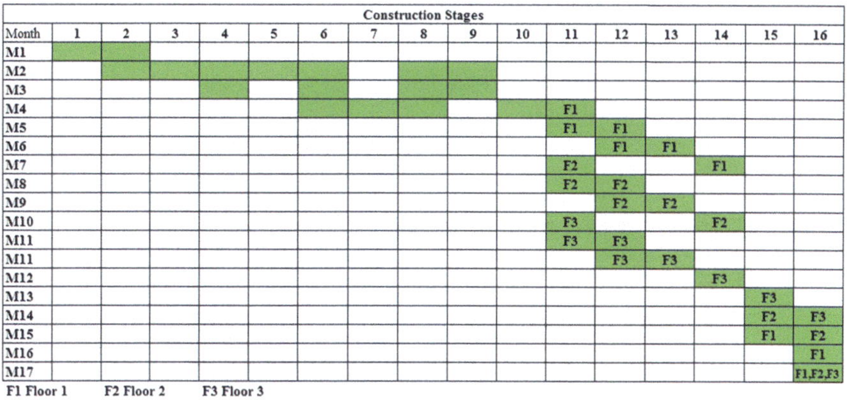

Fig. 5 Scheduling of construction activities. M represents the month number, F represents the floor, and digits from 1 to 16 represent the construction stage

Table 2 Water consumption in different construction stages

Construction stage (months)	Water consumption at the site (L)			
	Total water consumption	Waste water	Water recovered and reused	Runoff/Drained water
1	0	0	0	0
2	0	0	0	0
3	600	320	200	120
4	150	95	40	15
5	450	190	70	125
6	18,000	900	320	580
7	1000	750	700	50
8	21,000	700	350	350
9	14,000	5000	2700	2300
10	18,000	1000	400	600
11	21,000	4500	3000	1500
12	28,000	8000	7000	1000
13	1200	500	250	250
14	27,000	900	700	200
15	19,000	3000	800	2200
16	21,000	9000	6500	2500
Total	19,040	34,855	23,030	11,790

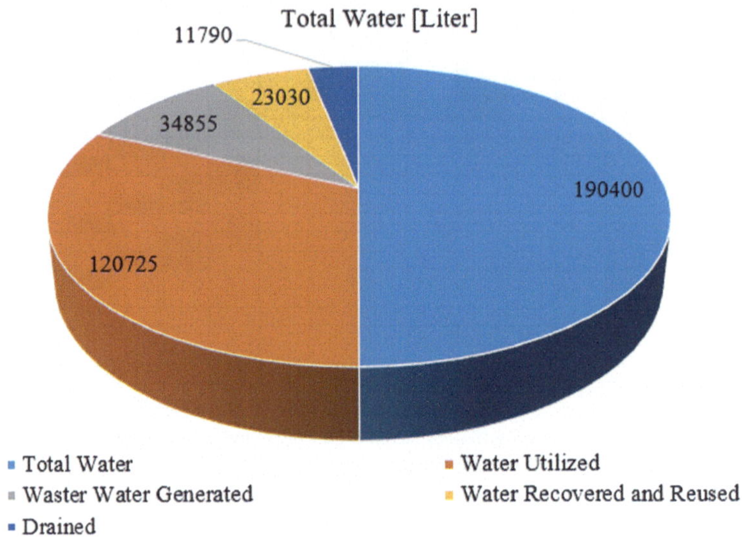

Fig. 6 Water consumption during the construction of the residential building

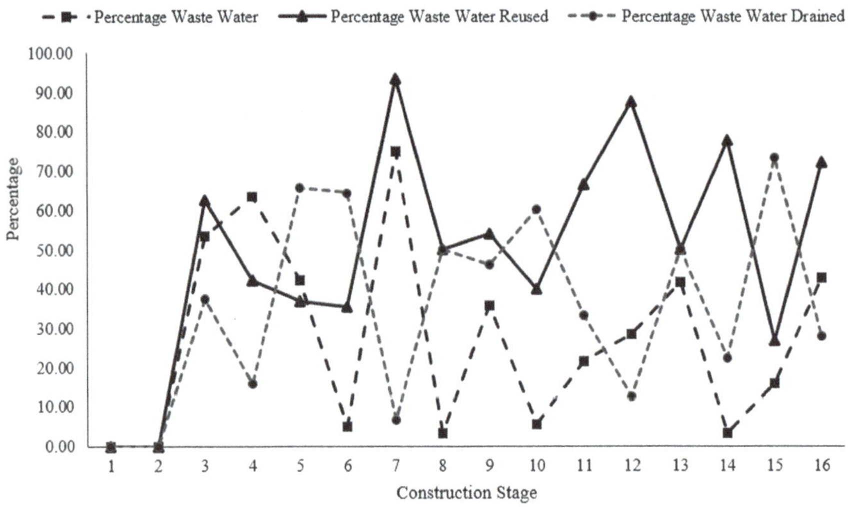

Fig. 7 Percentage of water consumed during different construction phases

4.5 *Water Consumption*

A large quantity of water consumed during the construction may be recovered and reused if the runoff is prevented. In the building understudy, a separate tank for collecting wastewater was made. Figure 6 shows the total amount of water consumed

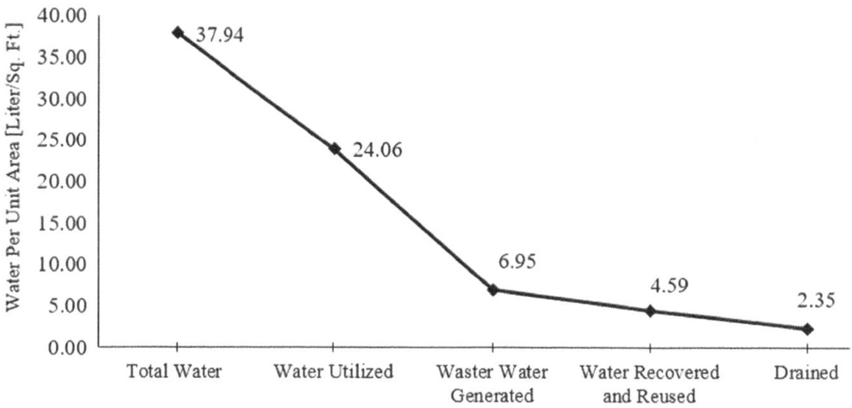

Fig. 8 Water consumption, usage, recovery, and drainage per unit covered area (L/ft^2)

during the construction of the building, the wastewater generated, the wastewater reused, and the water finally discarded in drains. Total water consumed was 190,400 L, the water absorbed by building material was 120,725 L, and the water drained was 11,790 L.

The fraction of wastewater, wastewater reutilized, and wastewater drained is plotted in Fig. 7 as a percentage during different construction stages. About 95% of the wastewater generated in stage 7 (sewage and wastewater drainage) was reused. About 85% of the wastewater generated at stage 12 (brickwork and roof slab) was reused, reducing the drained water to a minimum. The utilization of the wastewater in such large fraction not only conserves the water resources, but it reduces the water footprint of the building and water pollution resulting in the drainage of this water. The maximum fraction of drained water was in stages 5 (plot dressing), 6 (excavation), and 15 (flooring).

The water consumed per unit covered area (L/ft^2) is shown in Fig. 8. The water footprint of the building (without wastewater recovery) is 37.94 L/ft^2. The actual WFP of the building is 24.06 L/ft^2 (water consumed minus wastewater).

4.6 Water Wastage and Recycling [25, 26]

By wastewater recovery, the WFP of the building is reduced to 19.46 L/ft^2. Hence, recovery of 4.59 L/ft^2 of water has reduced the WFP of the building by 19%. The drained water is only 6.1% of the total water consumed. Such sustainable construction practices do not require any special equipment or investment or any compromise on the construction quality or strength of the building.

A schematic diagram of wastewater recovery at the construction site is represented in Fig. 9. The WFP of the RCC framed building may be reduced significantly by recovering the runoff or spilled water, and partial recovery of wastewater.

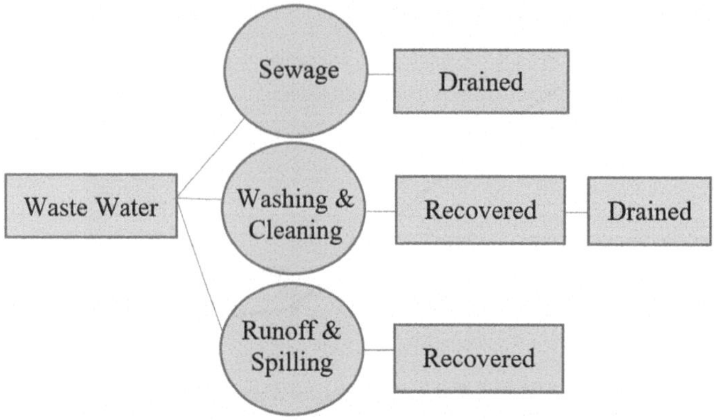

Fig. 9 Wastewater recovery and drainage

A line diagram of the water resource management at the construction site during different stages of building construction is presented in Fig. 10. Leakage prevention is the most important aspect of water conservation [25]. Spilled water is good enough to be reused in any construction activity. The wastewater generated shall be stored and reutilized in curing, Brick soaking, washing tools and equipment, and land settling [26]. The amount of wastewater may be significantly reduced during the finishing stage of the building construction. During the painting and polishing work, a huge amount of water is wasted in cleaning the tools and equipment. Such water is contaminated with oils, paints, and chemicals and is not fit for reuse, hence drained. If solvents are used in cleaning the tools and equipment a large amount of wastewater may be saved. The seasoning of construction activities, for example, land settlement in monsoon, roof slab casting in monsoon or in winters, plastering in winters, etc., may reduce the first-hand consumption of water to a great extent. The curing of roof slabs and wall plaster are water-intensive activities. If sprinklers are used for curing purposes, a considerable amount of water may be saved. The wastewater finally drained in the present study was about 2.35 L/ft^2. Such water may be effectively used for land settling.

Fig. 10 Water conservation and recovery during building construction

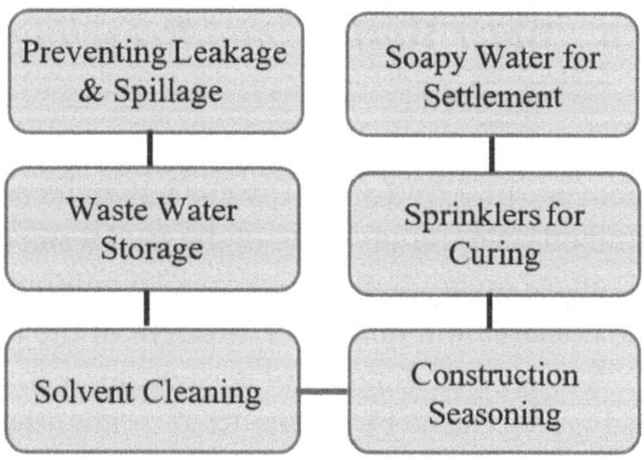

A few sustainable construction practices have been used to conserve the water resources in the building under study. Some additional practices (given in Fig. 10), which could not be used in the present building, may also be used to further reduce the WFP of the building.

5 Conclusions

The WFP of a three-floor residential building built on a plot of 199 m^2, having a covered area of 5018 ft^2, was estimated from the beginning of construction activities. The construction was divided into 16 stages and WFP of each stage was recorded. The following major conclusions are derived from the study.

1. Maximum water was consumed in Brickwork, Roof slab, plastering, foundation, flooring, excavation, and landfilling.
2. The total WFP of the building was 190,300 L, which is 37.94 L/ft^2.
3. By simple water conservation, recovery and reuse of wastewater, the WFP may be reduced by 4.59 L/ft^2.
4. A wastewater storage tank and controlled drainage of wastewater were effective to save 23,030 L of water. No spilled or runoff water was allowed to drain.
5. A few additional sustainable construction practices such as prevention of leakage, construction seasoning, use of sprinklers, solvent cleaning, etc., may further reduce the wastewater and hence, the WFP of the building.

Awareness and training of the public are required to a large extent to change the orthodox construction practices and construct low WFP buildings without any compromise on strength and quality. Water conservation at a construction site is possible with a minimum investment if it is considered during the planning and scheduling of construction stages. Unsustainable construction practices have to be controlled immediately to conserve the natural resources in the larger interest of the future generation.

6 Recommendations

The suggested water resource management strategy to reorder the construction stages in correlation with the ambient conditions to reduce the water consumption, recycle the water used for curing, and minimize the wastewater in washing and cleaning is expected to reduce the water footprint of building construction significantly. The suggested methods may be adopted irrespective of the building type or size. The state agencies and the architects should promote such sustainable construction practices to reduce the WFP of construction activities.

Acknowledgements The authors gratefully acknowledge resources and facilities given by the Mechanical Engineering Department, College of Engineering, King Khalid University, Asir-Abha KSA to conduct the study.

References

1. The Water Network (2020) What is blue green and grey water. https://thewaternetwork.com/question-0-y/what-is-blue-green-and-grey-water-6uuv13bt8lVovKyD7Andyw
2. Everblue (2013) The role of water efficiency in green building. https://everbluetraining.com/role-water-efficiency-green-building/
3. Water Saving Guide (2016) SFIC. https://ccsbestpractice.org.uk/wp-content/uploads/2016/11/Water-Saving-Guide.pdf
4. Sustainability (2020) Water conservation. https://sustainability.williams.edu/green-building-basics/water-conservation
5. Consulting Specifying Engineer (2012) 10 ways to save water in commercial buildings. https://www.csemag.com/articles/10-ways-to-save-water-in-commercial-buildings/
6. Inhabitat, GREEN BUILDING 101 (2014) Water efficiency, both inside and outside the home. https://inhabitat.com/green-building-101-water-efficiency/
7. Performance Services (2016) Benefits-of-water-reduction-in-the-building-process. https://www.performanceservices.com/benefits-of-water-reduction-in-the-building-process
8. Whole Building Design Guide (2018) A program for the National Institute of Building Sciences, the WBDG Sustainable Committee, Protect and Conserve Water WBDM. https://www.wbdg.org/design-objectives/sustainable/protect-conserve-water
9. Clean Leap (2015) Conserving the environment through sustainable construction. https://cleanleap.com/conserving-environment-through-sustainable-construction
10. Gerbens-Leenes PW, Hoekstra AY, Bosman R (2018) The blue and grey water footprint of construction materials: steel, cement and glass. Water Resour Ind 19:1–12
11. Bardhan S (2008) Baseline studies on embodied water foot-print of a RC frame constructed building in urban India. Int J Emerg Technol Adv Eng 5(6):171–174
12. Chen X, Li F, Li X, Hu Y, Hu P (2020) Evaluating and mapping water supply and demand for sustainable urban ecosystem management in Shenzhen, China. J Clean Prod 251:119754. https://doi.org/10.1016/j.jclepro.2019.119754
13. Mikulčić H, Baleta J, Klemeš JJ (2020) Sustainability through combined development of energy, water and environment systems. J Clean Prod 251. https://doi.org/10.1016/j.jclepro.2019.119727
14. Pham H, Kim SY, Van LuuT (2020) Managerial perceptions on barriers to sustainable construction in developing countries: Vietnam case. Environ Dev Sustain 22:2979–3003. https://doi.org/10.1007/s10668-019-00331-6
15. Kubba S (2012) Handbook of green building design and construction: LEED, BREEAM, and green globes. Butterworth-Heinemann, pp 361–384. https://doi.org/10.1016/b978-0-12-385128-4.00008-1
16. Das O, Bera P, Moulick S (2015) Water conservation aspects of green buildings. Int J Res Eng Technol 4:75–79
17. Al-Qawasmi J, Asif M, El Fattah AA, Babsail MO (2019) Water efficiency and management in sustainable building rating systems: examining variation in criteria usage. Sustainability 11:2416
18. EL-Nwsany RI, Maarouf I, El-Aal WA (2019) Water management as a vital factor for a sustainable school. Alex Eng J 58:303–313
19. Tirth V (2020) Nexus of water footprint with energy and GDP of Saudi Arabia and Solution for Sustainable Water Usage. In: Advances in water resources engineering and management. Springer, Singapore, pp 43–57

20. Treloar GJ, Crawford RH (2004) Assessing direct and indirect water requirements of construction. In: The 38th international conference of Architectural Science Association ANZAScA "Contexts of architecture", Launceston, Tasmania pp 186–191
21. Tirth V, Algarni S, Agarwal N, Saxena A (2019) Greenhouse gas emissions due to the construction of residential buildings in Moradabad, India. Appl Ecol Env Res 17:12111–12126
22. https://www.google.com/maps/place/28%C2%B051'56.8%22N+78%C2%B045'11.5%22E/@28.8656575,78.7527431,127m/data=!3m1!1e3!4m9!1m2!10m1!1e2!3m5!1s0x390af94de1b417cd:0x56d57aa31f7b5fb2!7e2!8m2!3d28.865775!4d78.7531831
23. CLIMATE-DATA.ORG (2020). https://en.climate-data.org/asia/india/uttar-pradesh/moradabad-5855/#climate-graph.
24. MapsofIndia (2020). https://www.mapsofindia.com/maps/india/climaticregions.htm
25. Hajji AM, Suprianto B, Ariestadi D (2018) Methods in water conservation as part of green building rating tools in Indonesia—case study. In: Design of integrated classrooms building in Universitas Negeri Malang, Indonesia. In: MATEC web of conferences, vol 204, p 04004
26. Sivasankar R, Sivasubramanian C, Malarvannan J, Jeganathan M, Balakumari M (2016) A study on water conservation aspects of green buildings. Life Sci Arch 2:832–836

Water Resources Availability and Its Teleconnection with Large Scale Climatic Oscillations Over Godavari River Basin

Jew Das and N. V. Umamahesh

Abstract The surface water resource, i.e., mostly streamflow/discharge, is the outcome of the complex hydrological and climatic interactions (directly and/or indirectly) which are modulated by the atmosphere–ocean circulation. In the present scenario of climate change, it is inevitable to study and analyse the surface water resources availability and their association with the low-frequency large scale climatic oscillations for better understanding and management of the water resources. In this context, the present chapter comprehensibly analyses the surface water resources availability over the second-largest basin, i.e., Godavari River basin in India, considering the 40 streamflow gauging stations. The surface water availability is assessed through constructing the Flow Duration Curve (FDC) for the monsoon flows over 40 different gauging stations. The variability and possible teleconnection of between monthly flow and Sea Surface Temperature (SST), Indian Ocean Dipole (IOD), and Indian Summer Monsoon Index (ISMI) are detected using wavelet coherence technique at 40 different stations. It is observed from the analysis that the flow during the monsoon periods varies greatly across the basin (from 0 to 26,700 m^3/s). From the wavelet analysis, it is observed that ISMI has significant association with the monthly streamflow at intra- and inter-annual scales over all the gauging stations. However, a uniform phase is not observed between the monthly streamflow and the three studied climatic oscillations. The present analysis enables hydrologists and water managers to understand the hydro-climatic process and their association with climatic oscillations along with the variability in the surface water resources over the Godavari Basin.

Keywords Flow duration curve · Godavari river basin · Indian summer monsoon index · Teleconnection · Wavelet analysis · Water availability

Supplementary Information The online version contains supplementary material available at (https://doi.org/10.1007/978-3-030-95786-5_15).

J. Das · N. V. Umamahesh (✉)
Department of Civil Engineering, National Institute of Technology Warangal, Warangal, Telangana 506004, India
e-mail: mahesh@nitw.ac.in

© Springer Nature Switzerland AG 2022

S. Yadav et al. (eds.), *Wastewater Assessment, Treatment, Reuse and Development in India*, Earth and Environmental Sciences Library,
https://doi.org/10.1007/978-3-030-95786-5_15

1 Introduction

In India, water has been recognised as one of the most essential natural sources for the sustainability of the ecosystems, development of economic status, recreational activities, etc. In particular, as mentioned by Ministry of Water Resources [1], different priority sectors for water allocations are (i) water for drinking, (ii) for irrigation, (iii) ecology, (iv) hydropower, (v) agro-industries ad non-agricultural industries, (vi) navigation. Therefore, integrated water resources management plays a vital role in managing and developing different natural resources in order to maximise the societal and economic aspects in an equitable manner. However, with the increasing pressure of climate change, population growth, industrialisation, and urbanisation, the demands to mitigate the requirements are increasing profoundly. Moreover, the uncontrolled exploitation/abstraction of water resources (both above and below ground) is leading to an imbalance between demand and supply.

Being a second most populated country [2], India accounts 18% of the total population but has only 4% of available water resources and 2.4% of land area (all values with respect to the global statistics). In addition, the water resources in India is intricately intertwined with the cultural, social and economic associations [3]. Unfortunately, India is undergoing a critical situation of acute water shortage in its history and millions of lives and livelihoods are under threat. With the terrifying figure, the present report by NITI Aayog [4] advocates that about 820 million population [5] over the twelve river basins have per capita water availability lower than 1000 m^3, nearly 82% of rural households don't have access to piped water supply, 70% of the available surface water is contaminated [6]. Hence, to plug the gap between the supply and demand by 2030, it is suggested an investment worth of $291 billion [7]. In addition, the future forecast of demand and supply of water in India is even more alarming. For instance, 40% of the total population will be deprived of accessing drinking water by 2030, the groundwater availability over 21 major cities will run out affecting 100 million people by 2020, and a loss of 6% of GDP as a result of water crisis by 2050 [8].

The intensifying pressure on the water sector in India is modulated by several crucial issues [9], such as (i) the significant variability in the precipitation over the country causes floods and droughts; (ii) uncontrolled groundwater exploitation; (iii) degradation of the quality of water as a result of water pollution; (iv) inefficient management of water resources; (v) poor water management law, the dispute among the states, growing financial crisis for development of water resources and safe drinking water; (vi) sea level rise and its intrusion to groundwater and inundation in coastal regions; (vii) implementation of water resources projects without considering the environmental sustainability, and holistic benefit to the people; (viii) improper maintenance of existing irrigation structure that decreases the irrigation efficiency; (ix) lack of awareness regarding the societal and economic value of water among the users; (x) a holistic and integrated approach to tackle the water related problems is lacking; (xi) the consequences due to the alteration of land use land cover modify the characteristics of surface water resources affecting the availability and distribution. In addition, inadequate institutional reforms and ineffective execution of present provisions also impact the efficiency of water service delivery. The present situation of the water crisis has led to conflicts amongst industrial, agriculture, and domestic sectors. As discussed earlier, these situations are likely to aggravate in the coming future and presently has become the most significant environmental and societal challenges in India.

In this light, much interest has been focused on water resources management and policy matters in India. In 1987, India adopted the first National Water Policy (NWP) to build and operate the multipurpose storage and diversion infrastructure to tackle drought and flood [10] and further revisions were adopted in 2002 and 2012 [9, 11]. The major provisions under the NWP are (i) recycling and planning of resources to ensure maximum availability; (ii) prioritise the project impacts on the environment and human settlement; (iii) regulation towards the groundwater overexploitation; (iv) setting up the water allocation priorities as drinking water, irrigation, hydropower, navigation, industrial and other uses; (v) water pricing keeping the benefits of small and marginal farmers in mind. Despite the limited available resources, India has taken steps to develop its water resources through large-scale investments in supporting and developing its water infrastructure. According to the World Bank [12] report, the storage capacity, irrigated area, and installed hydropower capacity in India are 200 BCM, 90 Mha, and 30,000 MW, respectively. Table 1 briefly explains the water sector in India. It should be noted that the datasets are collected from the Central Water Commission (CWC) [13].

The surface water resources in the country are classified as rivers, canals, tanks, ponds, reservoirs, lakes, brackish water, etc. In area-wise, the reservoirs have

Table 1 Water resources in India

Average annual precipitation (including snowfall)	4000 BCM
Geographical area	328.7 Million Hectare
Population in 2001	1028.74 Million
Population in 2011	1210.19 Million
Estimated annual precipitation	3669.35 BCM
Average annual water resources potential	1869 BCM
Per capita availability @ 2001	1816 Cubic meter
Per capita availability @ 2011	1544 Cubic meter
Utilisable surface water	690 BCM
Utilisable groundwater	431 BCM
Total Utilisable water	1121 BCM
Storage capacity (completed projects)	253.388 BCM
Storage capacity (under construction projects)	50.959 BCM

*BCM = Billion Cubic Meter = 10^9 m^3

maximum area of 2.9 Mha followed by tanks, lakes and ponds (about 2.4 Mha) (Source: Department of Animal Husbandry, Dairying and Fisheries; Ministry of Agriculture). The total length of the rivers flowing across the country is about 2 lakh km. Table 2 presents the river basin wise water resources availability along with the catchment area of each basin. It should be noted that the datasets are collected and reproduced from the Central Water Commission (CWC) [13].

It can be noted from the table that the average water resources potential is 1869 BCM with utilisable percentage is around 37% (i.e., 690 BCM). The Ganga–Brahmaputra–Meghna river basin is the largest basin based on the catchment area, i.e., about 11 lakh km^2. In addition, the catchment area of Godavari, Indus, Mahanadi, Krishna, and Narmada River basins' is ~one lakh km^2 or more. As far as the utilisable surface water resources are concerned, the ratio/proportion of the utilisable water to the average water resources is greater in the case of smaller river basins (catchment area wise). The minimum proportion is observed in case of Brahmaputra River basin. Out of 1869 BCM average available water resources, about 60% is available in Ganga–Brahmaputra–Meghna River basin followed by Godavari (5.9%), Krishna (4.2%), Indus (3.9%), Mahanadi (3.6%), Narmada (2.4%), and remaining 20.6% includes remaining river basins. Figures 1 and 2 represent the per capita availability (during 2010) and storage potential in India over different river basins, respectively.

From Fig. 2, it can be noted that Narmada has the highest per capita availability followed by Brahamani and Baitarani, Ganga, and Mahanadi. In 2010, the average water availability in the country was 1608 m^3 and it is projected that the per capita availability in the country will be 1140 m^3 in the year 2050, which is close to the water scarcity condition (less than 1000 m^3) according to the international agencies. The Cauvery, Pennar, Sabarmati and East Flowing Rivers are having per capita availability less than 500 m^3. Similarly, from Fig. 3, it is noticed that maximum storage capacity (including the projects under construction) lies in the Ganga River basin followed by Krishna, and Godavari. The storage capacity as a percentage of average annual flow is high for Pennar River basin (not presented in Fig. 3) and the percentage is more than 50% in case of Narmada, Tapi, and Krishna River basins.

The annual groundwater potential contributed through rainfall is about 342.43 km^3 that is 8.56% of the total annual rainfall of the country. In addition, canal irrigation contributed 89.46 km^3 making total replenishable groundwater resource to about 432 km^3. After providing 71 km^3 to domestic, industrial, and other uses, 361 km^3 is available for the irrigation purpose. The groundwater per capita availability over different river basins varies from ^{13,393} m^3 (Brahmaputra-Barak basin) to 300 m^3 (Sabarmati basin). Figure 3 represents the basin wise replenishable groundwater including the contribution from the canal irrigation and the datasets are obtained from Reddy [14]. Moreover, Fig. 4 describes the surface and groundwater resources availability in different states and the datasets are obtained from TERI [15].

From Fig. 3, it is observed that the Ganga River basin has the highest total replenishable groundwater resources followed by Godavari, Krishna, Brahmaputra and Indus. It can be noted from Fig. 4 that the percentage of surface water (groundwater) is highest in Odisha (West Bengal) over East, Maharashtra (Maharashtra) over West, Andhra Pradesh (Andhra Pradesh) over South, Uttar Pradesh (Uttar Pradesh) over

Table 2 Water resources potential over different river basins in India

River basin	Average water resources potential (BCM)	Utilisable surface water resources (BCM)	Catchment area (km^2)
Area of Inland drainage in Rajasthan	Negligible	–	–
Brahamani and Baitarni	28.48	18.3	51,822
Brahmaputra	537.24	24	194,413
Cauvery	21.36	19	81,155
East flowing rivers from Mahanadi to Pennar	22.52	13.1	86,643
East flowing rivers from Pennar to Kanyakumari	16.46	16.5	100,139
Ganga	525.02	250	861,452
Godavari	110.54	76.3	312,812
Indus	73.31	46	321,289
Krishna	78.12	58	258,948
Mahanadi	66.88	50	141,589
Mahi	11.02	3.1	34,842
Meghna	48.36	–	41,723
Minor river draining into Myanmar (Burma) and Bangladesh	31	–	36,302
Narmada	45.64	34.5	98,796
Pennar	6.32	6.9	55,213
Sabarmati	3.81	1.9	21,674
Subarnarekha	12.37	6.8	29,196
Tapi	14.88	14.5	65,145
West flowing rivers from Tadri to Kanyakumari	113.53	24.3	56,177
West flowing rivers from Tapi to Tadri	87.41	11.9	55,940
West flowing rivers of Kutch and Saurashtra including Luni	15.1	15	321,851
Total	1869.37	690.1	

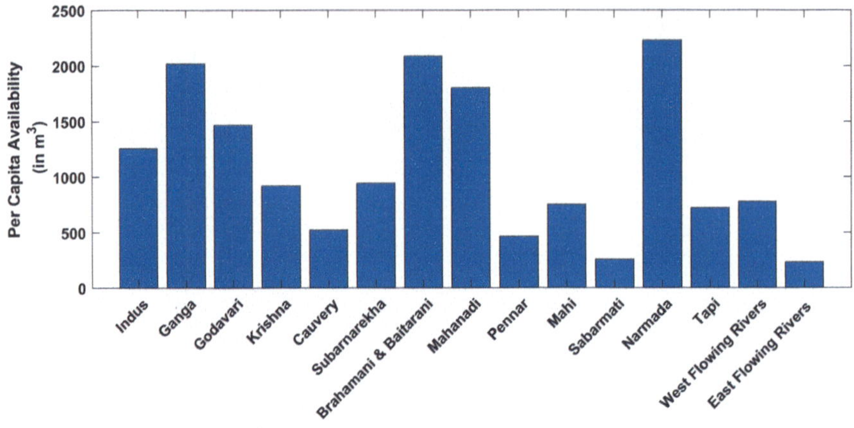

Fig. 1 Per capita water availability according to 2010 database

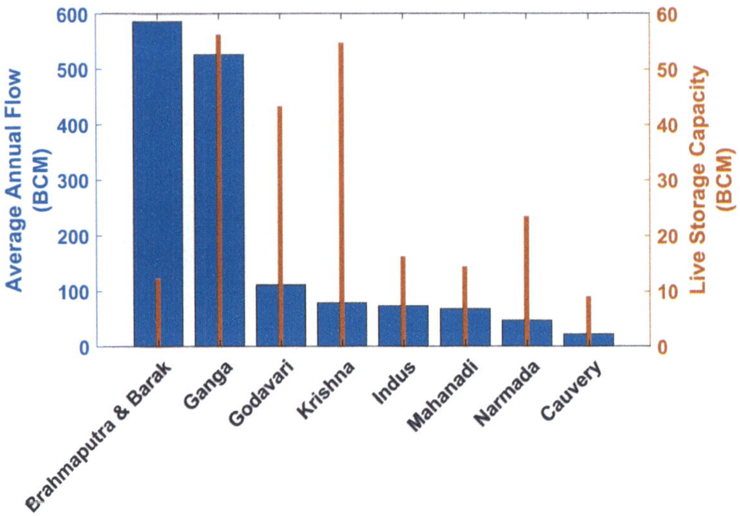

Fig. 2 Annual flow and storage potential over different river basins in India

North, Madhya Pradesh (Madhya Pradesh) over Central, and Assam (Assam) over North East region. In this light, it is well understood that the water resources planning decisions can have widespread significant impacts on environmental, social, and economic aspects. Therefore, it is vital to have the most relevant information before arriving at rational decisions to maximise the benefits. Moreover, in the climate change scenario, to understand the influence of external climatic factors on the available water resources enables to provide reliable information and can be a vital aid to

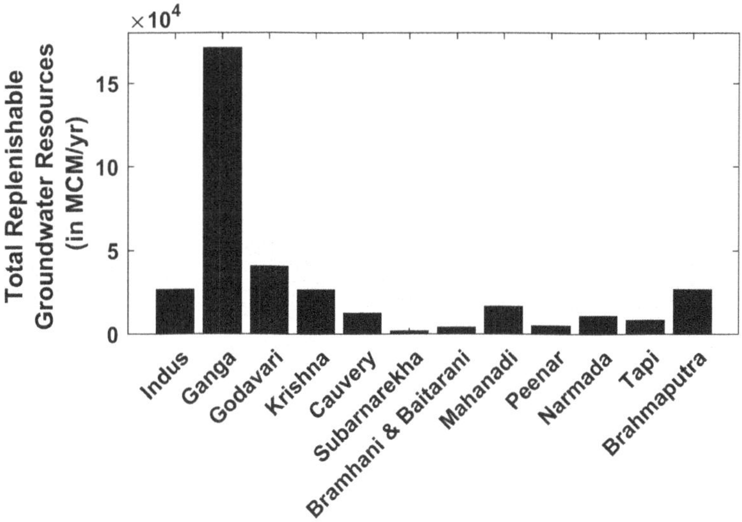

Fig. 3 Total replenishable groundwater resources over different river basins in India

the strategic management of the resource. The next section provides a brief discussion regarding the various external climatic factors and their association with the water resources (primarily surface water resources).

2 Streamflow and Large Scale Climatic Oscillation

The available/future availability of water resource (mostly surface water) is the outcome of the complex hydrological and climatic interactions (directly and/or indirectly) which are modulated by the atmosphere–ocean circulation. The availability of water resources is largely altered by the climate conditions at intra-annual, inter-annual and inter-decadal scale [16]. Therefore, characterising the large scale climatic oscillations and their possible connection with the hydrological variability improve forecasting of hydrological variables [17–19]. Moreover, it is evident from the previous study that the variability in the long record of hydrological series is associated with the variability in the climatic oscillations [20]. Analysing the association helps in the prediction of streamflow and provides guidelines to control the flood, management of agricultural activities, reservoir operation, etc.

The association of large-scale climatic oscillations to the streamflow is linked through the changes in precipitation, temperature, snow cover, and evapotranspiration [21]. With the increase in the population and urbanisation, the water resources availability and variability are tremendously affected [22, 23]. Moreover, the climatic variability in association with increasing greenhouse emissions results in modulating the global mean temperature [24]. Subsequently, the change in the evaporation rate

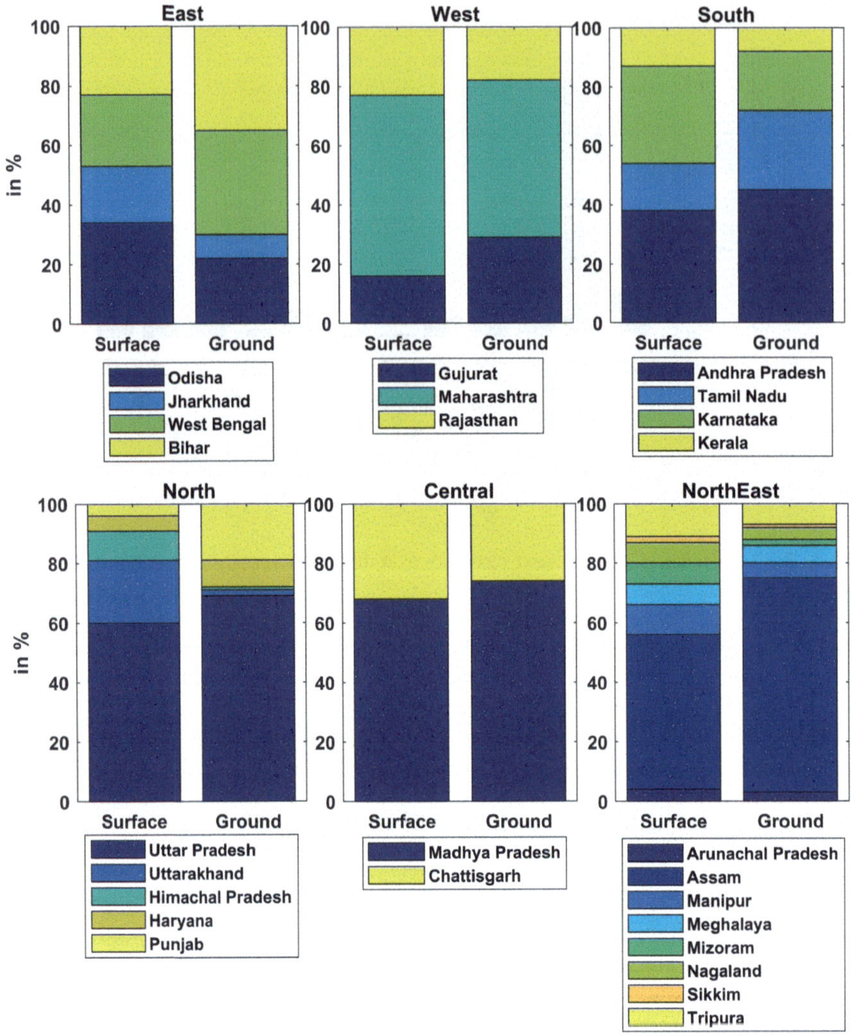

Fig. 4 Percentage of surface and groundwater resources availability over different provinces in India

and water vapour transport modify the hydrological cycle [25, 26]. Therefore, understanding spatio-temporal variability of streamflow under the influence of changeability in the climate is of paramount interest for sustainable management and planning of water resources [22, 27–29]. The association between the large-scale climatic variabilities and streamflow is expected to be highly complex and nonlinear in both space and time. Hydro-climatologists have made attempt to establish the association through teleconnection technique around the globe [30–33]. For instance, El Niño Southern Oscillation (ENSO), which is defined by the changes in the temperature of waters in the central and eastern tropical Pacific Ocean, was analysed to establish the relation with streamflow over Australia [34], over United States [35], over China [25], over Canada [36], over Central Europe [37], among others. Some studies found direct strong relationship between climate indices and streamflow, whereas, some studies tried to establish the connection through the association with the precipitation and temperature.

Similarly, the Indian Ocean Dipole (IOD) also plays a significant role in altering the water resources. In general, the IOD is characterised by Dipole Mode Index (DMI) and is derived from the difference in the sea surface temperature anomalies between western (50° E–70° E, 10° S–10° N) and eastern (90° E–110° E, 10° S–Equator) tropical Indian Ocean [38]. The positive phase of IOD brings precipitation over east Africa, India, Vietnam, and Bangladesh and at the same time droughts conditions are observed over Indonesia and Australia [38]. During the negative phase of IOD the opposite pattern is observed with increase in precipitation over Australia and Indonesia whereas droughts over Indian Subcontinent. Therefore, it is reasonable to expect the relationship between IOD and streamflow (directly and/or indirectly) [39]. The major part of the precipitation over India occurs during the monsoon season (i.e. from June to September), which is modulated by the Asian summer monsoon and particularly due to the Indian summer monsoon. Recent studies found significant teleconnections between the monsoon indices and precipitation over China [40] and India [41]. In this sense, Indian Summer Monsoon Index (ISMI) is a good indicator of the monsoon precipitation and can be helpful to analyse the variability in the streamflow as well. The ISMI is defined as the differences in the zonal winds at 850 hPa over (40° E–80° E, 5° N–15° N) and (70° E–90° E, 20° N–30° N) [42]. In addition, to these indices, studies have been performed to establish relationship between streamflow and the climate indices like Pacific Decadal Oscillation (PDO), Atlantic Multi-decadal Oscillation (AMO), Pacific North American Index, North Atlantic Oscillation (NAO) indices around the globe.

In this light, the present study analyses the streamflow teleconnection with the large scale climatic oscillations over the Godavari River basin in India. In this regard, Sea Surface Temperature (SST) as an indicator of ENSO, IOD, and ISMI are considered as climatic indices. Though the strength of the teleconnection decreases at smaller spatio-temporal scale, the influence at some particular geographical locations is still significant [43]. It is worth mentioning that the relation among the precipitation at upstream and the streamflow is highly non linear and the effect of precipitation on streamflow is obvious on a monthly scale [43]. Therefore, the association between the climatic indices and streamflow over the Godavari River basin is considered at monthly scale. The analysis is carried out at 40 different gauging stations over the Godavari. Moreover, for the reliability assessment of water availability, Flow Duration Curve (FDC) at 40 different stations are also analysed. The next section provides the detailed description about the study area and data used to perform the analysis.

3 Study Area and Data Used

In the present analysis, Godavari, the second largest river basin in India is considered as a study area to perform the teleconnection analysis with the large-scale climatic oscillations. The extent of the basin varies from 73°24′ E to 83°04′ E and 16°19′ N to 22°34′ N with a geographical area of 312,812 km^2. The river basin flows through 7 different states namely Maharashtra (48.7%), Andhra Pradesh (23.7%), Madhya Pradesh (7.8%), Odisha (5.7%), Karnataka (1.4%), Chhattisgarh (12.4%), and Puducherry (0.01%). The major part of the basin is covered with crop land. The average annual water potential over the basin is 110.54 BCM with utilizable surface water 76.30 BCM. The average annual runoff is 131 BCM over the basin. The basin has 8 sub-basins and 466 watersheds. The water resources projects over the basin include nearly 286 major and medium irrigation projects, and 15 hydro-electric projects. There are 1875 observation wells to assess the groundwater potential over the basin with a maximum number of 346 in the Wardha sub basin. The climate over the basin is tropical with evaporation losses varies from 1800 to 2400 mm over different parts of the basin [44]. The westernmost part of the basin is comparatively warmer than the eastern, central and northern parts. The most parts of the precipitation come during the monsoon months (i.e., from June to September). Figure 5 presents the location map of the Godavari River basin and the selected 40 hydrological stations across the basin. The exact location of the hydrological stations is presented in Table 3.

The streamflow datasets are taken from the Central Water Commission (CWC) online portal. For the analysis of precipitation variability, the daily precipitation data at a spatial resolution of 0.25° Latitude × 0.25° Longitude during 1901–2018 are procured from India Meteorological Department (IMD). Then, the daily precipitation series at each grid point is converted to the monthly series and the total length is divided into 2 parts, i.e., 1901–1950 and 1951–2018. As most of the precipitation occurs during the monsoon months, therefore, the spatio-temporal variability of precipitation is presented for the monsoon months only (Fig. 6).

From Fig. 6, it can be noted that the high value of precipitation is observed eastern part of the basin and the value decreases towards the west except a small portion over the Western Ghat region. Due to the spatial resolution of the dataset the Western Ghat region is not visible clearly. During the month of July, the spatial variability over the basin remains unchanged, whereas the precipitation amount is higher during 1901–1950 than 1951–2018. In June, there spatial variability of precipitation is significant over the basin. During 1901–1950, the higher precipitation depth is concentrated over the east and northeast parts of the basin. However, high precipitation amount is observed over the eastern part of the basin during 1951–2018. In the month of August, there is no significant spatial variability among the two durations; however, high precipitation depth is observed during 1951–2018. The precipitation depth during the month of September is high for the period 1901–1950. Therefore, the long-term variability in the precipitation during the monsoon months are going to affect the surface water variability and the precipitation variability can be attributed

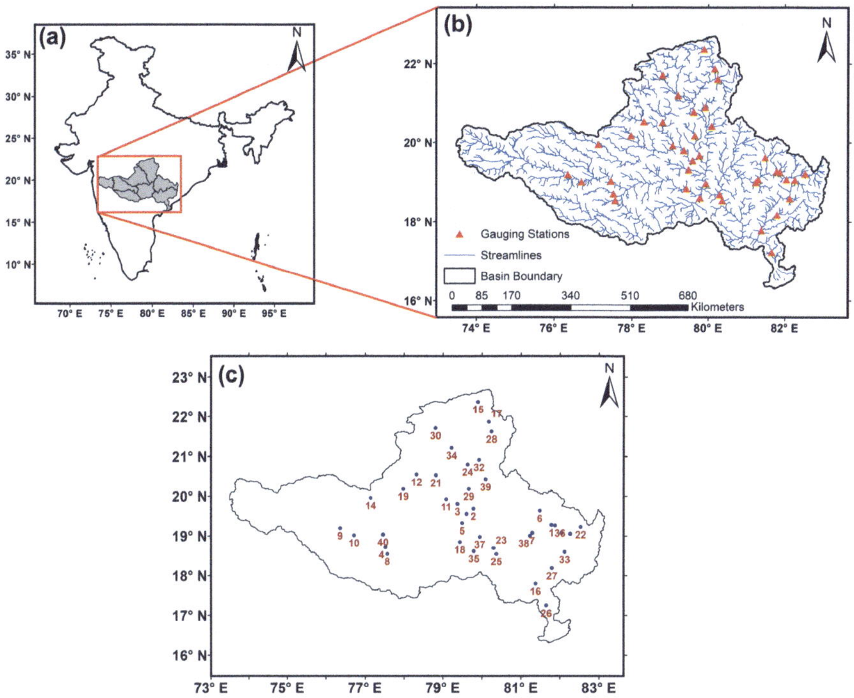

Fig. 5 **a** Location map of Godavari River basin and its sub-basin superimposed over India map; **b** Location of chosen 40 different streamflow gauging stations over Godavari River basin; **c** ID number of different streamflow gauging stations

duc to thc climatic and anthropogcnic changcability. In thc past study, thc prccipitation variability is analysed with respect to the large-scale climatic oscillations [41]; however, establishment of the direct connection between the streamflow variability and climatic oscillations is limited. In this sense, the next section discusses regarding the proposed methodology in the present analysis. The large-scale climatic oscillation datasets those are used to analyse the association with the streamflow are SST, IOD, and ISMI. The monthly oscillation datasets are downloaded from [45] for SST, [46] for IOD, and [47] for ISMI.

4 Methodology

To accomplish the case study, the methodology is presented in the form of a flow chart (Fig. 7).

The methodology includes the analysis of monsoon flow at 40 different selected stations to evaluate the water availability at different exceedance probability using the FDC. The large-scale climatic oscillations and their teleconnections with the

Table 3 Details of the streamflow gauging stations

ID. No.	Station name	Latitude	Longitude	Starting date	Ending date
1	Amabal	19.28	81.79	01-06-1993	31-12-2015
2	Asthi	19.69	79.79	01-06-1965	31-12-2015
3	Bamni	19.81	79.38	01-12-1965	31-12-2015
4	Betmogra	18.72	77.53	05-09-1986	31-12-2015
5	Bhatpalli	19.33	79.50	01-10-1986	31-12-2015
6	Cherribeda	19.64	81.49	01-06-1996	31-12-2015
7	Chindnar	19.08	81.30	06-06-1971	31-12-2015
8	Degloor	18.55	77.58	17-06-1987	31-12-2015
9	Dhalegaon	19.20	76.37	01-06-1965	31-12-2015
10	G.R.Bridge	19.02	76.72	17-06-1976	31-12-2015
11	Ghugus	19.93	79.09	01-12-1965	31-12-2015
12	Hivra	20.55	78.33	01-06-1987	31-12-2015
13	Jagdalpur	19.08	82.05	01-06-1965	31-12-2015
14	Kanhargaon	19.96	77.15	01-06-1992	31-12-2015
15	Keolari	22.38	79.90	01-11-1986	31-12-2015
16	Konta	17.80	81.38	01-12-1965	31-12-2015
17	Kumhari	21.88	80.18	01-06-1986	31-12-2015
18	Mancherial	18.85	79.44	01-06-1966	31-12-2015
19	Mangrul	20.19	77.99	01-11-1992	31-12-2015
20	Murthandi	19.06	82.27	01-12-1988	31-12-2015
21	Nandgaon	20.53	78.83	01-06-1986	31-12-2015
22	Nowrangpur	19.23	82.54	01-12-1965	31-12-2015
23	Pathagudem	18.70	80.30	01-06-1965	31-12-2015
24	Pauni	20.80	79.64	21-09-1964	31-12-2015
25	Perur	18.55	80.37	01-12-1965	31-12-2015
26	Polavaram	17.25	81.66	01-12-1965	31-12-2015
27	Potteru	18.19	81.80	01-06-1997	31-12-2015
28	Rajegaon	21.63	80.25	01-06-1986	31-12-2015
29	Rajoli	20.19	79.67	01-06-1986	31-12-2015
30	Ramakona	21.72	78.82	21-11-1986	31-12-2015
31	Sakmur_Sirpur	19.56	79.61	01-02-1968	31-12-2015
32	Salebardi	20.92	79.93	01-06-1986	31-12-2015
33	Saradaput	18.60	82.13	01-06-1970	31-12-2015
34	Satrapur	21.22	79.23	01-05-1986	31-12-2015
35	Somanpally	18.62	79.80	01-12-1966	31-12-2015
36	Sonarpal	19.27	81.88	01-12-1991	31-12-2015

(continued)

Table 3 (continued)

ID. No.	Station name	Latitude	Longitude	Starting date	Ending date
37	Tekra	18.98	79.95	01-06-1964	31-12-2015
38	Tumnar	19.01	81.24	01-12-1991	31-12-2015
39	Wairagarh	20.43	80.10	01-08-1992	31-12-2015
40	Yelli	19.04	77.47	01-06-1978	31-12-2015

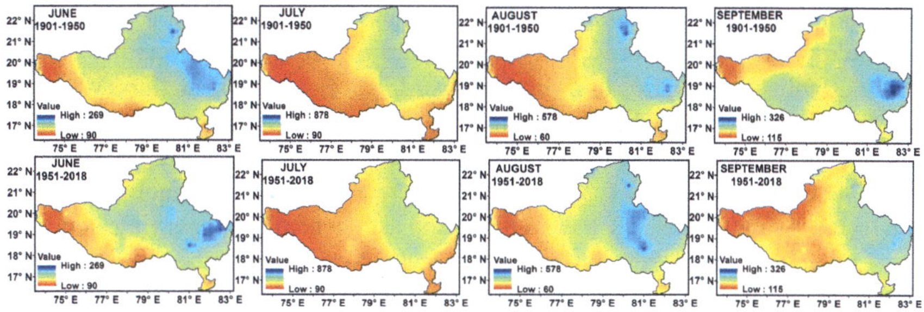

Fig. 6 Spatio-temporal variability of precipitation during different monsoon months

Fig. 7 Flow chart of the present study

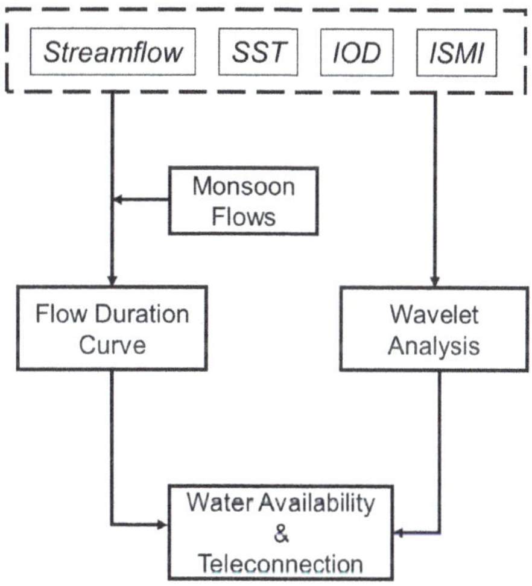

streamflow at different stations are performed through the wavelet analysis. The following sections describe the adopted methodology in details.

4.1 Wavelet Analysis: Background

For time series analysis and signal processing various methods have been adopted. In the present study, to perform the analysis and establish the relationship among the streamflow and climatic oscillations, wavelet analysis is used. In addition, FDC at each streamflow gauging point is plotted to analyse the water availability. The relationship between streamflow variability and large-scale climatic oscillations is more complex than a simple linear regression. The inter- and intra-annual variability of streamflow may be the possible reason for the complex association. Hence, the methodology that incorporates periodic attributes of streamflow variability can be used to establish the relationship. In this regard, wavelet transformation (WT) has been considered as one of the credible tools to analyse the signals/time series and is widely used in analysing the hydro-meteorological variables [41, 48–51]. Broadly, the WT is divided into two types, such as discrete WT and continuous WT. The scale factor in continuous WT has no limitation as it generates longer time series of $W(s, \tau)$ with τ and s define time shift and scale. Nevertheless, the scale factor in the discrete WT is dyadic [51]. More specifically, the scale factor in continuous WT follows arithmetic progression, whereas in case of discrete WT, it follows geometric progression. The transformation of the time series enables to detect signals and feature extraction [52]. Previous studies have revealed that the discrete WT has the capability to model the stationary time series over time than the non-stationary series with seasonality components [51]. The WT is a very powerful technique to model the time series and being widely used in time series analysis. The concept behind the WT is quite similar to the Fourier transforms (FT); however, WT provides much flexibility in capturing the all frequencies present in a time series [53].

The primary goal of the transformation is to decompose the original series into different time scales and the wavelet transformation can be defined as follow [54]

$$W(s, \tau) = \frac{1}{\sqrt{s}} \int_{-\infty}^{+\infty} x(t)\psi^*\left(\frac{t-\tau}{s}\right)dt \tag{1}$$

$$\psi_{(s \cdot \tau)}(t) - \frac{1}{\sqrt{s}}\psi\left(\frac{t-\tau}{s}\right) \tag{2}$$

where, $W(s, \tau)$ represents the wavelet transform with τ and s define time shift and scale, respectively. The wavelet function is presented by ψ and complex conjugate is denoted by ψ^*. With time shift 0 and scale 1, $\psi(t)$ denotes the basin or mother wavelet [55]. In the WT, the variability in scale enables to capture the long or short frequency events in a series and for scale more than 1, wavelet function corresponds to high frequency [55]. The larger value of $W(s, \tau)$ represents strong correlation between the time series and ψ [49]. In WT method, the phase describes the relative displacement among two signals (for e.g., in the present study streamflow and climatic oscillations) and often called as phase difference. Usually the phase difference is denoted by degree. The in-phase signals will have no phase difference among them;

whereas the phase difference of 180° suggests anti-phase among the time series. In hydrological studies, the use of Morlet wavelet (φ_0) is extensive and can be defined as [56]

$$\varphi_0(\theta) = \pi^{-1/4} e^{i\omega_0\theta} e^{-\theta^2/2} \tag{3}$$

where, ω_0 and θ represent dimensionless frequency and time, respectively. The frequency controls the resolution of scale and time, i.e. with high frequency, resolution of time decreases and at the same time scale increases. To identify the long term fluctuation in the time series, it is recommended to use lower resolution and vice versa [50]. As Morlet wavelet is localised in scale, high resolution in the frequency can be obtained [55]. In addition, application of Morlet wavelet enables to segregate the phase and amplitude in any time series [50, 57].

4.2 Wavelet Coherence Analysis

To evaluate the linkage between the two time series within the time–frequency space, wavelet coherence is used and evaluated through the correlation coefficient that varies between 0 and 1. The wavelet coherence evaluates the degree of coherence of cross wavelet transform over time frequency space [40]. The wavelet coherence coefficient ($R^2(s, \tau)$) can be computed using the Eq. 4.

$$R^2(s, \tau) = \frac{\left|S\left(s^{-1} W_{xy}(s, \tau)\right)\right|^2}{S\left(s^{-1}|W_x(s, \tau)|^2\right) \cdot S\left(s^{-1}\left|W_y(s, \tau)\right|^2\right)} \tag{4}$$

$W_{xy}(s, \tau)$ defines the cross wavelet transform among the two series, S presents the smoothing operator (Eq. 5) [40]. The footprint of smoothing operator is similar as the wavelet used and in the present study, for the Morlet wavelet the smoothing operator is defined in Eq. 6 [58].

$$S(W) = S_{scale}(S_{time}(W(s, \tau))) \tag{5}$$

$$\begin{aligned} S_{time}(W)|_s &= \left(W(s, \tau) \cdot c_1^{\frac{-t^2}{2s^2}}\right)|_s \\ S_{scale}(W)|_\tau &= \left(W(s, \tau) \cdot c_2 \prod(0.6s)\right)|_\tau \end{aligned} \tag{6}$$

The smoothing along scale axis is represented by S_{scale} and time is denoted by S_{time}. The rectangular function is defined by $\prod$ and c_1, c_2 are normalisation constants. The value 0.6 is the scale decorrelation length determined empirically [58].

In the present study, the wavelet coherence is accomplished using the Monte Carlo simulations using 1000 ensemble surrogate pairs with the lag-1 autoregressive coefficients as the input datasets. It is established that the resolution has the substantial impact on computing the scale smoothing and significance level. Therefore, the number of scale per octave should be carefully chosen to capture the rectangular shape of the scale smoothing operator with minimum computation time [41]. With reference to the past studies [40, 58], 12 scales per octave are chosen and the analysis is carried out at a significance level of 5%. The matlab source code to perform the analysis can be obtained from http://noc.ac.uk/using-science/crosswavelet-wavelet-coherence. It is worth mentioning that the analysis is performed over all the 40 gauging stations across Godavari River basin.

4.3 Flow Duration Curve

In addition to the wavelet analysis, flow duration curves (FDC) for all the gauging stations are plotted to understand the water availability during the observed period as mentioned in Table 3. The FDC is a cumulative frequency curve that presents the percent of time a given discharge is equalled or exceeded during a given period [59]. Precisely, FDC illustrates the relationship between frequency and streamflow magnitude [60]. In the present study, monsoon flow is considered at each gauging point to develop FDC. Initially, the monsoon streamflow values are arranged in descending order and each value is given rank (M), i.e. starting 1 for the highest and n to the lowest value for a data series of length n. Then, the probability of exceedance (P) at each streamflow value is computed as $P = \left[M/n + 1\right] * 100$. The shape of the FDC is significant in characterising the basin and stream characteristics. In high flow region, the shape of FDC specifies the type of flood flow regime over the basin. However, over low flow region, the shape of the FDC indicate the ability of the basin to sustain during the dry seasons.

5 Results and Discussion

The precipitation anomalies associated with the climatic oscillations could influence the hydrology of river basin [39]. The climatic oscillations and its non-linear large-scale climate forcing at large spatial extent directly effects the streamflow over river basin. Therefore, teleconnections between the monthly precipitation variability and the selected three large-scale climatic oscillations are performed. For brevity, 8 out of 40 gauging stations and their association with the selected oscillations is presented in Fig. 8. However, the teleconnection of remaining 32 stations are present in the supplementary file.

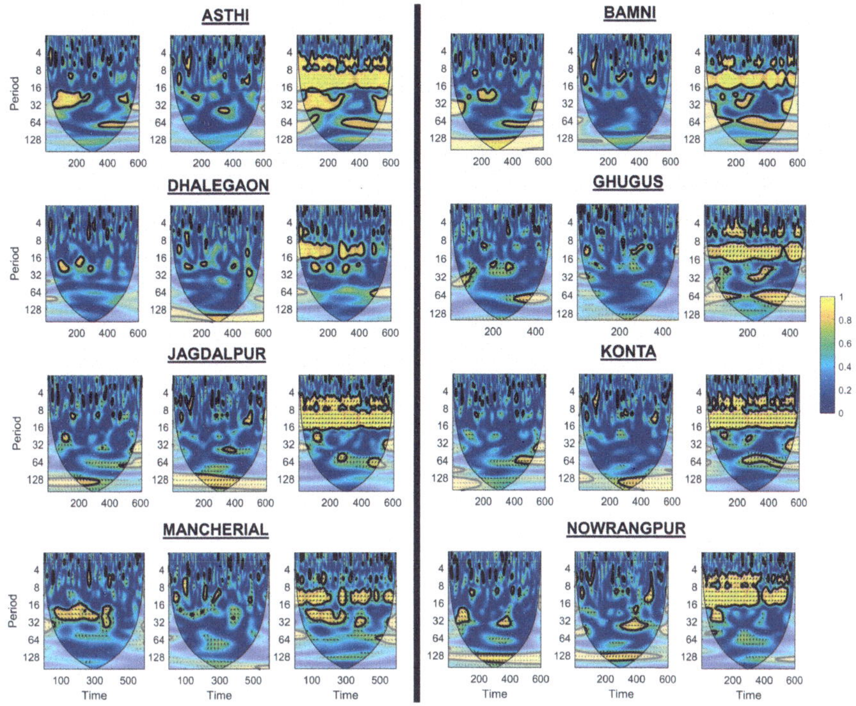

Fig. 8 Wavelet coherence between monthly streamflow and selected large-scale climatic oscillations at 8 different streamflow gauging stations. The order from left towards right specifies SST, IOD, and ISMI. The significant level is presented as thick contour

5.1 Association Between Monthly Streamflow and SST

In Fig. 8, the wavelet coherence plot of SST with the streamflow of a particular station is presented towards left. Similarly, the centre and the right plots belong to the IOD and ISMI, respectively. The abscissa of the plot represents the monthly time scale and the ordinate presents the different period (temporal scale) at which the correlation between the climatic oscillations and streamflow is evaluated. The abscissa range at different stations vary because the length of the dataset of streamflow at each station is not same. It should be noted that the above-said pattern is also followed in all the figures presented in the supplementary material. From Fig. 8, it can be noted that for the stations Asthi, and Mancherial the association between SST and streamflow is significant at inter-annual scale (16–32 months). However, at other stations (in Fig. 8), the teleconnection is scattered over the time scale. Over Bamni, and Asthi, intermittency of other scales is also noticed during different years, while high coherence is noticed at different time scales in different years. However, stations like Jagdalpur, Konta, and Ghugus, the coherence is not that much significant at any time scale for the total length of datasets. From Fig. S1 (refer to the

Supplementary Material), Pauni, Polavaram, Somanpally, and Perur stations have significant correlation with SST at a time scale of 16–32 months. In addition, the coherence structure at Peruru, Tekra, and Polavaram follows the same pattern over the entire length of dataset. Pathagudem and Chindnar show no significant association with SST. Similarly, from Fig. S2, except Yelli station, the monthly streamflow at rest of the stations show in significant teleconnection with SST. In Fig. S3, significant correlation at 64 months is observed over Wairagarh, Bhatpalli, and Hivra over the complete length of the dataset. However, at Degloor station, the teleconnection is noticed during 16–64 months. It can be observed form Fig. S4 that the gauging stations namely Kumhari, Salebardi, and Ramakona don't have significant correlation with SST at any time period. However, Nandagaon and Rajoli stations show significant association with SST during 32–64 months. In a recent study by Das et al. [41], it is observed that the association between SST and monthly precipitation over West Central India (WCI) is significant at 20–54 months during 1951–2015. In the present analysis, it is observed that the association of SST with streamflow at most of the stations come within the range of teleconnection in case of precipitation. This suggests that, though not completely (as streamflow is also modulated by catchment characteristics, manmade structures, potential evapotranspiration, etc.), the SST teleconnection with the streamflow can be attributed through the connection with precipitation.

5.2 Association Between Monthly Streamflow and IOD

Like SST, association between IOD and streamflow is also examined over 40 different stations over Godavari River basin. For instance, IOD over 8 different hydrological stations are presented in Fig. 8 (i.e. the middle one for each station). It can be noticed from the fig. that except Jagdalpur, Nowrangpur stations there is no significant association between IOD and streamflow is observed. For the two stations, high significant coherence is observed at inter-decadal scale (128 months). Similarly, in Fig. S1, the stations such as Perur, Tekra, Pauni show insignificant association with IOD. However, for Pathagudem (128 months), Sakmur-Sirpur (64 months), and Chindnar (128 months) linkage with IOD is significant. In case of Somanpally, the significant association with IOD is highly scattered in nature. In Fig. S2, the significant association is scattered in nature at different time periods is observed over Yelli, Cherribeda, and Amabal stations. However, over Saradaput, and Kanhargaon stations, the link between IOD and streamflow is significant at 128 and 64 months, respectively. In Fig. S3, over all the 8 gauging stations, intermittency of other scales is noticed over during different years, while high coherence is noticed at different time scales in different years and over different stations (however, over all association with IOD seems to be insignificant). In Fig. S4, Murthandi, Rajegaon, and Ramakona have significant association at 32 month scale during the initial years. However, other 5 gauging points have no significant linkage with the monthly streamflow.

5.3 Association Between Monthly Streamflow and ISMI

As discussed, the availability of water resources is highly depending on the precipitation during the monsoon seasons. Moreover, the monsoon rainfall is modulated by the ISMI. Therefore, ISMI will also have the influence on the streamflow availability. In this light, the teleconnection between the streamflow and ISMI is carried out at 40 different gauging stations over Godavari River basin. From Fig. 8, it can be noted that Asthi, Bamni, Mancherial, Ghugus, and Konta stations have intra as well as inter-annual linkage with ISMI. As monsoon precipitation has seasonal behaviour and hence it is presumed that ISMI will have significant association with streamflow at intra-annual scale. The recent study by Das et al. [41] showed the significant coherence of ISMI with monthly precipitation over different homogeneous meteorological regions in India. In the present analysis, it is also found that (including all the supplementary figures) all the gauging stations have significant teleconnection at intra-annual scale. However, some of the station exhibits association at inter-annual scale also. It is also noticed that the pattern of the association is not scattered over the total time length; however, in most of the cases the coherence is continuous over the entire time scale.

5.4 Water Availability Over Godavari River

The water availability is at each station is computed using the FDC and is presented in Fig. 9. The FDC is plotted for the monsoon season as most of the precipitation occurs during that period.

It is noted from the figure that the flow during the monsoon periods varies greatly across the basin (from 0 to 26,700 m^3/s). FDC is an important technique to analyse the water availability for different purposes like irrigation, hydropower generation, etc. In general, Q75 (i.e., flow that is equalled or exceeded for 75% of the flow record) is used to analyse the irrigation potential based on the water availability. Similarly, Q90 and Q95 are significant to represent the low flow parameter to assess the river water quality and hydropower generation. However, the very low discharge is highly sensitive to the proportional variability between the natural flow and the alteration due to man-made water infrastructures. Therefore, Q75, Q90, and Q95 are extracted from the FDC for each gauging station. According to the water availability at Q75, the top four stations with maximum Q75 among 40 stations are Polavaram, Perur, Tekra, and Pathagudem (presented in descending order). Moreover, the same order is also observed in case of Q90 and Q95 during the monsoon season. The percentage of station for different ranges of streamflow during the monsoon season is also plotted and presented in Fig. 10.

It is noticed from the figure that the percentage of station below 50 m^3/s is high for Q75, Q90, and Q95 flows. In particular, more than equal to 50% of the total stations have streamflow 50 m^3/s. However, only 10% (for Q75), 5% (for Q90), and

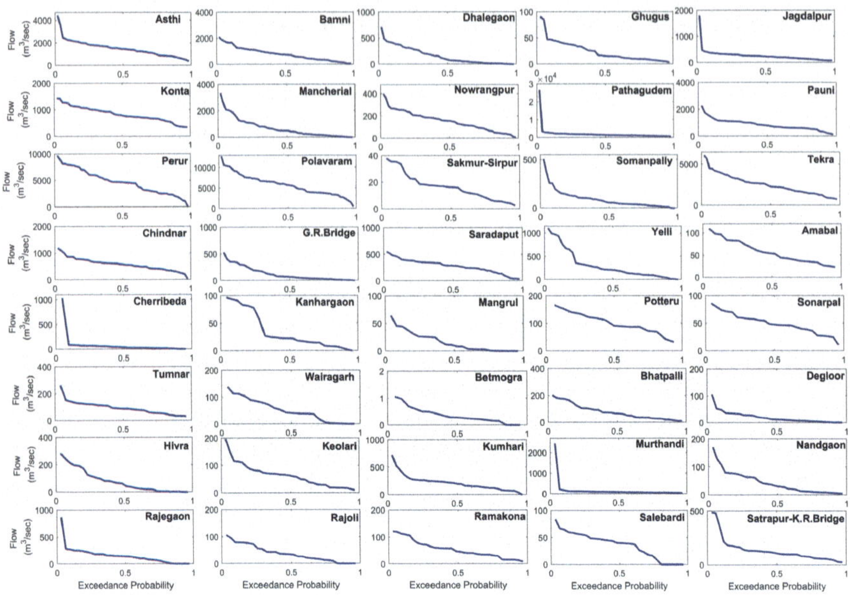

Fig. 9 Flow duration curve for the 40 selected stations over Godavari River basin

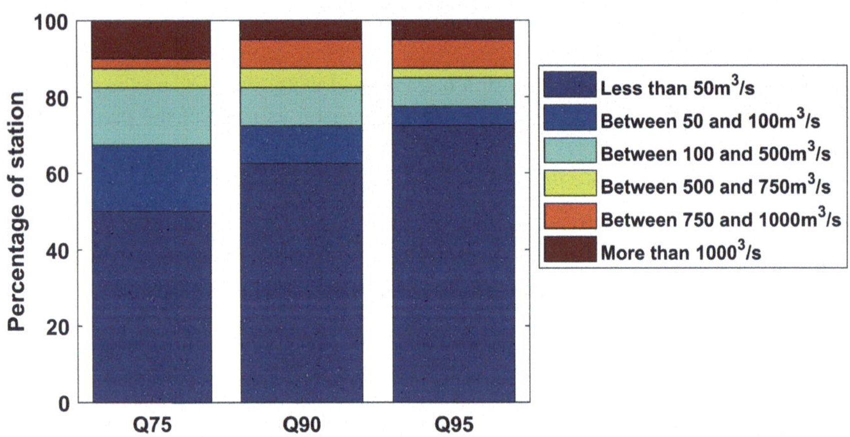

Fig. 10 Percentage of stations for different ranges of streamflow during the monsoon season for Q75, Q90, and Q95 flows

5% (for Q95) stations have flow more than 1000 m^3/s. Therefore, it is a matter of great concern for the river basin management point of view to build a sustainable and resilience water infrastructure to assure the water availability in the changing scenario of climate change and fulfil the demand of growing population.

6 Concluding Remarks

The present chapter, discusses various aspects of large-scale climatic variability on streamflow. Moreover, the water availability over second largest river basin in India is analysed using the FDC. It is observed from the analysis that the flow during the monsoon periods varies greatly across the basin (from 0 to 26,700 m^3/s). From the analysis, it is found that mostly the ISMI has significant teleconnection with the streamflow over all the stations in the basin. However, a uniform phase is not observed between the monthly streamflow and the three studied climatic oscillations. The present analysis enables hydrologists and water managers to understand the hydro-climatic process and their association with climatic oscillations along with the variability in the surface water resources over the Godavari Basin.

Water availability and its sustainable management in the climate change and population growth scenarios require proper adaptation. Here, the adaptation refers to the adjustment in socio-economic and ecological system in response to the present and projected climate change effects in order to alleviate the adverse impact and take advantages of new opportunities [61]. The adjustment towards the adaptation can be protective or opportunistic. Therefore, adaptation can include actions by individuals and communities, from a simple water harvesting structure in a community to new coastal infrastructure that can accommodate future sea level rise. Neil Adger et al. [62] advocated that successful implementation of adaptation strategy incorporates certain criteria like effectiveness in adaptation, efficiency of adaptation, equity and legitimacy in adaptation. India with an agrarian economy has severe threat on the water resources in terms of drought, flood, water quality, groundwater recharge etc. Despite of implementation of different policies by the government at various scales, water laws continues to remain inadequate and inconsistent in twenty-first century [63]. The reason is attributed to the various elements at local, regional, national, and international levels and their implementation issues.

7 Recommendations

As a paramount interest, the water resources and its variability in the context of large-scale climatic variabilities should be assessed over different river basins to devise sustainable water resources management plans. In addition to the present study, a greater number of large-scale climatic variabilities can be selected to assess the variabilities of not only streamflow but also other hydro-climatic variables around the globe.

In order to combat the climate change impact on water resources, development in the technical aspects, improvement in the understanding and computational facilities, the estimation in the probable variability in the foreseeable future need considerable research activities for robust adaptation strategies and policies. The multidisciplinary integrated approach will enable to make scientific decisions on water and clime change policy [64]. In addition, to improve the adaptation strategy and policy, the researchers should make significant improvement in climate change assessment and identify the suitable and robust approaches by involving key stakeholder organisations and local communities. In this light, significant steps should be taken to avail the quality datasets, analysis of uncertainty, multi criteria based management practices and strategy, assessment of vulnerability and most essentially, establishing the linkage between the academician and policy makers will enable to develop a climate resilient nation as climate change impacts are highly underutilized in decision and policy making.

References

1. Ministry of Water Resources (2002) National water policy
2. Census of India (2011) Size, growth rate and distribution of population
3. UNICEF (2013) Water in India: situation and prospects. UNICEF, FAO Saci-WATERs, New Delhi, India
4. NITI Aayog (2019) Composite water management index
5. Central Water Statistics (2015) Water and related statistics
6. Water Aid (2018) The water gap: the state of the world's water 2018
7. ASSOCHAM (2017) Investments worth $291 bn needed to plug water demand-supply gap in India: study
8. NITI Aayog (2018) Composite water management index
9. Government of India (2012) National water policy. New Delhi
10. Government of India (1987) National water policy. New Delhi
11. Government of India (2002) National water policy. New Delhi
12. World Bank (2005) India's water economy: bracing for a turbulent future. Rep No 34750-IN
13. Central Water Commission (2013) Water and related statistics
14. Reddy MS (1998) Five decades of water resource development in India
15. TERI (2017) Study of assessment of water foot prints of India's long term energy scenarios. New Delhi
16. Ionita M, Lohmann G, Rimbu N, Chelcea S (2012) Interannual variability of Rhine river streamflow and its relationship with large-scale anomaly patterns in spring and autumn. J Hydrometeorol 13:172–188. https://doi.org/10.1175/JHM-D-11-063.1
17. Croley TE, Luukkonen CL (2003) Potential effects of climate change on ground water in Lansing, Michigan. J Am Water Resour Assoc 39:149–163. https://doi.org/10.1111/j.1752-1688.2003.tb01568.x
18. Souza Filho FA, Lall U (2003) Seasonal to interannual ensemble streamflow forecasts for Ceara, Brazil: applications of a multivariate, semiparametric algorithm. Water Resour Res 39. https://doi.org/10.1029/2002WR001373
19. Das J, Jha S, Goyal MK (2020) Non-stationary and copula-based approach to assess the drought characteristics encompassing climate indices over the Himalayan states in India. J Hydrol 580:124356. https://doi.org/10.1016/j.jhydrol.2019.124356
20. Arnell NW, Krasovskaia I, Gottschalk L (1993) River flow regimes in Europe
21. Schulte JA, Najjar RG, Li M (2016) The influence of climate modes on streamflow in the Mid-Atlantic region of the United States. J Hydrol Reg Stud 5:80–99. https://doi.org/10.1016/j.ejrh.2015.11.003
22. Das J, Umamahesh NV (2018) Spatio-temporal variation of water availability in a river basin under CORDEX simulated future projections. Water Resour Manag 32:1399–1419. https://doi.org/10.1007/s11269-017-1876-2

23. Das J, Umamahesh NV (2017) Uncertainty and nonstationarity in streamflow extremes under climate change scenarios over a river basin. J Hydrol Eng 22.https://doi.org/10.1061/(ASCE)HE.1943-5584.0001571
24. IPCC (2007) Climate change 2007: the physical science basis. Contribution of Working Group I to the Fourth Assessment Report of the Intergovernmental Panel on Climate Change
25. Zhang Q, Xu C, Jiang T, Wu Y (2007) Possible influence of ENSO on annual maximum streamflow of the Yangtze River, China. J Hydrol 333:265–274. https://doi.org/10.1016/j.jhydrol.2006.08.010
26. Menzel L, Bürger G (2002) Climate change scenarios and runoff response in the Mulde catchment (Southern Elbe, Germany). J Hydrol 267:53–64. https://doi.org/10.1016/S0022-1694(02)00139-7
27. Qaiser K, Ahmad S, Johnson W, Batista JR (2013) Evaluating water conservation and reuse policies using a dynamic water balance model. Environ Manag 51:449–458. https://doi.org/10.1007/s00267-012-9965-8
28. Dawadi S, Ahmad S (2013) Evaluating the impact of demand-side management on water resources under changing climatic conditions and increasing population. J Environ Manag 114:261–275. https://doi.org/10.1016/j.jenvman.2012.10.015
29. Wu G, Li L, Ahmad S et al (2013) A dynamic model for vulnerability assessment of regional water resources in arid areas: a case study of Bayingolin, China. Water Resour Manag 27:3085–3101. https://doi.org/10.1007/s11269-013-0334-z
30. Kahya E, Dracup JA (1994) The influences of Type 1 El Nino and La Nina events on streamflows in the Pacific southwest of the United States. J Clim 7:965–976. https://doi.org/10.1175/1520-0442(1994)007%3c0965:TIOTEN%3e2.0.CO;2
31. Rogers JC, Coleman JSM (2003) Interactions between the Atlantic Multidecadal Oscillation, El Niño/La Niña, and the PNA in winter Mississippi Valley stream flow. Geophys Res Lett 30. https://doi.org/10.1029/2003gl017216
32. Kalra A, Li L, Li X, Ahmad S (2013) Improving streamflow forecast lead time using oceanic-atmospheric oscillations for Kaidu River basin, Xinjiang, China. J Hydrol Eng 18:1031–1040. https://doi.org/10.1061/(ASCE)HE.1943-5584.0000707
33. Sagarika S, Kalra A, Ahmad S (2015) Interconnections between oceanic-atmospheric indices and variability in the U.S. streamflow. J Hydrol 525:724–736. https://doi.org/10.1016/j.jhydrol.2015.04.020
34. Chiew FHS, Piechota TC, Dracup JA, McMahon TA (1998) El Nino/Southern oscillation and Australian rainfall, streamflow and drought: links and potential for forecasting. J Hydrol 204:138–149. https://doi.org/10.1016/S0022-1694(97)00121-2
35. Barlow M, Nigam S, Berbery EH (2001) ENSO, pacific decadal variability, and U.S. summertime precipitation, drought, and stream flow. J Clim 14:2105–2128. https://doi.org/10.1175/1520-0442(2001)014%3c2105:EPDVAU%3e2.0.CO;2
36. Nalley D, Adamowski J, Biswas A et al (2019) A multiscale and multivariate analysis of precipitation and streamflow variability in relation to ENSO, NAO and PDO. J Hydrol 574:288–307. https://doi.org/10.1016/j.jhydrol.2019.04.024
37. Steirou E, Gerlitz L, Apel H, Merz B (2017) Links between large-scale circulation patterns and streamflow in Central Europe: a review. J Hydrol 549:484–500. https://doi.org/10.1016/j.jhydrol.2017.04.003
38. Saji NH, Goswami BN, Vinayachandran PN, Yamagata T (1999) A dipole mode in the tropical Indian Ocean. Nature 401:360–363. https://doi.org/10.1038/43854
39. Sahu N, Behera SK, Yamashiki Y et al (2012) IOD and ENSO impacts on the extreme streamflows of Citarum river in Indonesia. Clim Dyn 39:1673–1680. https://doi.org/10.1007/s00382-011-1158-2
40. Chang X, Wang B, Yan Y et al (2019) Characterizing effects of monsoons and climate teleconnections on precipitation in China using wavelet coherence and global coherence. Clim Dyn 52:5213–5228. https://doi.org/10.1007/s00382-018-4439-1
41. Das J, Jha S, Goyal MK (2020) On the relationship of climatic and monsoon teleconnections with monthly precipitation over meteorologically homogenous regions in India:

wavelet & global coherence approaches. Atmos Res 238:104889. https://doi.org/10.1016/j.atmosres.2020.104889
42. Wang B (2000) Commentary and analysis: comments on "choice of south Asian summer monsoon indices." Bull Am Meteorol Soc 81:821–822. https://doi.org/10.1175/1520-0477(2000)081%3c0821:CAA%3e2.3.CO;2
43. Maity R, Nagesh Kumar D (2008) Basin-scale stream-flow forecasting using the information of large-scale atmospheric circulation phenomena. Hydrol Process 22:643–650. https://doi.org/10.1002/hyp.6630
44. CWC (1999) Reassessment of water resources potential of India
45. SST Data http://www.esrl.noaa.gov/psd/gcos_wgsp/Timeseries/Data/nino34.long.anom.data
46. IOD Data http://www.jamstec.go.jp/frcgc/research/d1/iod/DATA/dmi.monthly.txt
47. ISMI Data http://apdrc.soest.hawaii.edu/projects/monsoon/seasonal-monidx.html
48. Araghi A, Martinez CJ, Adamowski J, Olesen JE (2019) Associations between large-scale climate oscillations and land surface phenology in Iran. Agric For Meteorol 278:107682. https://doi.org/10.1016/j.agrformet.2019.107682
49. Adamowski J, Sun K (2010) Development of a coupled wavelet transform and neural network method for flow forecasting of non-perennial rivers in semi-arid watersheds. J Hydrol 390:85–91. https://doi.org/10.1016/j.jhydrol.2010.06.033
50. Araghi A, Mousavi-Baygi M, Adamowski J, Martinez C (2017) Association between three prominent climatic teleconnections and precipitation in Iran using wavelet coherence. Int J Climatol 37:2809–2830. https://doi.org/10.1002/joc.4881
51. Araghi A, Mousavi Baygi M, Adamowski J et al (2015) Using wavelet transforms to estimate surface temperature trends and dominant periodicities in Iran based on gridded reanalysis data. Atmos Res 155:52–72. https://doi.org/10.1016/j.atmosres.2014.11.016
52. Kreyszig E (2011) Advanced engineering mathematics. Wiley
53. Percival DB, Walden AT (2000) Wavelet methods for time series analysis. Cambridge University Press
54. Adamowski K, Prokoph A, Adamowski J (2009) Development of a new method of wavelet aided trend detection and estimation. Hydrol Process 23:2686–2696. https://doi.org/10.1002/hyp.7260
55. Mallat S (2009) A wavelet tour of signal processing. Elsevier
56. Maraun D, Kurths J (2004) Cross wavelet analysis: significance testing and pitfalls. Nonlinear Process Geophys 11:505–514. https://doi.org/10.5194/npg-11-505-2004
57. Grinsted A, Moore JC, Jevrejeva S (2004) Application of the cross wavelet transform and wavelet coherence to geophysical time series. Nonlinear Process Geophys 11:561–566. https://doi.org/10.5194/npg-11-561-2004
58. Torrence C, Compo GP (1998) A practical guide to wavelet analysis. Bull Am Meteorol Soc 79:61–78. https://doi.org/10.1175/1520-0477(1998)079%3c0061:APGTWA%3e2.0.CO;2
59. Searcy JK (1959) Flow-duration curves. US Government Printing Office
60. Vogel R, Fennessey N (1995) Flow duration curves II: a review of applications in water resources planning. J Am Water Resour Assoc 31:1029–1039. https://doi.org/10.1111/jawr.1995.31.issue-6
61. McCarthy JJ, Canziani OF, Leary NA et al (2001) Climate change 2001: impacts, adaptation, and vulnerability: contribution of Working Group II to the third assessment report of the Intergovernmental Panel on Climate Change. Cambridge University Press
62. Neil Adger W, Arnell NW, Tompkins EL (2005) Successful adaptation to climate change across scales. Glob Environ Chang 15:77–86. https://doi.org/10.1016/j.gloenvcha.2004.12.005
63. Kumar SV, Bharat GK (2014) Perspectives on a water resource policy for India. Energy Resour Institute. www.teriin.org
64. Goyal MK, Surampalli RY (2018) Impact of climate change on water resources in India. J Environ Eng 144:04018054. https://doi.org/10.1061/(ASCE)EE.1943-7870.0001394

Conclusions and Recommendations

Update, Conclusions and Recommendations for "Wastewater Assessment, Treatment, Reuse and Development in India"

Abdelazim M. Negm, El-Sayed E. Omran, Shalini Yadav, and Ram Narayan Yadava

Abstract Update and conclusions chapter covers and sheds light on the key conclusions and recommendations of the 16 chapters that provided in the book titled "Wastewater Assessment, Treatment, Reuse and Development in India". In addition, some observations from a few recently published analytical work relevant to water quality. Therefore, this chapter provides details on water pollution management in India, its assessment, treatment and remediation. Furthermore, a set of guidelines for future research operation is highlighted to direct future work towards environmental conservation and sustainability, which under Indian circumstances is the key subject of strategic significance.

Keywords Environment · Assessment · Wastewater · Management · Water pollution · Water reuse · Treatment · Development · Sustainability · India

A. M. Negm (✉)
Water and Water Structures Engineering Department, Faculty of Engineering, Zagazig University, Zagazig 44519, Egypt
e-mail: amnegm@zu.edu.eg

E.-S. E. Omran
Soil and Water Department, Faculty of Agriculture, Suez Canal University, Ismailia 41522, Egypt

Institute of African Research and Studies and Nile Basin Countries, Aswan University, Aswan, Egypt

S. Yadav
Centre of Excellence in Advanced Water and Environmental Research, Rabindranath Tagore University, Bhopal, Madhya Pradesh, India

R. N. Yadava
Research and International Affairs, Madhyanchal Professional University, Bhadbhada Road, Ratibad 462944 (M. P.), Bhopal, India

© Springer Nature Switzerland AG 2022
S. Yadav et al. (eds.), *Wastewater Assessment, Treatment, Reuse and Development in India*, Earth and Environmental Sciences Library,
https://doi.org/10.1007/978-3-030-95786-5_16

1 Background

Pollution of water is an important environmental issue in India. Untreated waste is India's largest source of water pollution. Many emissions sources include farm runoff and unregulated small-scale industry. Many of India's rivers, reservoirs, and surface water are contaminated by agriculture, untreated wastewater, and industrial waste. Untreated sewage discharge is the single most significant source of surface and groundwater pollution in India. There is a major difference between domestic wastewater generation and treatment in India. The problem is not only that India lacks adequate capacity for treatment but also that the current sewage treatment plants do not work and be not maintained.

Throughout India, the surface water quality is higher than the groundwater. Contamination of water not only impacts water quality but also endangers human health, economic development and social stability. It is vital that it is monitored on a regular basis to maintain that the water quality is being preserved or restored at the level desired.

Therefore, this chapter provides general findings of the assessment, treatment and reuse of water in India. In planning sustainable water supply management, it is necessary to give due consideration to the various resources used which make the resulting production system unsustainable. The major issue of this book is that it sheds light on water pollution management in India. So, the purpose of the book is to strengthen and discuss the following key theme:

- Monitoring and assessment of polluted water
- Wastewater treatment and reuse
- Wastewater remediation
- Waste and sustainability

The next section presents a brief of the important findings of some of the recent(updated) published studies on water pollution management in India, then the main conclusions of the book chapters in addition to the main recommendations for researchers and decision-makers. The update, conclusions, and recommendations presented in this chapter come from the data presented in this book.

2 Update

The following are the major update for the book project based on the main book theme.

2.1 Monitoring and Assessment of Polluted Water

Three monitoring and assessment tools for polluted water in India were identified. The first tool is "Coastal Reservoirs-a technology that can quench Indian thirst". In terms of natural resources such as fresh water, the world is facing the crucial problems. According to the World Health Organization, at home today 2.1 billion people lack safe drinking water [1]. Despite India becoming the world's second-most populous country, water demand is projected to surpass all existing supply capability and the country is predicted to become water-scarce by 2025 [2]. Roughly 70% of irrigation and 80% of domestic water currently come from the rapidly depleting groundwater [3]. To maintain towards this scarcity India needs stepping further on introducing new strategies with alternative resources and water management systems. As India is the 9th country by total renewable water resources [4] developing such strategies at earliest can minimize water scarcity in future.

The second tool is "Heavy metal accumulation in dominant edible fish species of River Kolab in Koraput District of Odisha." Rivers play a vital role in connecting habitats and their values to plants and animals that extends far beyond the surface area they cover [5]. India is one of the world's mega biodiversity countries and in terms of freshwater mega biodiversity it occupies the ninth position. Due to their non-biodegradable nature, accumulating properties and their long biological half-lives, the study of toxic and trace metals in the environment is more important compared to the other pollutants. Metal accumulation in fish is a part of the absorption and excretion. Uptake is assumed passive, involving gradients of diffusion generated by adsorption or binding of the metal to the tissue and cell surfaces. There are three possible ways by which metals may enter the body: (i) the body surface, (ii) the gills, and (iii) the alimentary tract. Yet little is known about heavy metals being taken up by the skin. No extensive research has been done on the fish found at Kolab River in Koraput Odisha district.

The third tool is "Essentials of water purification in mitigating water demand in explosive global population: The changing patterns of cake and membrane filtration." As the fundamental demand of the life processes, its purification, recycling and reuse are essential for sustainable water management. One sustainable resource management will be Zero Discharge, where 80% of the water used would be recycled and used primarily for various industrial purposes. In this case origin and used water nature is essential. MBST (Membrane Based Separation Technology) is applied to purify the said used water for drinking purpose as a major breakthrough in the history of civilization. In addition to increasing water demands, the degree of contamination may also increase (i.e. people usually use water from different areas where population and waste are dense, which may also help group activities to create water born diseases).It is possible to present the problem through technological innovation specifically through MBST (Membrane Based Separation Technology) and jute pulp cake filtration [6]. Cake filtration and MBST as integrated phenomena will be cost-effective, environmentally friendly and suitable technologies to use for human survivals.

2.2 Wastewater Treatment and Reuse

Five potential approaches were identified for Wastewater Treatment and Reuse in India: First possible approach is "Potential for Use of Treated Wastewater for Industrial Reuse in India." There is plenty of water on earth to sustain life, but keeping in mind overuse and increased rates of pollution, we must do what we can to conserve and protect water resources [7]. The water we actually waste (i.e. wastewater) must also be treated and reused. It has been recorded that the largest population still has no access to improved water supplies and adequate sanitation [8]. The diseases caused due to contaminated water are the major cause of mortality and morbidity worldwide [8]. The countries that are still making much of the investments in water and sanitation have emphasized out-of-household water supply systems, wastewater treatment, and main building facilities. As such, these services have no positive impact on the lives of disadvantaged urban residents [9]. The key factors that affect the distribution process and the continuous supply of water systems are the quality of the water and the scheme is the quality of the water and the schemes given to the citizens [10, 11]. On the list of priorities, wastewater drainage, sanitation and treatment tend to rank higher [12].

Second possible approach is "Understanding sustainable arsenic mitigation technology application in the Indian Subcontinent." Often road to hell is paved with good intentions and this is the case in groundwater problem with the origin of arsenic. Trying to save the world from waterborne diseases and encouraging groundwater as a protected supply without checking for the presence of arsenic is this century's grave unparalleled tragedy. Because of its widespread occurrence, high groundwater arsenic combined with aquifers with strongly reducing conditions presents the most serious threat from a public health perspective. In Bangladesh, India, Vietnam, Cambodia, and Pakistan, for example, aquifers with strongly reducing conditions exist. The most commonly known mechanism for releasing arsenic into anoxic groundwater is reductive oxidation of iron oxides in the aquifer mediated by microbes. The iron oxides in the aquifers allow for arsenic binding sites to remain in the solid phase. The reductive dissolution of iron oxides decreases the binding sites for arsenic which causes the mobilization of arsenic into groundwater [13]. MOSPI, GoI. Subsurface arsenic removal (SAR) is a relatively new treatment method that has been tested in Bangladesh and West Bengal [14–16]. SAR has the potential to be a cost-effective option to provide safe drinking water in rural areas of Bangladesh and West Bengal as existing shallow tube wells can be modified to perform SAR [15].

Third possible approach is "Sequential anaerobic-aerobic co-treatment of three herbicides mixture in water: A comprehensive study on bio-transformation". The used herbicides and their derivatives have become a major concern in the field of environmental engineering as they join freshwater sources by agricultural runoff and then enter the downstream water bodies and increase its toxicity. The use of herbicides before runoff has contributed to a rise in herbicide load on the downstream bodies of water [17]. The origins of herbicides include pesticide processing plants, soil leaching, surface runoffs, accidental spills, improper disposal, etc. Herbicide removal using pure cultures isolated from algae, fungi, and bacteria have also been used to remove herbicides efficiently [18, 19]. But their adequacy is restricted to a specific type of herbicides, and basically it is difficult to keep in pure form on a large scale at field conditions. The bio-degradation of herbicides mainly due to breaking up of the bond between the benzene ring and the substituent group by methanogens in the existence of electron donor microbes [20]. They increase the rate of reaction by shuttling electrons from primary electron donors or from bulk electron donors to the electron-accepting organic compounds.

Fourth possible approach is "Wastewater management and treatment technologies with recycling and reuse issues in India leading to Zero Liquid Discharge (ZLD)." A very low amount of household and industrial wastewater is currently treated in India. The wastewater created by the industrial and municipal sectors must be properly handled before it is discharged. The need for fresh water in India has increased steadily at a rapid rate. The reducing per capita supply of water and declining water quality has forced the country to search for safe and efficient water technologies to provide clean and reliable water. Wastewater treatment and its disposal problems are seen as one of the biggest obstacles in today's perspective leading to sustainable practices in water management. To meet the regulatory discharge limits set by Central Pollution Control Boards (CPCB) in recent month, State and Local governments and to protect the quality of Nation's water, proper technology solutions must be provided not only to treat these polluted and sometimes toxic wastewaters to meet the recent stringent norms, the wastewater recycling and reuse issues at the same time need to be think of for protection of our fresh water resources on long term basis. In addition, Central Board has also introduced electronic monitoring of effluent quality levels for close monitoring of wastewater related issues, and at the same time helps to protect our fresh water supplies nationwide.

Further possible approach is "Wastewater and its impacts in India". The primary sources of wastewater production can be considered loosely divided into both domestic and industrial sources. More than 50% of the sewage in India are generated from the states of Maharashtra, Tamil Nadu, Uttar Pradesh, Delhi and Gujarat Besides the domestic sources. A further source of wastewater is from the large number of small-scale factories that are haphazardly situated in crowded places like residential areas. In the context of industrial wastewater treatment, around 5,500 MLD generated by small-scale industries are left untreated [21]. This industry is unable to develop its own treatment facilities because of a lack of sufficient financial capital. Therefore, in order to reduce the polluting load of the receiving water bodies, there is a need for specific effluent treatment plants (CETPs) built across the country for these industries. Conventional centralized sewage treatment technologies, currently widely used in India, are costly, require complex technology and, most importantly, are incapable of meeting the demand for treatment of all the wastewater produced. Now the emphasis will turn to the implementation of advanced treatment systems to treat and reuse wastewater locally. Decentralized wastewater treatment network is one of the possible approaches for sustainable treatment technologies. In a decentralized wastewater system; besides lower operation and maintenance cost the capital investment requirement is also lower in comparison with conventional centralized wastewater system.

2.3 Wastewater Remediation

Five methods were identified for Wastewater remediation as a potential way for sustainability. First is "Nutrient loading impact on remediation of hydrocarbon polluted groundwater using constructed wetland". Rapid urbanization and population growth have created an unparalleled global burden on the surface and subsurface water supplies. Shallow regions of Indian aquifer, particularly petrochemical refineries, have been identified to be the most vulnerable to hydrocarbon pollutants in the severe industrial growth. The water soluble compound like toluene, starts dissolving to underlying groundwater resources and finally form a dissolved plume [22]. As toluene is carcinogenic, the use of this polluted groundwater for drinking beverage will cause major negative impacts on micro biota and human health. For this purpose, more attention is required from environmental scientists and geochemists to remediate these contaminants from groundwater systems to meet the drinking water (demand of that population and to reduce the effect of emissions from hydrocarbons. Several researchers have developed numerous approaches to bioremediation in recent decades, with plant-assisted bioremediation/constructed wetlands among the most cost-effective techniques documented [23]. Nevertheless, very little attention has been given to the performance evaluation of plant supported bioremediation and developed wetland for the management of groundwater supplies polluted with hydrocarbons. Some of these studies concentrate on storm water management and wastewater management that include many contaminants including hydrocarbons [24]. The results/results of these studies have limitations to be used specifically to develop a bioremediation program for groundwater contaminated with hydrocarbons.

Second is "Municipal Wastewater: A remedy for water stress in India". Total sewage generation in India is 61754 MLD, out of which installed sewage treatment capacity is only 22963 MLD. However, with respect to installed sewage treatment capacity, India has only 12,000 MLD. The difference between sewage generation and installed sewage treatment capacity remains around 26000 MLD. This difference is around 68.4% by percentage and even in some of the cities the current treatment capacity remains underused [25]. While the population continues to grow, competition and water shortages become unmanageable. If India is estimated to rise to 1.6 billion by 2051, the water supply per capita is projected to decline to 1174 m^3 per year [13]. Therefore, ensuring 100% municipal wastewater treatments and enabling the reuse of water, which can be one of India's remedies for water stress, becomes important for the country. Through reusing water, the country will benefit greatly through making use of all available water sources to increase its effectiveness.

Third is "Biosorptive removal of a Herbicide, Glyphosate using waste activated sludge of tannery industry". Biosorption process is an efficient, eco-friendly, cost effective and attractive option for removal of synthetic organic matter, as well as, toxic metals from aqueous solution. The mechanism also occurs through physico-chemical interactions such as adsorption, exchange of ions, electrostatic attraction, binding or complex formation between ions and functional groups present on the surface of the biosorbents. Adsorptive glyphosate reduction has focused on the use of different materials, such as waste industrial residues such as alum sludge [26], clay based substances, and activated carbon [27]. The waste activated tannery sludge was effectively applied for elimination of Ni(II), Co(II), Zn(II) and Cd(II) ions from aqueous environment both as single and multi-metal removal system [28, 29]. The potential of multi-metal laden spent biosorbent on removing glyphosate has inspired us to exploit this kind of toxic sludge as adsorbents in wastewater for herbicide remediation.

Fourth is "Wastewater treatment processes with special reference to activated sludge process in Indian conditions". Wastewater treatment is a procedure that requires wastewater treatment before being released into the natural world, so that it does not pose a hazard to human or animal health after disposal. Preliminary treatment is the first step in treating wastewater. The primary aim is to eliminate coarse solids and other large materials that are often present in raw wastes. The aim of primary treatment is to eliminate larger floating particles and suspended solids. The primary treatment's main aim is to create a typically homogeneous liquid that can be biologically treated, and a sludge that can be handled separately. Secondary treatment is the additional application of the main application effluent to remove organic residuals and suspended solids. In urban wastewater treatment plants, the advanced wastewater treatment method also known as tertiary treatment is used when the higher quality efficiency is needed.

Fifth is "Assessment of water resources in development of Rajasthan." In context of India, over the ages a relative share of water consumption is observed changing as people are moving forward urbanization. Water consumption may increase more rapidly in industrial and residential areas. A preference for water usage should be determined on the basis of available water supplies. This is carried out at district level with the help of data accessible to the water market. Residential-industrial agricultural demand data is obtained from the respective sources along with population data. Business forecasts should be made wise. There has to be a detailed calculation method involved. Policy makers give significant attention to the question of urban residential water scarcity.

2.4 Waste and Sustainability

Three potential methodologies were identified for waste and sustainability issue. The first approach is "Evaluation of factors influencing surface water quality in a coalfield area of Damodar Valley, India: A Sustainable Use Approach." At this point of view, assessing the surface water quality is necessary to determine the appropriateness of water for a specific use. Scarcity of clean and potable water has emerged in recent years as one of the most serious developmental issues in many parts of West Bengal,

Jharkahnd, Orissa, Western Uttar Pradesh, Andhra Pradesh, Rajasthan and Punjab [30]. The surface water quality problems are more severe in densely populated and heavily developed mining areas. Coal mining is believed to be linked to the liberation of Fe, Mn, Se and SO_4^{-2} in the environment [31]. Likewise, concentrations of soil and surface water are also large of coal mining areas where toxic waste and waste water from mining fields and coal power plants are disposed of to the atmosphere. Also all three metals are known as being related to vehicle emissions.

The second approach is "Sustainable construction practices for residential buildings to reduce the water footprint." Water is indispensable for life on the planet to survive. Humanity, animals, and plants are a key requirement. Gray Water is wastewater polluted with sand, soil, soap and oily substances but does not come into contact with fecal matter [32]. It is obtained from bathing, dishwashing, laundry, and agricultural runoff, which may include fertilizers and pesticides. Such water can be recovered and reused in different means including for agriculture [33]. It can be absorbed by the soil, and filtered to water tables during its percolation cycle. The fecal matter contaminates the Black Water. This is the water that is collected by rinsing excreta from humans or livestock. This has large fecal coliform and other pathogenic compounds. Gray Water does not come into contact with the Black water. The primary impurities which contaminate Process water are toxins and heavy metals [34]. Therefore all forms of water can be recycled by proper treatment. Wastewater regeneration is necessary for the humanity's survival. Direct water consumption (or WFP) is the volume of water absorbed, for example, during mixing, binding, curing, hydration, and dilution during the construction process, etc. [35, 36]. The indirect water intake is the WFP as the building material is manufactured, packed, and transported to the construction site. The overall WFP of the material or process is a combination of direct and indirect WFP [37].

The third approach is "Water resources availability and its teleconnection with large scale climatic oscillations over Godavari River Basin." Sadly, India is facing a crucial situation in its past of severe water shortages and threatening millions of lives and livelihoods. With the terrifying figure, the present report by NITI Aayog [38] advocates that about 820 million population [39] over the twelve river basins have per capita water availability lower than 1000 m^3, nearly 82% of rural house holds doesn't have access to piped water supply, 70% of the available surface water is contaminated [40]. So, to close the gap between supply and demand by 2030, an investment worth $291 billion is suggested [41]. On top of that, India's potential demand and water availability projection is even more worrying. For example, by 2030 40% of the global population will be deprived of access to drinking water, the supply of groundwater in 21 major cities will run out affecting 100 million people by 2020 and a loss of 6% of GDP as a result of the water crisis by 2050 [42]. The association between the large-scale climatic variability and stream flow is expected to be highly complex and nonlinear in both space and time. Hydro-climatologists have made an attempt to establish the association through teleconnection technique around the globe [43]. By changing precipitation, temperature, snow cover and evapotranspiration, the interaction of large-scale climatic oscillations with the stream flow [44].

3 Conclusions

In the course of the current book project the editorial teams took some conclusions from this book. The chapter draws valuable lessons from the cases in the book, particularly the positive characteristics of water quality, assessment, and management in India, in addition to methodological observations. Such conclusions are significant for India's environmental management. The concluding remark for this chapter is to highlight the following:

3.1 Monitoring and Assessment of Polluted Water

Coastal reservoirs could effectively mitigate flood disasters, and convert the disastrous floodwater into water resources in dry seasons. Once local runoff is developed, downstream water demand is met and flood disasters are mitigated, the upstream water crisis could be solved, e.g., Chennai and Bangalore, thus India will have sufficient freshwater from the sea without desalination. Water pollution is another problem for India. The wetland pre-treatment can be developed to purify the river water prior to storage. The total runoff in India is about 1887 km^3/year (the average runoff coefficient is about 0.47), among it 1033 km^3/year of water could be developed by the inland strategy and at least 854 km^3/year of floodwater reaches at the sea, for which coastal reservoirs can harvest in the sea. It is enough for India to quench its thirst, or India is not running out of water, but water is running out of India.

The fish diversity study of river Kolab exhibits the presence of nineteen (19) finfish species. Site 1 is more congenial compared to site 2 as confirmed by high Shannon-Weiner index value and low Index of Dominance value. Pb accumulation occurred in the intestine of *Catlacatla* by crossing the barrier of the safe limit as prescribed by WHO, 1984. The statistical panorama suggests the negative impact of pH and temperature on the magnitude of bioaccumulation. However, turbidity exhibits a positive correlation which may be due to adsorption of the metal species on suspended particulate matter. Management policy advocates are liming of the effluents released from the industries to the river to trigger the alkaline condition of the water body so that the heavy metal species get compartmented in the sediment bed instead of entering in the body tissues of aquatic organisms. The dose of liming has to be optimized depending on the pH of the waste water generated from these industries.

MBST has been considered as efficient water purification system. Jute pulp cake filtration is an alternative innovation, where the cake is formed on the surface of jute pulp, attached to clarifier basin, rotating surface and the said cake is used to separate pollutants up to 2 microns (μm). Cake filtration and MBST are as integrated process. Membrane must be used to intensify water purifications. It is cost-effective, eco-friendly and appropriate to maintain water quality, quantity of tertiary water reservoir. Water treating in primary and secondary clarifies can be made by various chemical and biological system. Jute pulp cake filtration can be made before primary and secondary treatment to remove sediment particles and slurry from water. Ganga water can be treated in such integrated filtration process. MBST is used also for desalination and for removals of heavy metals including As, Cr, and Phosphate. The selection of membrane properties, and their modules, plate, hollow fiber and wickle are important. Water treatments by microbes (Bacteria, fungi and Algae) are also essential, amongst them algae took an important role in clarifications of waste water and to use algal biomass for the fuel, food and fertilizer an alternative solution to mitigate energy and pollution.

3.2 Wastewater Treatment and Reuse

The long-term study was conducted using two anaerobic reactors namely R1 (anaerobic control with no herbicide) and R2 (anaerobic reactor fed with 2,4-d + ametryn + dicamba mixtures). The effect of increased herbicides concentration on the anaerobic–aerobic reactor during 400 days of the treatment period was evaluated. Two aerobic reactors were operated simultaneously to give post-treatment to the anaerobic effluent. The reactor performance was evaluated by monitoring herbicide removal efficiency of chemical oxygen demand (COD) and biogas production. The reactor's stability parameters pH, alkalinity, volatile fatty acids (VFA) and oxidation–reduction potential (ORP) were monitored on a daily basis. Both the anaerobic reactors were stabilized using 2 g/L of starch with total organic loading rate (OLR) of 0.21–0.215 kg-COD/m^3/d during 48 days, and aerobic reactors were stabilized in 14 days using anaerobic effluent as feed having OLR of 0.02 to 0.038 kg-COD/m^3/d. Sequential anaerobic–aerobic removal efficiency for A2 reactor was found to be >85%. Addition of 5–10 mg/L of anthraquinone-2,6-disulphonate (AQS) as a redox mediator enhanced the herbicides removal efficiency in the anaerobic reactors (R2) by 5–10%. The samples were analyzed using GC-HRMS to identify the intermediate compounds.

Wastewater Management with recycle and reuse must be taken as a key performance indicator to protect our fresh water resources. Looking into Industry, it is the second largest user of water after agriculture. The amount of water used varies widely from one industry type to another. Many businesses, notably the Textile, food, beverages, and pharmaceutical sectors are highly water intensive and consume water by using it as an ingredient in processing finished products. Many businesses discharge of wastewater or wash water into natural fresh water ecosystems. Each liter of water moving through a system represents a significant energy cost. Water losses in the form of inefficient treatment, delivery leakages, theft, and consumer waste all directly affect the amount of energy required to deliver water for its proper use. Wastage of water directly leads to a waste of energy.

The treatment of wastewater is an essential process for the conservation of our environment and needs to be encouraged in India. The sector has been characterized with a huge gap between the infrastructure facilities and the amount of waste being generated, thereby leaving behind a significant portion of wastewater left untreated. The failure to ensure a holistic wastewater management could be attributed to the key challenges being faced by the sector. Some of the key challenges worth mentioning are lack of budgetary resources, practice of treating only the wastewater conveyed through underground sewers leading to low collection efficiency, preference for centralized system which entails a huge capital investment, lack of interest to adopt innovative sustainable treatment technologies.

3.3 Wastewater Remediation

The wetlands are the self-sustaining green aquatic ecosystems and are preferred over conventional physio-chemical treatment methods. The biological processes of periphyton (microorganism community of wetland ecosystem) along with physical removal of pollutants by plant root biomass are responsible for pollutant removal in constructed wetlands. Indeed the viability of any treatment wetland depends on factors like type of plants, porous media, and characteristics of polluted water along with the prevailing environmental conditions. Presence of plants to the treatment wetland system drastically improved the decontamination efficiency from 68% to >90% in the current case study which fortifies the crucial role of plants on bioremediation.

Biosorption efficiency of waste tannery industry sludge could be increased from 26.18% to >92% for glyphosate removal after incorporation of multi-metal loading on the sludge. Binding with biosorbent surface functional groups, physical adsorption, vanderwaal forces, chelation, complexation and ligand formation were established in the studied biosorption process. Feasibility study for actual effluent treatment showed 89% glyphosate removal efficiency in simulated municipal wastewater.

Three main steps in wastewater treatment are primary treatment, secondary treatment and tertiary treatment. Activated sludge process is an important type of secondary treatment, which aims to clean the wastewater by separating it into effluent and sludge. The effluent can then be discharged into any water body while the sludge can be used as manure or incinerated as per the requirement.

The rise in number of population values in Rajasthan is now just a normal phenomenon with no exception. The population rise in Rajasthan follows a rate of 3.08% per annum compared to the overall population. With the increasing number of inhabitants the throat cut demand for the sources also increases which results in degradation of the surroundings and deterioration of the quality of life. The reason behind the shortage of resources is not only because of increasing population but one of the major causes of this is improper and insufficient management and usage of them. It is extremely essential to study the various aspects of consumption and management of various natural resources predominantly water.

3.4 Waste and Sustainability

The overall study indicated the effect of mining and allied activities on the quality of the surface water resources of the East Bokaro coalfield area. The geochemistry indicated that dominance of rock weathering is controlling the surface water chemistry of the study area. Silicate weathering, carbonate weathering, and ion exchange have been identified as the hydrogeochemical processes controlling the chemical composition of water resources in the study area. The contamination is predicted to be more along the Damodar River. Quality assessment for irrigation use shows that surface water of the area is good to permissible quality and can be used for irrigation purposes. The conservation measures that can be suggested for the enhancement of quality of the water resources of the study area are rainwater harvesting, surface water reclamation and treatment of the available water resources.

The sustainable construction practices primarily include three aspects namely; the conservation and optimization of resources, waste management, and reduction of energy and water footprint. Maximum water was consumed in Brickwork, Roof slab, plastering, foundation, flooring, excavation, and landfilling. The total WFP of the building was 190,300 L, which is 37.94 L/ft^2. By simple water conservation, recovery and reuse of wastewater, the WFP may be reduced by 4.59 ^{L}/ft^2. A wastewater storage tank and controlled drainage of wastewater were effective to save 23,030 L of water. No spilled or runoff water was allowed to drain.

The variability and possible teleconnection of between monthly flow and Sea Surface Temperature (SST), Indian Ocean Dipole (IOD), and Indian Summer Monsoon Index (ISMI) are detected using wavelet coherence technique at 40 different stations. It is observed from the analysis that the flow during the monsoon periods varies greatly across the basin (from 0 m^3/sec to ^{26,700} m^3/sec). From the wavelet analysis, it is observed that ISMI has significant association with the monthly stream flow at intra- and inter-annual scales over all the gauging stations. However, a uniform phase is not observed between the monthly stream flow and the three studied climatic oscillations. The present analysis enables hydrologists and water managers to understand the hydro-climatic process and their association with climatic oscillations along with the variability in the surface water resources over the Godavari Basin.

4 Recommendations

Throughout this book project the editorial teams recognized other areas that could be explored to further develop. Based on the observations and conclusions of the authors, this section contains a number of recommendations which provide suggestions for potential researchers to go beyond this book's scope. The following suggestions are derived primarily from the chapters set out in this volume:

- It is beneficial that India has such a large stretch of coastline that provides the country with possibly the best solution for its water supply shortages in coastal reservoirs. Coastal reservoirs would manage to capture and hold large amounts of usable water that would otherwise go to waste. As the population of India continues to increase, the water crisis will embellish so as action needs be taken more drastically as time goes on, coastal reservoirs will more prominently become the most viable option.

- The present study area needs constant monitoring of bioaccumulation pattern in consumable aquatic fauna because of the presence of major industries in and around its vicinity. Conservation of natural resources in the river Kolab needs to be taken up through awareness programme involving the industrialists, policy makers, fisherman and people of ranks of the society of Koraput in the loop.
- The cake filtration models are needed to be changed at secondary treatments. MF (Microfiltration), UF (Ultra-filtation), NF (Nano-filtration) and RO (Reverse Osmosis) membranes are used according to the decreasing pore sizes from 1 micron (m) till 0.001 μm for optimizing DWQ (Drinking Water Quality). For treatments of WW (waste water) combined jute pulp cake filtration and membrane could be effective. Heavy metals like Arsenic (arsenic As^{3+}/As^{5+}), Chromium (Cr^{6+}), Copper (Cu^{2+}), Cd (Cadmium), and lead (Pb) originated from industrial and domestic effluents can also be treated by microbes (Pseudomonas putida, Aspegillusnieger, Candida species) by designing various flora and by fauna cultures, like Water Hyansinth to grow in a reserved low WLC (Wet land Cultivation) by bio-augmenting and bio-remediation to develop a healthy socio-economic environment.
- Studies may be conducted to quantify the concentrations of intermediate compounds formed due to herbicides biotransformation. Studies may be conducted to treat high concentration of different herbicides using anaerobic–aerobic system. Research may be extended to find out the type of bacteria present in the reactor sludge by polymer chain reaction (PCR) amplification and 16 S rRNA sequencing. Studies can be conducted to reduce the treatment time required by modifying the existing anaerobic–aerobic method.
- Industries and municipalities must focused on effective water uses by reducing demand, maximizing recycling opportunities; minimizing water losses due to breakdowns and leakages; and investing in new water storage infrastructure. For Sustainable development optimal management of water and wastewater leading to zero discharge is now the another challenge before the industry. Finally, in terms of working methodology, it may be looked as a-Systematic approach of Identifying, Measuring, Monitoring and Reducing the Water Consumption by different activities in an Industry and exercise of stewardship of water resources through deployment of cost effective, performance oriented appropriate/advance technology with a focus on recycle and reusing the wastewater for the greatest good of society and the environment on sustainable basis.
- There is a need for paradigm shift in treatment of wastewater in India. The focus should be to embrace new sustainable technologies for treatment in place of the conventional treatment process. Practices of opting conventional centralized wastewater system also need to be relooked in the light of the need to build in flexibility in the system and making it cost effectiveness. The feasibility of opting

decentralized wastewater system over the conventional system should be explored with the objective to overcome the challenges being faced by the sector.

- The efficiency of bioremediation technique can be amplified by controlling the prevailing environmental conditions and by modifying the constituents using plants, termed as biostimulation. Biodegradation rate of toluene was found to be augmented by the addition of the nutrient in planted wetland. A high removal efficiency of toluene is attributed to biostimulation of the periphyton in the root zone. Efficiency can further be enhanced by addition of the nutrients which will increase the microbial metabolic activities. Further studies are required to quantify the nutrient and pollutant uptake by plant root biomass to their subsequent translocation to shoot biomass and to incorporate evapotranspiration losses.
- Mechanical drying of sludge can be carried out which will decrease the amount of space required for conventional dewatering process and lessen the ground water contamination possibilities. OSTP does not have the tertiary level treatment process which can be considered there. Incinerators can also be installed which will help in reducing excess sludge and even help in the disposal of screenings and grit. Energy consumption can be reduced by substituting aerators with fine bubble aeration system.
- To make water economy, it is essential to follow certain steps: Development of technology is focusing on research and development. Make each and every citizen aware of the importance of water and this should be done by seeking their Corporation. Proper pricing policies and law and order. To strictly follow the concept of recycling of water. Effective implementation of the awareness programs, stating how to use water economically. Proper maintenance of all the pipeline, service lines and feeder lines, canal and the catchment areas should be executed time to time.
- A few additional sustainable construction practices such as prevention of leakage, construction seasoning, use of sprinklers, solvent cleaning, etc., may further reduce the wastewater and hence, the WFP of the building. Awareness and training of the public are required to a large extent to change the orthodox construction practices and construct low WFP buildings without any compromise on strength and quality. Water conservation at a construction site is possible with a minimum investment if it is considered during the planning and scheduling of construction stages. Unsustainable construction practices have to be controlled immediately to conserve the natural resources in the larger interest of the future generation.
- In order to combat the climate change impact on water resources, development in the technical aspects, improvement in the understanding and computational facilities, the estimation in the probable variability in the foreseeable future need considerable research activities for robust adaptation strategies and policies. In addition, to improve the adaptation strategy and policy, the researchers should make significant improvement in the climate change assessment and identify the suitable and robust approaches by involving key stakeholder organisations and local communities.

References

1. World Health Organization (2018) 2.1 billion people lack safe drinking water at home, more than twice as many lack safe sanitation. [online] Available at: http://www.who.int/news-room/detail/12-07-2017-2-1-billion-people-lack-safe-drinking-water-at-home-more-than-twice-as-many-lack-safe-sanitation. Accessed 10 April 2020
2. Amarasinghe UA, Shah T, Turral H, Anand BK (2007) India's water future to 2025–2050: business-as-usual scenario and deviations. International Water Management Institute, Colombo, Sri Lanka, 47 p. (IWMI Research Report 123)
3. The Times of India (2018) India set to become water scarce country by 2025: report—Times of India. [online] Available at: https://timesofindia.indiatimes.com/home/environment/developmental-issues/India-set-to-become-water-scarce-country-by-2025-Report/articleshow/47403894.cms. Accessed 10 April 2020
4. En.wikipedia.org (2018) List of countries by total renewable water resources. [online] Available at: https://en.wikipedia.org/wiki/List_of_countries_by_total_renewable_water_resources. Accessed 10 April 2020
5. Alam MS, Hossain MS, Monwar MM, Enamul M (2013) Assessment of fish distribution and Biodiversity Status in Upper Halda River, Chittagong, Bangaladesh. Int J Biodivers Conserv 5(6):349–357
6. Brahma NK (2008) Zero discharge and global warming: role of MBST, O3, UV and ME (water recycling). Chem Bus 22(6):18–24. ISSN: 0970–3136
7. Desai SS (2017) Recycling, reuse and economics of treated wastewater in urban India. Int J Eng Res Mech Civil Eng (IJERMCE), 2:12. ISSN (Online) 2456–1290
8. UNICEF/WHO (2017) Progress on household drinking water, sanitation and hygiene, special focus of inequalities
9. Philip L, Ramprasad C, Krithik D (2018) Sustainable wastewater management through decentralized systems: case studies. Appl Environ Sci Eng Sustain Future. https://doi.org/10.1007/978-3-319-75199-3_2
10. Anastasios IZ, Ioannis AK (2019) Recent advances in water and waste water treatment with emphasis in membrane treatment operations. Water 11:45.https://doi.org/10.3390/w11010045. www.mdpi.com/journal/water
11. Kulesha O, Maletsky Z, Ratnaweera H (2018) Multivariate chemometric analysis of membrane fouling patters in biofilm ceramic membrane bioreactor. Water 10:982
12. Alfredo VR, Elsa NAG, Gloria TE (2019) Potential uses of treated municipal wastewater in a semiarid region of Mexico. Sustain Open Access J 11(8):1–23
13. MOSPI, GoI (2018) EnviStats India 2018 (supplement of environmental accounts). Retrieved from http://www.mospi.gov.in/publication/envistats-india-2018. Accessed on 12 April 2020
14. Van Halem D, Bakker SA, Amy GL, Van Dijk JC (2009) Arsenic in drinking water: a worldwide water quality concern for water supply companies. Drinking Water Eng Sci 2. http://resolver.tudelft.nl/uuid:ac069d55-cb87-474b-9af6-421a07225339. Accessed 10 April 2020
15. Van Halem D, Heijman SGJ, Johnston R et al. (2010a) Subsurface iron and arsenic removal: low-cost technology for community-based water supply in Bangladesh. Water Sci Technol 62(11). Scopus: 2702–2709
16. Van Halem D, Olivero S, de Vet WWJM et al. (2010b) Subsurface iron and arsenic removal for shallow tube well drinking water supply in rural Bangladesh. Water Res 44(19). Scopus: 5761–5769
17. Conte LO, Schenone AV et al (2016) Photo-Fenton degradation of the herbicide 2, 4-D in aqueous medium at pH conditions close to neutrality. J E M 170:60–69
18. Bhaskar S, Manu B, Sreenivasa MY (2019) Bacteriological synthesis of iron hydroxysulfate using an isolated Acidithiobacillusferrooxidans strain and its application in ametryn degradation by Fenton's oxidation process. J EnvManag 232:236–242
19. Szewczyk R, Kuśmierska A, Bernat P (2018) Ametryn removal by Metarhiziumbrunneum: biodegradation pathway proposal and metabolic background revealed. Chemos 190:174–183

20. Ghattas AK, Fischer F, Wick A, Ternes TA (2017) Anaerobic biodegradation of (emerging) organic contaminants in the aquatic environment. Water Res 116:268–295
21. ENVIS (2017) Newsletter [Online]. ENVIS, CPCB, New Delhi. Available: http://cpcbenvis.nic.in/cpcb_newsletter/sewagepollution.pdf. Accessed 10 April 2020
22. Gupta PK, Yadav B, Yadav BK (2019) Assessment of LNAPL in subsurface under fluctuating groundwater table using 2D sand tank experiments. ASCE J Environ Eng
23. Basu S, Yadav BK, Mathur S (2015) Enhanced bioremediation of BTEX contaminated groundwater in pot-scale wetlands. Environ SciPollut Res 22:20041
24. Mustapha HI, Gupta PK, Yadav BK, van Bruggen JJA, Lens PNL (2018) Performance evaluation of duplex constructed wetlands for the treatment of diesel contaminated wastewater. Chemosphere 205:166–177
25. MoEF&CC, GoI (2019) National status of waste water generation & treatment, ENVIS Centre on hygiene, sanitation, sewage treatment systems and technology. Retrieved from http://www.sulabhenvis.nic.in/Database/STST_wastewater_2090.aspx. Accessed on 12 April 2020
26. Hu Y, Zhao Y, Sorohan B (2011) Removal of glyphosate from aqueous environment by adsorption using water industrial residual. Desalination 271:150–156
27. Mayakaduwa SS, Kumarathilaka P, Herath I, Ahmad M, Al-Wabel M, Ok YS, Usman A, Abduljabbar A, Vithanage M (2016) Equilibrium and kinetic mechanisms of woody biochar on aqueous glyphosate removal. Chemosphere 144:2516–2521
28. Ramrakhiani L, Halder A, Majumder A, Mandal AK, Majumdar S, Ghosh S (2017) Industrial waste derived biosorbent for toxic metal remediation: Mechanism studies and spent biosorbent management. J Chem Eng 308:1048–1064
29. Ramrakhiani L, Majumdar S, Sarkar S, Ghosh S (2016) Heavy metal biosorption in multi component system on dried activated sludge: Investigation of adsorption mechanism by surface characterization. Mater Today Proceedings 3:3538–3552
30. Tiwari AK, Singh AK (2014) Hydrogeochemical investigation and groundwater quality assessment of Pratapgarh district, Uttar Pradesh. J GeolSoc India 83(3):329–343
31. Sun L, Gui H, Peng W (2014) Heavy metals in groundwater from the Wolonghu coal mine, northern Anhui Province, China and their hydrological implications. Water Prac Technol 9(1):79–87
32. Water Saving Guide 2016 (2016) SFIC. Available on https://ccsbestpractice.org.uk/wp-content/uploads/2016/11/Water-Saving-Guide.pdf. Accessed on 22 Jan 2020
33. Performance Services (2016) Benefits-of-water-reduction-in-the-building-process. Available on https://www.performanceservices.com/benefits-of-water-reduction-in-the-building-process. Accessed on 22 Jan 2020
34. Gerbens-Leenes PW, Hoekstra AY, Bosman R (2018) The blue and grey water footprint of construction materials: Steel, cement and glass. Water ResourInd 19:1–12. https://doi.org/10.1016/j.wri.2017.11.002
35. EL-Nwsany RI, Maarouf I, El-Aal WA (2019) Water management as a vital factor for a sustainable school. Alexandria Eng J 58(1):303–313
36. Tirth V (2020) Nexus of water footprint with energy and GDP of Saudi Arabia and solution for sustainable water usage. In: Advances in water resources engineering and management. Springer, Singapore, pp 43–57
37. Tirth V, Algarni S, Agarwal N, Saxena A (2019) Greenhouse gas emissions due to the construction of residential buildings in Moradabad, India. Appl Ecol Environ Res 17(5):12111–12126
38. NITI Aayog (2018) Composite water management index
39. Central Water Statistics (2015) http://www.indiaenvironmentportal.org.in/content/412064/water-and-related-statistics-2015/, p 31
40. Water Aid (2018) The water gap: the state of the world's water 2018
41. ASSOCHAM (2017) Investments worth $291 bn needed to plug water demand-supply gap in India: study
42. NITI Aayog (2019) Composite water management index

43. Sagarika S, Kalra A, Ahmad S (2015) Interconnections between oceanic-atmospheric indices and variability in the U.S. streamflow. J Hydrol 525:724–736. https://doi.org/10.1016/j.jhydrol.2015.04.020
44. Schulte JA, Najjar RG, Li M (2016) The influence of climate modes on streamflow in the Mid-Atlantic region of the United States. J Hydrol Reg Stud 5:80–99. https://doi.org/10.1016/j.ejrh.2015.11.003

The manufacturer's authorised representative in the EU is Springer Nature Customer Service Centre GmbH, Europaplatz 3, 69115 Heidelberg, Germany. If you have any concerns regarding our products, please contact ProductSafety@springernature.com

Printed and bound by CPI Group (UK) Ltd, Croydon, CR0 4YY
30/09/2026
02225498-0012